Shramila Yadav, Meenakshi Gupta (Eds.)
Polymers, Colloids, and Surface Chemistry

Also of interest

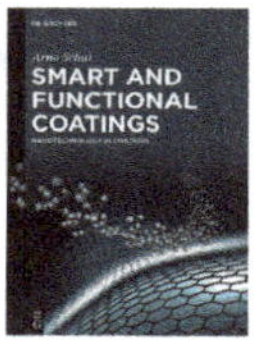

Smart and Functional Coatings.
Nanotechnology in Coatings
Arno Schut, 2025
ISBN 978-3-11-132644-3, e-ISBN (PDF) 978-3-11-132649-8

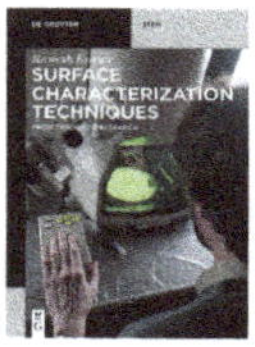

Surface Characterization Techniques.
From Theory to Research
Rawesh Kumar, 2022
ISBN 978-3-11-065599-5, e-ISBN (PDF) 978-3-11-065648-0

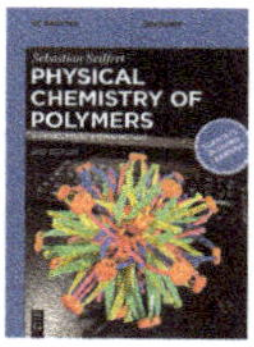

Physical Chemistry of Polymers.
A Conceptual Introduction
Sebastian Seiffert, 2023
ISBN 978-3-11-071327-5, e-ISBN (PDF) 978-3-11-071326-8

Polymers.
Chemistry, Morphology, Characterization, Processing, Technology and Recycling
Mohamed Elzagheid, 2025
ISBN: 978-3-11-158565-9, e-ISBN 978-3-11-158573-4

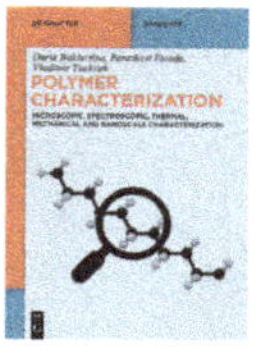

Polymer Characterization.
Microscopic, Spectroscopic, Thermal, Mechanical and Nanoscale Characterization
Daria Bukharina, Paraskevi Flouda and Vladimir Tsukruk, 2025
ISBN: 978-3-11-134536-9, e-ISBN 978-3-11-134574-1

Polymers, Colloids, and Surface Chemistry

Polymerization, Suspensions, Emulsions, and Adsorption Fundamentals

Edited by
Shramila Yadav and Meenakshi Gupta

DE GRUYTER

Editors
Dr Shramila Yadav
Department of Chemistry
Rajdhani College
University of Delhi
Raja Garden, Mahatma Gandhi Road
110015 New Delhi, India
syadav@rajdhani.du.ac.in

Dr Meenakshi Gupta
Department of Chemistry
Atma Ram Sanatan Dharma College
University of Delhi
Chemistry Room No. 14
110021 New Delhi, India
mgupta@arsd.du.ac.in

ISBN 978-3-11-914664-7
e-ISBN (PDF) 978-3-11-220823-6
e-ISBN (EPUB) 978-3-11-220840-3

Library of Congress Control Number: 2026930371

Bibliographic information published by the Deutsche Nationalbibliothek
The Deutsche Nationalbibliothek lists this publication in the Deutsche Nationalbibliografie; detailed bibliographic data are available on the Internet at http://dnb.dnb.de.

Cover image: selvanegra/iStock/Getty Images Plus
Typesetting: Integra Software Services Pvt. Ltd.

www.degruyterbrill.com
Questions about General Product Safety Regulation:
productsafety@degruyterbrill.com

Preface

"Nothing in life is to be feared, it is only to be understood."

Marie Curie

Guided by this profound thought, we present this book as an effort to simplify and illuminate the complex yet fascinating domains of polymers, colloids, and surface chemistry. These three areas form the backbone of countless phenomena and technologies that shape our everyday lives. From the plastics and fibers we use daily to the emulsions in foods, pharmaceuticals, and the catalytic surfaces in industry, these subjects are not only intellectually engaging but also practically indispensable. Their interconnection provides a rich field of study for understanding how matter behaves, interacts, and functions across different scales.

The present volume has been written to provide a clear, comprehensive, and integrated account of these three domains. While each of these areas has been extensively studied individually, students often encounter them in a fragmented manner. This book brings them together in a single text, emphasizing their relationships and combined relevance to modern science and industry.

The chapters are organized systematically to support progressive learning. Beginning with the foundations of polymer science and the chemistry of polymerization, the text advances through polymer solutions, the essentials of colloid chemistry, and the fundamental properties of colloidal systems. The role of surface chemistry in colloids, practical applications of colloids, and adsorption phenomena are then discussed in depth. Each chapter is written in simple and accessible language, making the concepts approachable while retaining scientific rigor.

This book is designed specifically for undergraduate students pursuing B.Sc. (Hons) chemistry, B.Sc. (Prog.) physical sciences, and B.Sc. (Prog.) life sciences under the University of Delhi curriculum, particularly for the Discipline Specific Elective Courses (DSE-06 and DSE-09). It will also serve as a useful reference for students of analytical chemistry, industrial chemistry, and related disciplines.

We are deeply indebted to Prof. Gurmeet Singh for his invaluable guidance, constant encouragement, and inspiration throughout our academic journey. A special mention must also be made of his wife, Prof. Gurmeet Kaur, whose unwavering support has been a source of motivation for us. We extend our heartfelt gratitude to Prof. Darshan Pandey, Principal, Rajdhani College, and Prof. Gyantosh Jha, Principal, Atma Ram Sanatan Dharma College, University of Delhi, for their encouragement and support. We express our sincere gratitude to Prof. Yudhvir Sharma and Dr Shikha Kaushik for believing in us and for their constant encouragement, which greatly motivated us in bringing this book to fruition.

Our deepest love and appreciation go to our superstars Moulik, Aarav, and Bhavya, whose joy, energy, and smiles have been a source of strength during the writing of this book. We are forever grateful to our parents, who laid the foundation of every-

https://doi.org/10.1515/9783112208236-202

thing we are today. We extend our heartfelt gratitude to Mr Mukesh Yadav and Mr Vaibhav Gupta, whose unwavering support, understanding, and companionship have been a constant source of strength and inspiration. We extend our heartfelt gratitude to our family and friends, whose unwavering support and understanding have been a constant source of strength and inspiration, and whose love and blessings have always guided us.

We are equally grateful to all the contributing authors, whose insights and scholarly contributions have greatly enriched the quality and depth of this work. We also wish to acknowledge our worthy colleagues, whose constant support and encouragement have been invaluable throughout the preparation of this book.

In preparing this text, our primary objective has been to bridge the gap between theory and application, enabling students to appreciate not only the scientific principles but also their real-world significance. We hope that this book will not only serve as an academic resource but also inspire curiosity and critical thinking in young learners.

Dr Shramila Yadav
Dr Meenakshi Gupta

Contents

About the Editors

Dr Shramila Yadav is an associate professor in the Department of Chemistry at Rajdhani College, University of Delhi. She completed her graduation, post-graduation, M.Phil., and Ph.D. from the University of Delhi. With over 15 years of teaching experience at the undergraduate level, she has taught students across courses, including B.Sc. (H) chemistry, B.Sc. (physical science), and B.Sc. (applied physical science). Her specialization is in physical chemistry and she had completed her Ph.D. in corrosion chemistry. Her research interests include electrochemistry, surface chemistry, and nanomaterials. She has authored numerous research articles in reputed international peer-reviewed journals and has contributed several chapters to academic books, demonstrating her dedication to advancing research and sharing knowledge in her field.

Dr Meenakshi Gupta has been serving as an associate professor in the Department of Chemistry at Atma Ram Sanatan Dharma (ARSD) College, University of Delhi, for the past 17 years. She has taught undergraduate courses including B.Sc. (H) chemistry, B.Sc. (physical science), and B.Sc. (applied physical science). She completed her graduation, post-graduation, and Ph.D. at Delhi University, earning gold medals in both graduation and post-graduation. Her specialization is in physical chemistry and she had pursued her Ph.D. in corrosion chemistry. Her research interests include surface chemistry, electrochemistry, and computational chemistry. She has guided several DBT-funded projects with undergraduate students and has authored research articles in reputed international peer-reviewed journals. Additionally, she contributed two chapters to the edited book *Analytical Methods in Chemical Analysis* published by De Gruyter, demonstrating her commitment to academic writing and knowledge dissemination.

https://doi.org/10.1515/9783112208236-204

List of Contributors

Chanchal Singh
Department of Chemistry
D.S.N. (P.G.) College
C.S.M. University
A.B. Nagar
Unnao 209801
Uttar Pradesh, India
premchanchal@gmail.com
Chapter 1

Jaspreet Kaur
Hindu College
University Enclave
North Campus
Delhi 110007, India
Jaspreet.kutaal@gmail.com
Chapter 2

Mukesh
Keshav Mahavidyalya
Near Sainik Vihar, Zone H-45
Co-operative Group Housing Societies
Pitampura
Delhi 110034, India
mukeshgupta.chem@gmail.com
Chapter 2

Mohan Kumar
M.M.H. College
Ghaziabad 201009
Uttar Pradesh, India
kumar.mohan46@gmsail.com
Chapter 3

Ravi Kumar Thakur
Department of Chemistry
Maharshi Vishwamitra College
Buxar 802101
Bihar, India
ravikumarthakr@gmail.com
Chapter 3

Madhuri Chaurasia
Daulat Ram College
University of Delhi
Delhi 110007, India
madhurichaurasia2011@gmail.com
Chapter 3

Vishnu Kumawat
Department of Chemistry
Atma Ram Sanatan Dharma College
University of Delhi
Dhaula Kuan
Delhi 110021, India
vkumawat@arsd.du.ac.in
Chapter 4

Subash Chandra Mohapatra
Department of Chemistry
Atma Ram Sanatan Dharma College
University of Delhi
Dhaula Kuan
Delhi 110021, India
scmohapatra@arsd.du.ac.in
Chapter 4

Richa Tyagi
Department of Chemistry
Shyam Lal College
University of Delhi
Shahdara
Delhi 110032, India
dr.richa.tyagi5@gmail.com
Chapter 5

Rajni Grover
Department of Chemistry
Rajdhani College
University of Delhi
Raja Garden
Delhi 110015, India
rajni.grover@rajdhani.du.ac.in
Chapter 5

https://doi.org/10.1515/9783112208236-205

Sujata Kumari
School of Basic and Applied Sciences
K.R. Mangalam University
Sohna 122103
Haryana, India
sujata.kumari@krmangalam.edu.in
Chapter 6

Pratibha Sharma
School of Basic and Applied Sciences
K.R. Mangalam University
Sohna 122103
Haryana, India
pratibha.s@krmangalam.edu.in
Chapter 6

Deepak Yadav
School of Basic and Applied Sciences
K.R. Mangalam University
Sohna 122103
Haryana, India
deepak.y@krmangalam.edu.in
Chapters 6 and 7

Harsh Kumar Rai
Netaji Subhas University of Technology
Sector-3, Dwarka
Delhi 110078, India
harshkumarrai02@gmail.com
Chapter 7

Nilesh Singh
Netaji Subhas University of Technology
Sector-3, Dwarka
Delhi 110078, India
singhnilesh2121@gmail.com
Chapter 7

Rashika
Netaji Subhas University of Technology
Sector-3, Dwarka
Delhi 110078, India
rashika.phd23@nsut.ac.in
Chapter 7

Tannavi Badhan
Netaji Subhas University of Technology
Sector-3, Dwarka
Delhi 110078, India
tannavi.badhan.phd23@nsut.ac.in
Chapter 7

Sushmita
Netaji Subhas University of Technology
Sector-3, Dwarka
Delhi 110078, India
yadavsushmita60@gmail.com
Chapter 7

Neeta Azad
Department of Chemistry
Atma Ram Sanatan Dharma College
University of Delhi
Dhaula Kuan
Delhi 110021, India
neetaazad@aesd.du.ac.in
Chapter 8

Namita Khandpur Johar
Department of Chemistry
Maharaja Agarsen Institute of Technology
GGSIPU
Rohini
Delhi 110086, India
namitajohar1979@gmail.com
Chapter 8

Abbreviations

ABS	Acrylonitrile butadiene styrene
AFM	Atomic force microscopy
API	American Petroleum Institute
CCC	Critical coagulation concentration
CMC	Critical micelle concentration
$\overline{D_p}$	Degree of polymerization
DLVO	Derjaguin-Landau-Verwey-Overbeek
DLS	Dynamic light scattering
DMSO	Dimethyl sulfoxide
DMDHEU	Dimethyloldihydroxyethyleneurea
DNA	Deoxyribonucleic acid
EDL	Electric double layer
ELS	Electrophoretic light scattering
EOF	Electrostatic flow
EOR	Enhanced oil recovery
ESP	Electrostatic precipitators
H-Bonding	Hydrogen bonding
HDPE	High-density polyethylene
HEC	Hydroxyethyl cellulose
HEUR	Hydrophobically modified ethoxylated urethane
HLB	Hydrophile-lipophile balance
IEP	Isoelectric point
IHP	Inner Helmholtz plane
LB film	Langmuir-Blodgett film
LCST	Lower critical solution temperature
LDPE	Low-density polyethylene
Melmac	Melamine formaldehyde resin
M.W.	Molecular weight
O/W	Oil-in-water
O/W/O	Oil-water-oil
OLED	Organic light-emitting diodes
OHP	Outer Helmholtz plane
PDI	Polydispersity index
PE	Polyethylene
PEDOT:PSS	Poly(3,4-ethylenedioxythiophene): polystyrene sulfonate
PEEK	Polyether ether ketone
PEG	Polyethylene glycol
PFG-NM	Pulsed field gradient nuclear magnetic resonance
PGA	Polyglycolic acid
PLA	Polylactic acid
PLGA	Poly(lactic-co-glycolic acid)
PNIPAM	Poly(*N*-isopropylacrylamide)
PP	Polypropylene
PS	Polystyrene
PTFE	Polytetrafluoroethylene
PU	Polyurethanes
PVA	Polyvinyl alcohol

https://doi.org/10.1515/9783112208236-206

PVC	Polyvinyl chloride
QCM-D	Quartz crystal microbalance with dissipation
SAMs	Self-assembled monolayers
SANS	Small-angle neutron scattering
SAXS	Small-angle X-ray scattering
SBR	Styrene butadiene rubber
SLS	Static light scattering
SPR	Surface plasmon resonance
TSS	Total suspended solids
UCST	Upper critical solution temperature
UHMWPE	Ultra-high-molecular-weight polyethylene
VOC	Volatile organic compound
W/O	Water-in-oil
W/O/W	Water-oil-water
ZNC	Ziegler-Natta catalyst

Symbols

K_b	Boltzmann constant
ΔH	Change in enthalpy
ΔS	Change in entropy
μ	Coefficient of friction
C	Concentration of the polymer
ξ	Correlation length
Φ_c	Critical composition
T_c	Critical temperature
κ	Debye length
D	Diffusion coefficient
ΔS_{mix}	Entropy of mixing
χ	Flory-Huggins interaction parameter
G	Gibbs free energy
H	Helmholtz free energy
R_H	Hydrodynamic radius
yow	Interfacial tension of oil
ysl	Interfacial tension of solid
r	Intermolecular separation
U	Internal energy
KB	Kirkwood-Buff
KBI	Kirkwood-Buff integral
Z	Lattice coordinate
p	Mode index
µm	Micrometer
M	Molecular weight
r_n	Motion associated with the *n*th bead
nm	Nanometer
N	Number of monomers
n_P	Number of polymer molecules
n_S	Number of solvent molecules
Π	Osmotic pressure
c^*	Overlap concentration
RDF	Radial distribution function
s	Radius of gyration
f_n	Random force
R_Θ	Rayleigh ratio or refractive index increment
τ_p	Relaxation time
τ	Reptation time
s_{rms}	Root mean square of speed
γ	Surface tension
π	Surface pressure
R_s	Stokes' radius
T	Temperature
L	Tube length
μ_{tube}	Tube mobility
R	Universal gas constant
v	Velocity of the polymer chain

https://doi.org/10.1515/9783112208236-207

B_2 and B_3	Virial coefficients
H	Viscosity
η_s	Viscosity of solvent
Φ	Volume fraction of the polymer
$(1-\Phi)$	Volume fraction of the solvent
V_{site}	Volume per site
λ	Wavelength
ζ	Zeta potential

Chanchal Singh

1 Introduction to Polymers

Abstract: Polymers are described as very large molecules composed of numerous repeating units called monomers. The term "poly" means many, and "mers" means units, thus polymers. The process of joining several monomeric units is called polymerization. Since they consist of many units, polymers are substances of high molecular weight. Polymers occur both naturally and synthetically. The examples of naturally occurring polymers are rubber, wool, and silk, while polyethylene, nylon, polyvinyl chloride, and Bakelite are synthetic polymers. Polymers find their applications in various fields, such as fabrics for clothing, furniture, packaging, transportation, insulation, and storage. Polymers may be classified by their origin, by structure, by the nature of monomers, and by method of synthesis.

1.1 Introduction

A polymer is a high-molecular-weight organic molecule fabricated by smaller units called monomers. These monomeric units are covalently bonded to each other in a repetitive pattern all over the chain to form a macromolecule, often measured in the hundreds or thousands. In addition to strong covalent bonds within the polymer chain, macromolecules may also exhibit intermolecular interactions such as hydrogen bonding and weak van der Waals forces [1]. One of the most famous examples of a polymer is polythene, which is manufactured using ethene as a monomer:

$$\text{Ethene} \xrightarrow{\text{many units}} \text{Polyethene}$$

$$\text{Monomer} \xrightarrow{\text{many units}} \text{Polymer}$$

Polymer means many repeat units ("mer" from the Greek word meros, meaning part). The most naturally occurring constituents, such as cellulose, rubber, silk, and cotton, have been employed in tools and textiles since ancient times. Naturally occurring polymers often have limitations in terms of stability and strength. For example, natural rubber has poor resistance to heat and reacts easily with ozone due to the presence of double bonds in its backbone. To overcome these drawbacks, Charles Goodyear developed the process of vulcanization in 1839, which involves adding sulfur to rubber. This treatment enhances its durability and improves resistance to heat. The history of synthetic polymers began with the invention of Bakelite in 1907 by Leo Baekeland, the first fully synthetic plastic [2]. Later, in 1935, the discovery of nylon revolutionized the textile industry by providing an alternative to traditional natural fibers. A major breakthrough came with the discovery of the Ziegler-Natta catalyst by chemists Karl

https://doi.org/10.1515/9783112208236-001

Ziegler and Giulio Natta, which enabled the controlled synthesis of synthetic polymers and earned them the Nobel Prize in Chemistry [1].

Since then, the polymer industry has developed into a full-fledged branch of science called polymer science, which is interdisciplinary than most sciences, combining chemistry, chemical engineering, and materials science to transform raw materials into versatile polymers. These synthesized macromolecules are superior in strength, flexible, and lightweight over natural polymers.

1.2 Classification of Polymers

Polymers can be categorized in several ways depending on their source, structure, monomer type, and method of synthesis.

1.2.1 Classification Based on Origin/Source

According to their origin, polymers are broadly divided into two categories:

1. Natural polymers: There are natural organic polymers and natural inorganic polymers, here, only organic polymers will be discussed. Natural polymers are found in both plants and animals. Most natural polymers are proteins, DNA, and RNA, which are essential for life processes, like keratin in hair and collagen in skin, whereas cellulose in plants provides the mechanical basis, and starch serves as a food source. One of the well-known natural polymers is natural rubber (polymer of isoprene) harvested from the plant *Hevea brasiliensis* [2]. The latex collected from the plant is a white colloid that is sticky and milky in nature. It has a wide range of uses in refrigeration seals, air conditioning units, cooling systems, automotive gaskets, and weather stripping.

2. Synthetic polymers: These polymers are synthesized in a laboratory, for example, plastic is a man-made synthetic polymer. They have various advantages, such as durability, flexibility, lightweight, chemical resistance, water resistance, thermal insulation, electrical insulation, versatility, and cost-effectiveness. Examples include nylon, polyvinyl chloride (PVC), polystyrene, Teflon, polyesters, and polyethylene [2].

1.2.2 Classification Based on Structure

Based on structural differences, polymers are classified into three types:

1. Linear polymers: These polymers contain an uninterrupted straight and long chain joined together end to end by van der Waals interaction and H-bonding, as represented in Figure 1.1. A linear polymer can have side groups called pendant groups, which are not considered as side chains. Examples include Teflon, polypropylene (PP), and polystyrene.

Figure 1.1: Linear arrangement of monomeric units.

The arrangement of pendant groups along the backbone of a polymer chain is described by the concept of tacticity, which refers to the stereochemical regularity of a polymer chain. Depending on how the side groups (pendant groups) are oriented, polymers can exhibit different degrees of order [1]. The three main types of tacticity are:

(i) Isotactic polymers: All pendant groups are located on the same side of the polymer backbone. Such polymers generally exhibit semicrystalline behavior, leading to higher strength and rigidity:

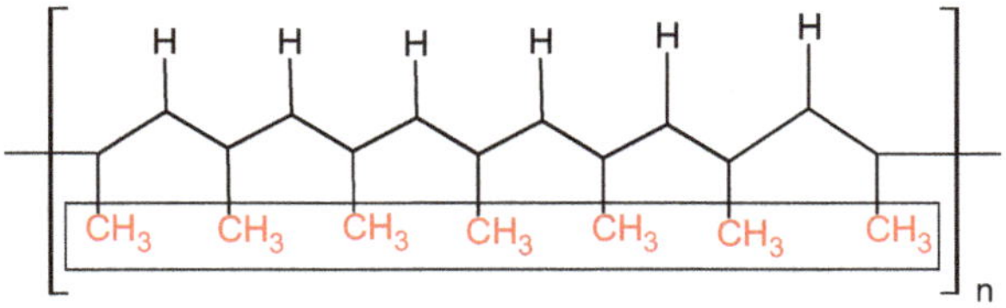

(ii) Syndiotactic polymers: Pendant groups are arranged in an alternating pattern along the chain. These polymers are often crystalline, making them strong and thermally stable:

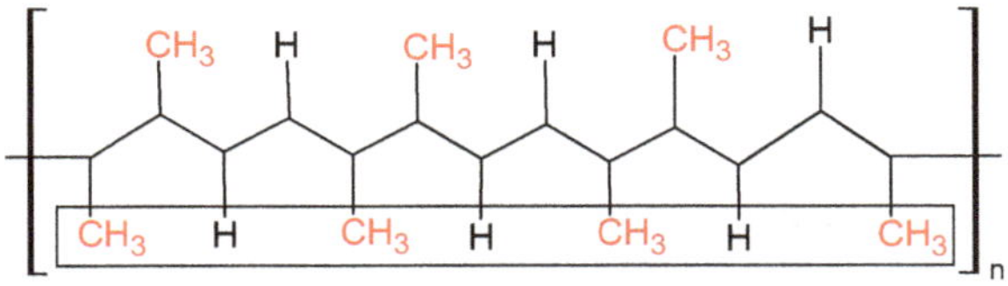

(iii) Atactic polymers: Pendant groups are oriented in a random manner without any regular arrangement. Atactic polymers are typically amorphous, soft, and flexible in nature:

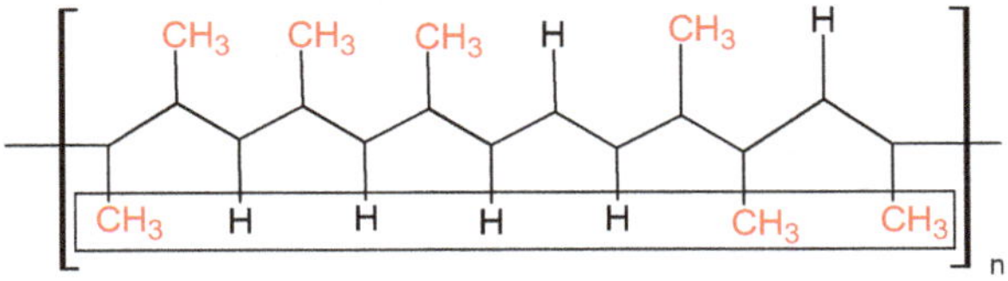

2. Branched-chain polymers: These polymers contain one main linear chain with many side branches attached to the main chain, as depicted in Figure 1.2. Branching in a polymer chain affects the properties, for example, the shorter branched chains interfere with efficient packing, thus making them less dense and decreasing melting and boiling points than linear polymers. Branching in a polymer chain allows the polymer to be a thermoplastic.

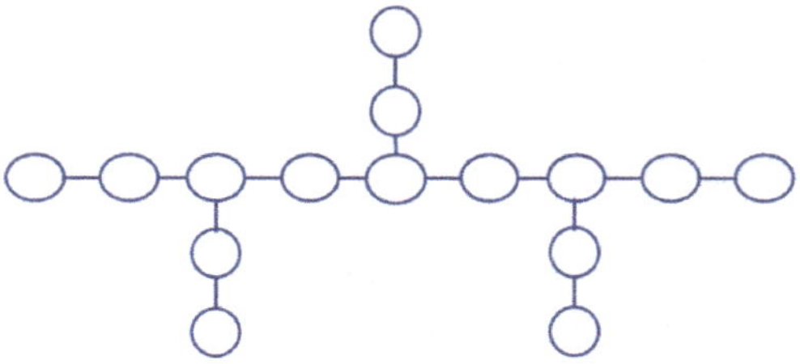

Figure 1.2: Side branches attached to the main chain.

3. Cross-linked polymers: Cross-linked polymers, shown in Figure 1.3, are formed when many linear polymer chains are interconnected through covalent bonds, known as cross-links. Monomers used in the manufacturing of these polymers are bi-functional and trifunctional in nature; the added functionality produces the cross-links. Most cross-linked polymers are thermosetting, with only a few exceptions.

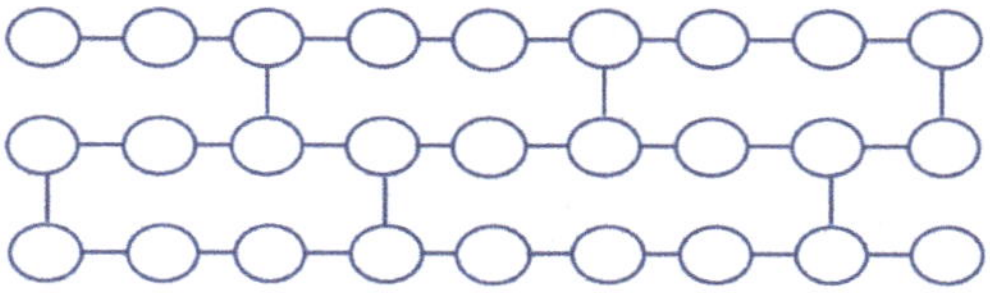

Figure 1.3: Cross-linking of many linear chains.

1.2.3 Classification Based on the Nature of Monomers

Based on the monomer used, polymers are classified into three types:
1. Homopolymers: These polymers contain only one type of monomer, e.g., polyethene, PVC, and nylon-6, etc. They are represented as $-(A-)_n$ and shown in Figure 1.4.

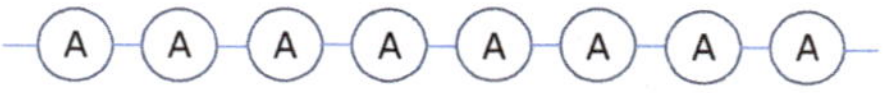

Figure 1.4: Single monomeric unit in a linear chain.

2. Copolymers: This polymer has more than one type of monomer bonded together by covalent bonds in a long, strong polymer chain. Copolymers exhibit different arrangements of monomers in the chain, and they can be differentiated based on monomer placement in the polymer chain with respect to each other [3]. These are:

(i) Alternate/periodic copolymers: When monomeric units are placed in alternating order (ABABAB), a regular order is followed throughout the chain, and exhibits an accurate sequence of monomer units. They are represented in Figure 1.5 as –(A–B)$_n$.

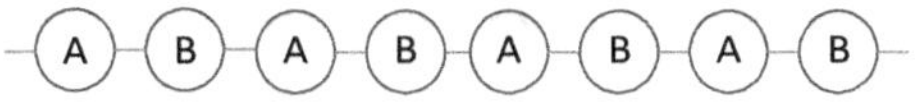

Figure 1.5: Two monomeric units in a linear chain.

(ii) Block copolymers: When a block of monomeric units is placed in alternating order (AAAABBBB), each block consists of a series of repeating monomer units different from the neighboring blocks. A regular order is followed throughout the chain. They are represented in Figure 1.6.

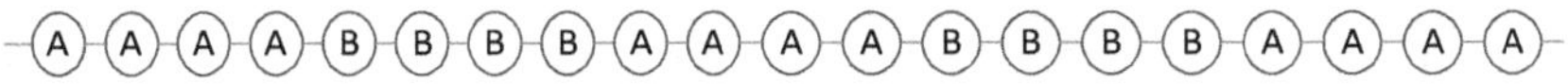

Figure 1.6: Repeated arrangement of two monomeric units in a linear chain.

(iii) Graft copolymers: One monomeric unit forms the main chain while another monomeric unit forms the branches. The chemical properties and structure of these branches are different from the main chain. Graft copolymer is always branched in structure. They are represented in Figure 1.7.

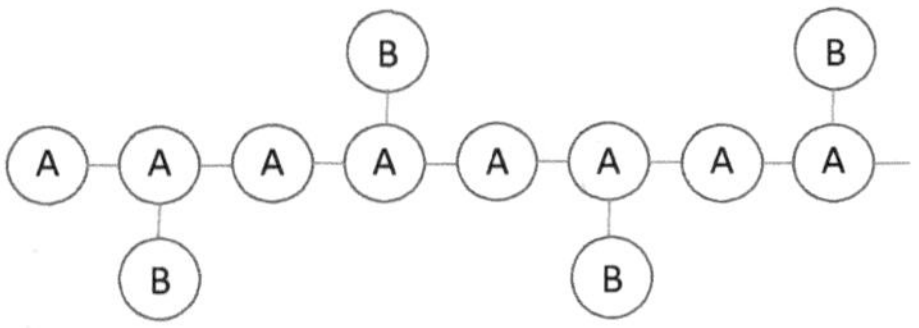

Figure 1.7: Side chain of different monomeric units attached to the main chain.

(iv) Random copolymers: In random copolymers, there is no regular or systematic arrangement of the monomer units along the polymer chain, which can be understood in Figure 1.8. Instead, the different monomers are distributed in a completely random manner, without any fixed sequence. Such polymers are usually formed when two or more monomers with comparable reactivity ratios undergo polymerization.

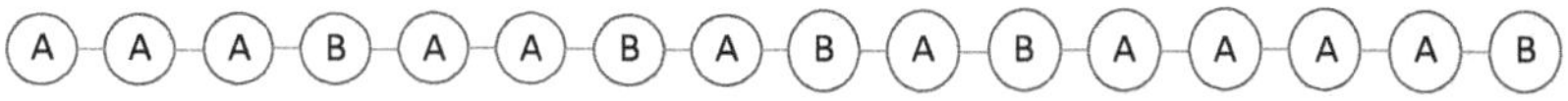

Figure 1.8: Random arrangement of two monomeric units.

1.2.4 Classification Based on Thermal Behavior

Polymers show different responses when exposed to heat, and this forms the basis of their classification by thermal behavior. Such classification helps in understanding their processing, stability, and suitability for various applications.

They are classified as:

(i) Thermoplastic polymers
(ii) Thermosetting polymers

(i) Thermoplastic polymers: They can be shaped into various forms, as they are soft and moldable without significant chemical changes when heated. A thermoplastic polymer comprises individual chains that are bound together by van der Waals forces. Thermoplastic polymers have both ordered crystalline regions and amorphous noncrystalline regions (Figure 1.9). They offer many advantages, such as being meltable, chemically resistant in nature, have a high strength-to-weight ratio, are transparent or translucent, and provide electrical insulation. Examples include polyethylene (low-density polyethylene [LDPE] and high-density polyethylene [HDPE]), PVC, polyethylene terephthalate, polyamide, and polystyrene [4].

(ii) Thermosetting polymers: They contain strong cross-links between the polymer chains (Figure 1.10), which are caused by chemical reactions during the curing process, creating a rigid and durable material. Thermosetting polymers owe their strength to strong covalent linkages (cross-linking) between polymer chains, which results in a more brittle permanent shape. They cannot be recycled after they are hardened, can endure high temperatures without melting or softening, have superior chemical resistance, and excellent dimensional stability compared to thermoplastics [5]. Examples are epoxy resin, phenolic resin, polyurethane (PU), melamine formaldehyde resin, and Bakelite (phenolic formaldehyde).

1.2.5 Classification Based on Synthesis

Copolymers are synthesized by using two or more different monomers. Addition and condensation polymerization are two major methods used for the synthesis of a variety of polymers [1, 6].

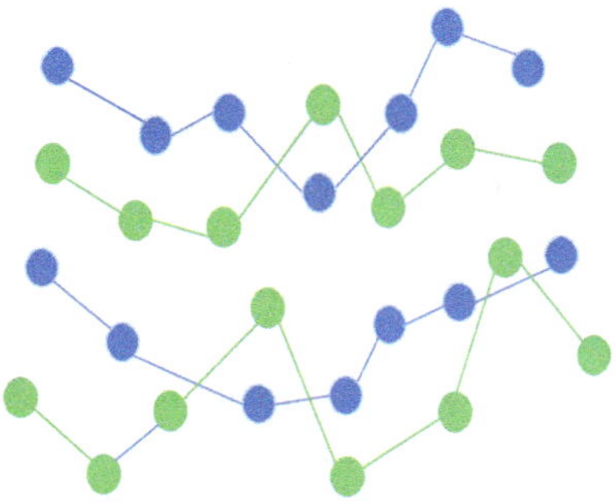

Figure 1.9: Polymer chain arrangement in a thermoplastic polymer.

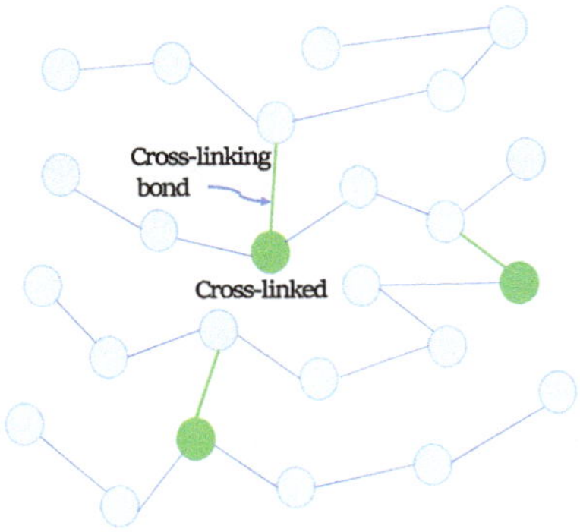

Figure 1.10: Cross-linking of polymer chains in a thermosetting polymer.

1.2.5.1 Addition Polymerization

Monomers with carbon-carbon double bonds are connected to each other to form a chain. It involves cationic, anionic, and free radical polymerization. Initiators are commonly used to start the polymerization process. Polymers produced by an addition reaction are called addition polymers. Monomers containing one or more double bonds are added in a sequential order to each other without the elimination of any by-products. For example, the formation of polythene and polyacrylonitrile:

Ethene Polythene

Polyacrylonitrile

Addition polymerization involves three steps:

First step is chain initiation, followed by chain propagation, and lastly, chain termination step takes place. Addition polymerization occurs via three different types of mechanisms: (i) free radical mechanism, (ii) cationic mechanism, and (iii) anionic mechanism:

(i) Free radical mechanism: It is a type of chain growth polymerization in which polymer chains are formed through the successive addition of monomer units containing carbon-carbon double bonds. Since the process primarily involves unsaturated monomers, such as alkenes, it is also commonly referred to as vinyl polymerization [1]. Polythene, PVC, and polystyrene are some of the common examples. Polymers obtained by the addition growth process of repeating monomeric units have a molecular weight of multiple monomeric units added to the polymer chain. The three steps involved are:

Step 1: Initiation

The reaction is initiated by light, heat, or by free radical-inducing species such as benzoyl peroxide and azobisisobutyronitrile. Benzoyl peroxide breaks down into benzene free radical:

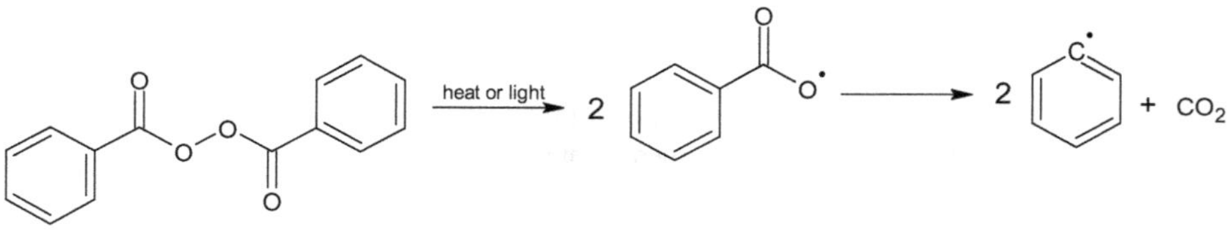

Azobisisobutyronitrile breaks down to 3^0 free radicals:

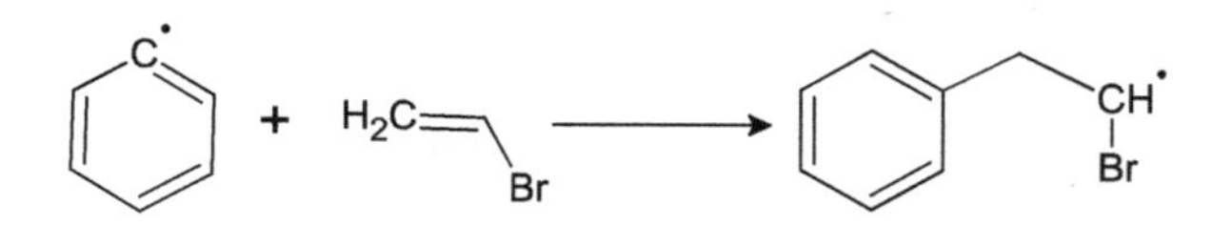

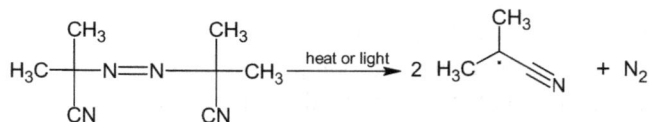

In the first step, the initiator is dissociated into a free radical, and in the second step, it adds to the monomeric unit to give monomeric free radical species:

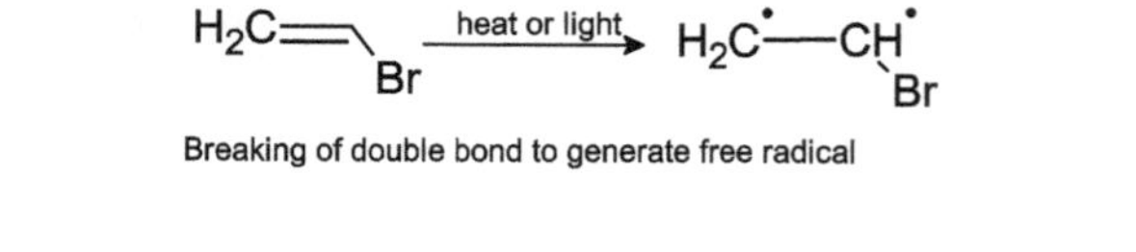

addition of free radical initiator to monomeric unit

Breaking of double bond to generate free radical

Step 2: Propagation

It involves the continuous addition of free radical species to another monomeric unit without the loss of any small molecule to propagate the active free radical center for the chain growth process. The free radical always attacks on methylene carbon to give a secondary free radical:

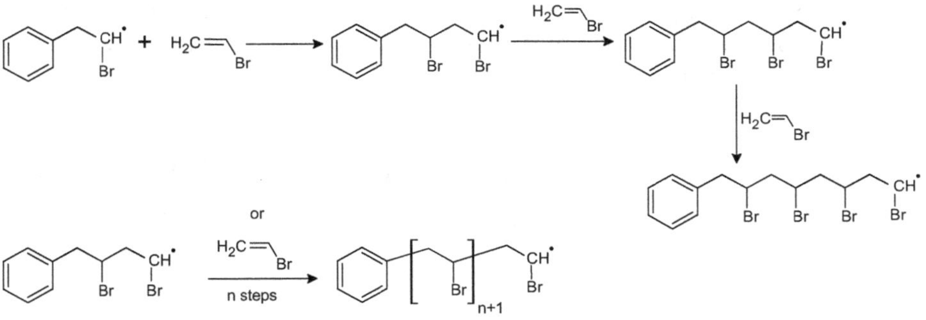

Similarly, a double free radical after breaking of double bond can propagate from both sides.

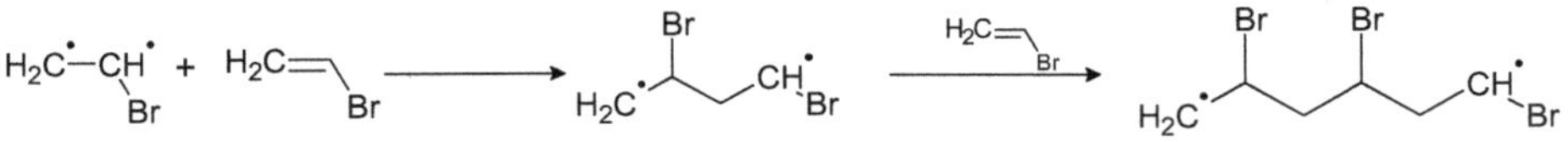

Step 3: Termination

The termination step in free radical polymerization usually occurs when two active radical species combine, thereby neutralizing each other and stopping further chain growth. This can take place through:

(i) Coupling (combination): Two growing radical chains join together to form a single, longer polymer chain.

(ii) Disproportionation: A hydrogen atom is transferred from one radical chain to another, producing two stable polymer chains without radicals.

In some cases, the presence of impurities or additives in the monomer system can act as chain transfer agents. These species interact with the growing radical chain, effectively terminating its growth. As a result, the overall molecular weight of the polymer is reduced, and a new radical species may be generated that can initiate another chain:

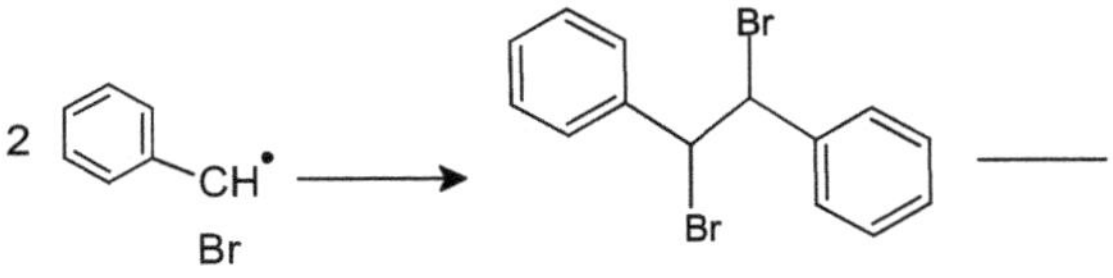 (i)

Coupling of free radical which terminate the reaction

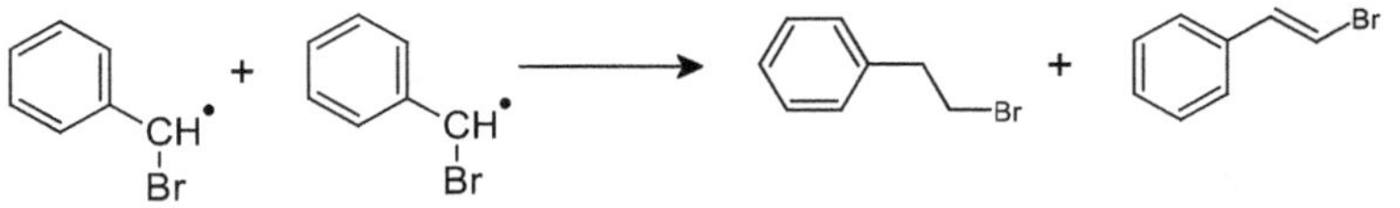

(ii)

Disproprotionation of free radical to give one satutated and one unsaturated unit

(iii) Chain transfer: A growing polymer abstracts an atom from RSH (chain transfer agent) to terminate the growth process and produce a new free radical, which can initiate a new polymerization chain:

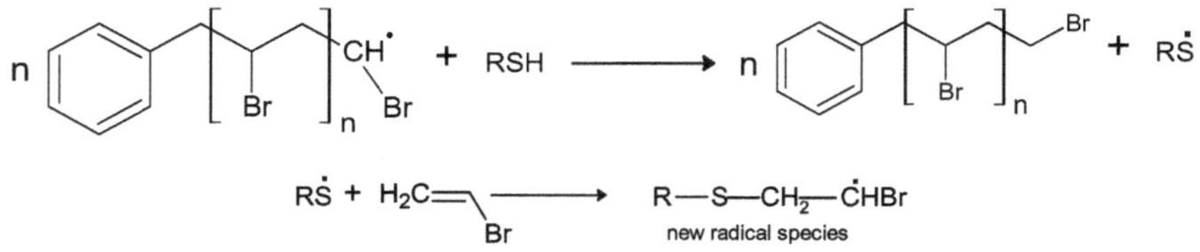

A monomer may also act as a chain transfer agent. The reaction below clearly shows how a monomer gives rise to a dead polymer and a new free radical. This new radical can initiate the formation of a fresh polymer chain:

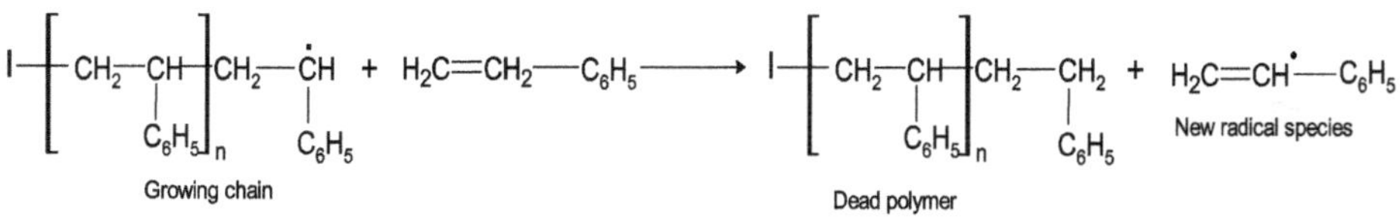

Branching due to chain transfer reaction, the new radical generated gives rise to a branched chain polymer:

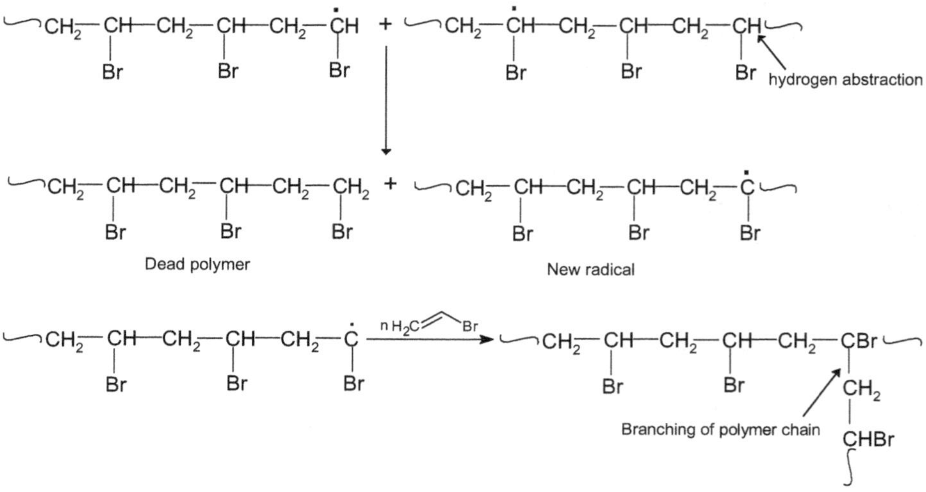

Hydrogen abstraction from different molecules (intermolecular) leads to a long-branched chain polymer. However, hydrogen abstraction from the same molecule (intramolecular) leads to a short-branched chain polymer and cross-linking:

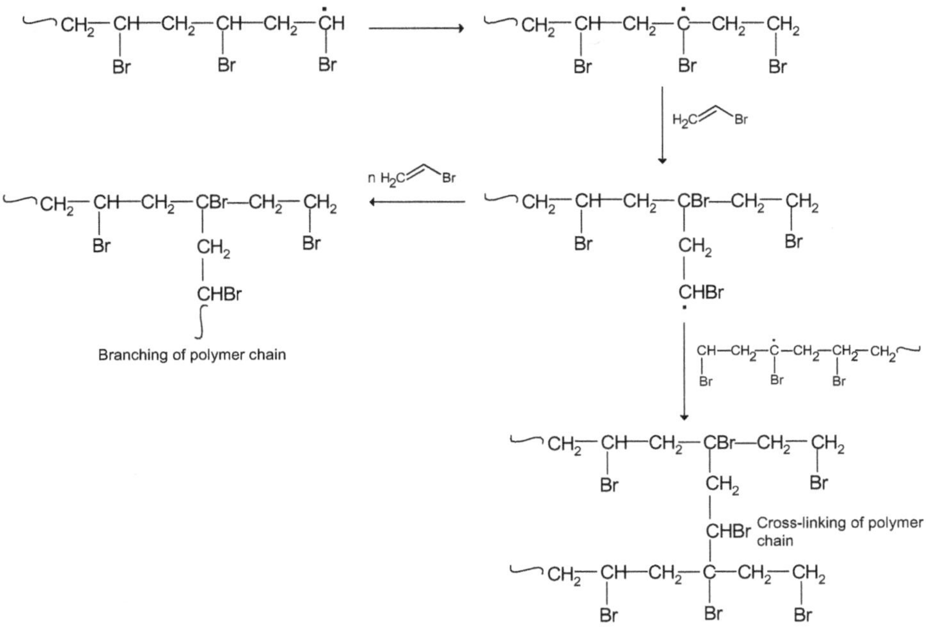

(ii) Cationic polymerization mechanism: Cationic polymerization comprises three steps, which involve a carbocation as an intermediate. The reaction is initiated by an electrophile and terminated by a nucleophile. It can be summarized as follows: the initiator is an electrophile, the intermediate is a carbocation, and the inhibitor (terminator) is a nucleophile. Lewis acids such as BF_3, $TiCl_4$, $AlCl_3$, and $SnCl_4$ are used as initiators. Usually, the Lewis acids require water or methanol as a cocatalyst, suggesting that the reactive initiator is actually a protic acid. Protonic acids, such as H_2SO_4 and $HClO_4$, which are used as initiators and involve the addition of a proton (H) to monomer [1, 2]. The mechanism can be shown as:

Step 1: Initiation

$$BF_3 + H_2O \rightleftharpoons BF_3.H_2O$$

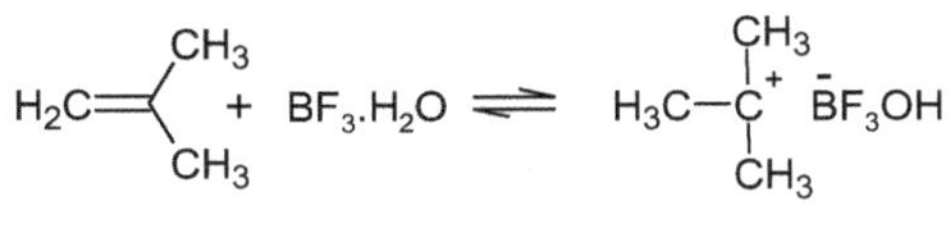

Step 2: Propagation

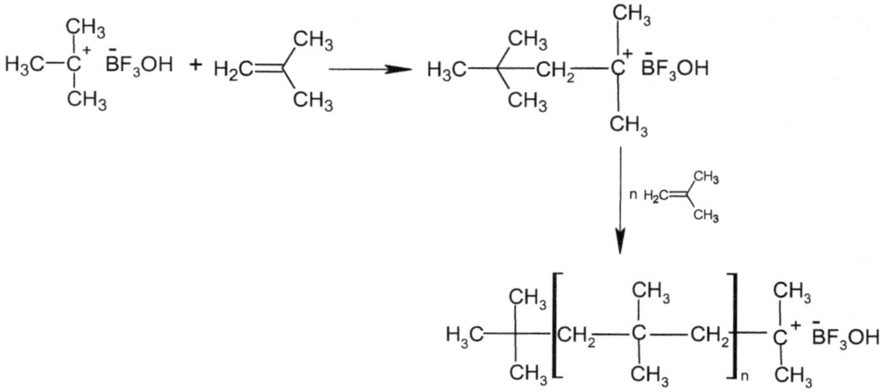

Step 3: Termination

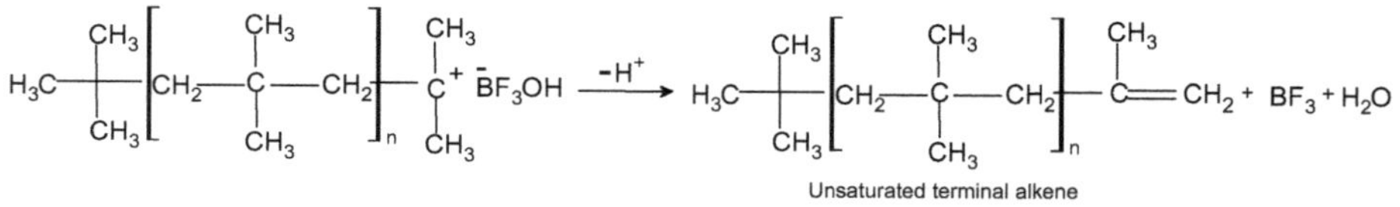

Termination of the polymeric chain occurs via the rearrangement and formation of an unsaturated terminal (double bond) unit.

(iii) Anionic polymerization mechanism: Anionic polymerization involves the repetitive addition of monomers to a propagating anionic species. Strong bases, such as *n*-butyllithium and lithium amide, act as initiators, they both are strong nucleophiles, which is a requisite condition to attack the electron-rich olefins. Electron-withdrawing substituents on the olefin bond increase the rate of addition. The anionic polymerization is very fast.

Step 1: Initiation

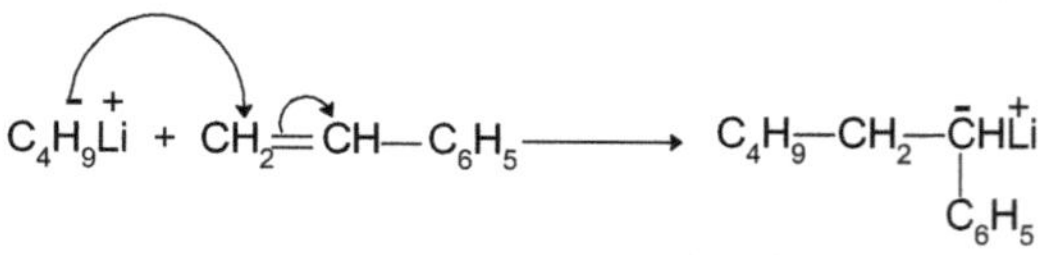

Step 2: Propagation

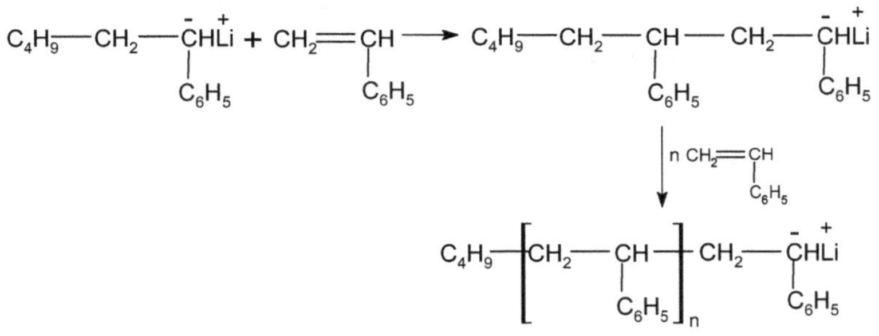

Step 3: Termination
Termination occurs by a halide transfer by solvent medium or by reaction of generated nucleophiles with impurity:

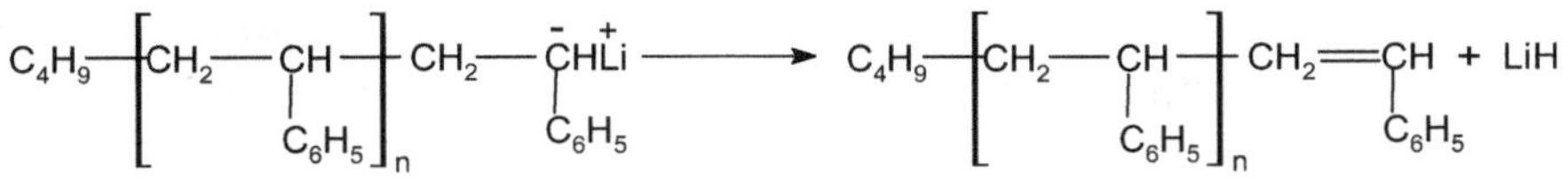

1.2.5.2 Condensation Polymerization

Polymers that are generated by the condensation of two monomer units comprising two functional groups, with the loss of small molecules such as H_2O, HCl, and NH_3. It is also known as step-growth polymers. Condensation polymerization reaction is slower than the addition polymerization process, it occurs in the presence of heat and a catalyst, and due to the loss of small molecules, they have low molecular weight. Polar functional groups present on the chains give rise to hydrogen bonding, thereby increasing crystallinity and tensile strength [6].

1.2.5.2.1 Examples of Condensation Polymer

(a) Polyesters (ethylene terephthalate) or Dacron-Mylar: Coupling of dimethyl esters of terephthalic acid with ethylene glycol by the transesterification process gives rise to a polyester molecule [7]:

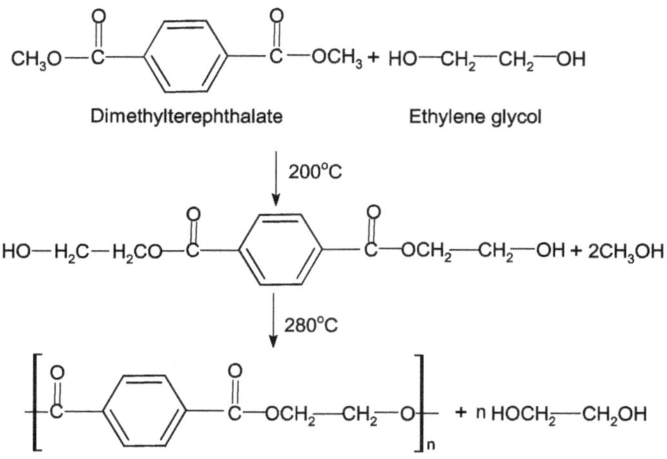

Polyesters are most commonly synthesized through a condensation reaction in which a single molecule is repeated thousands of times to form a long polymeric chain. Polyester is used in products such as disposable enamel paints, soft-drink bottles, rubber tires, compact discs, etc. Dacron and Mylar have almost the same composition, which finds its usage in clothing, packaging, magnetic recording, tape, etc.:

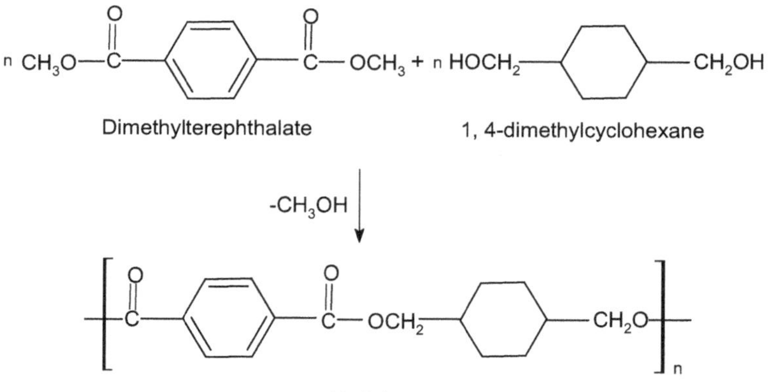

Kodel

(b) Poly(ethylene phthalate) alkyld resin: Coupling of poly (ethylene phthalate) and phthalic anhydride with condensation reaction results in alkyld resin, which is linear in structure, used in paints and lacquers.

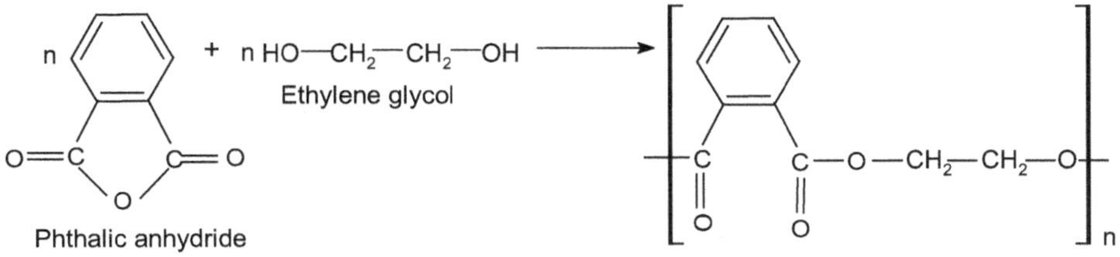

(c) Polyamides: Polyamides contain an amide group, formed by the condensation polymerization of a carboxylic group and an amino group [8]. A long, repeated chain of amide (-CO-NH-) networks gives rise to a polyamide polymer. Polyamides are classified as natural polyamides, such as proteins, wool, and silk, whereas synthetic polyamides comprise nylons, aramids, and sodium polyaspartate. Polyamides can be distinguished into three families: (i) aliphatic polyamides, (ii) semiaromatic polyamides, and aromatic polyamides (aramids). Polyamides find their application in a number of industries like textiles, 3D printing, engineering, automobiles, electronics, etc.

(i) Reaction of diamine and dicarboxylic acid:

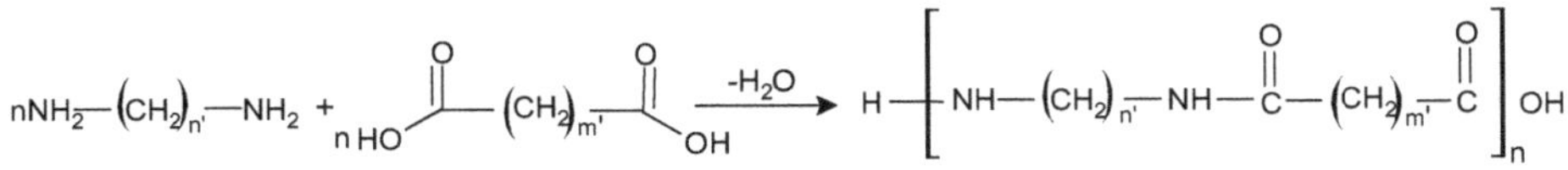

(ii) Self-condensation of amino carboxylic acid:

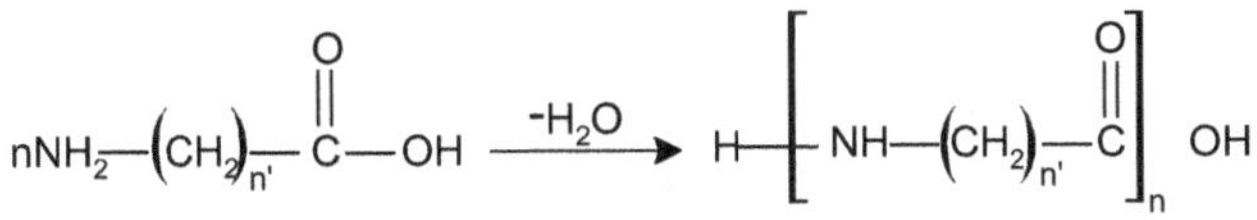

(iii) Ring-opening polymerization of a lactam:

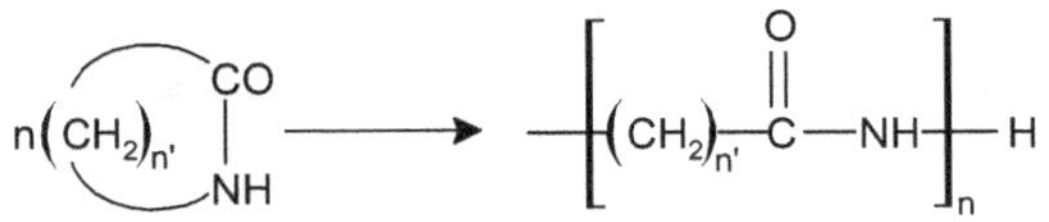

(d) Nylon 6: When caprolactam is heated to about 250 °C in the presence of water and acetic acid, it undergoes a ring-opening polymerization reaction. Caprolactam, which contains a six-carbon ring, opens up and links together to form long polymer chains of nylon 6. Nylon has good thermal stability, and can withstand elevated temperatures without undergoing significant decomposition. Nylon 6 is often used in zippers, fishing nets, parachute cords, plastic gears, and technical parts in machinery [9]:

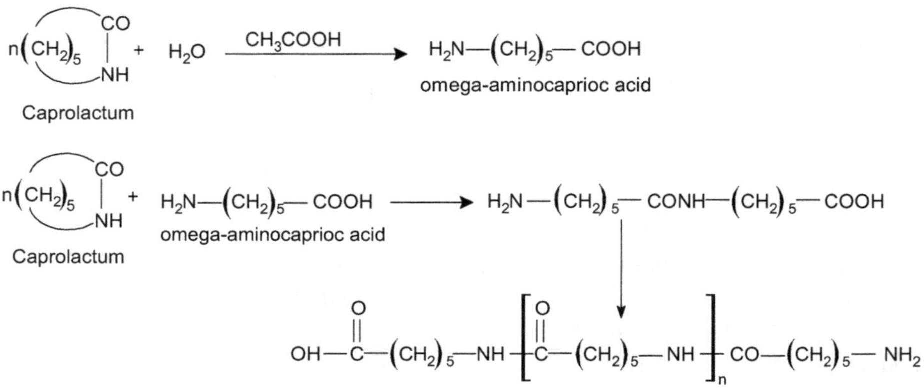

(e) Nylon 6, 6: Coupling of hexamethylene diamine and adipic acid by a condensation reaction yields nylon 6, 6. The nylon fibers are stretched into long lengths to produce a strong, elastic filament. One of the major advantages of nylon is its thermoplasticity, which permits everlasting waviness and texturing of the fibers, resulting in bulk and stretch properties [10]:

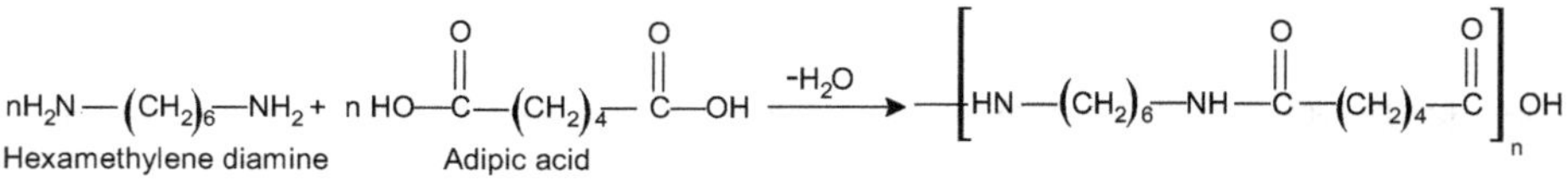

(f) Nylon 6, 10: Nylon 6, 10 is made of two monomers, one containing six carbon atoms and the other containing 10 carbon atoms. The six-membered monomer is hexamethylenenediamine and 10-membered is 1,10-decandioic acid. It has various advantages, such as low moisture absorption and good chemical resistance, and is used to make toothbrush bristles, engine covers, air intakes, and radiators [1]:

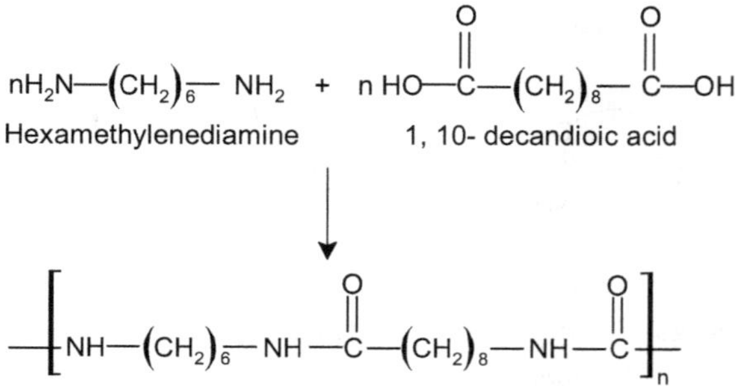

(g) Polyurethanes: Polyurethanes are plastic polymers made by combining diisocyanates and alcohols. Polyurethane is considered a polymer composed of a large number of urethanes. The urethane monomer is produced by the reaction between an isocyanate and an alcohol. When diisocyanates react with organic compounds that acts as prepolymer containing OH end groups (MW 1,000–2000), such as polyesters containing carboxyl groups, the result is the formation of foamed polyurethanes. During this process, bubbles of carbon dioxide are released and remain trapped within the material, giving rise to its characteristic cellular structure. Two types of polyurethane are thermosetting (heat-resistant) and thermoplastic (heat-sensitive in nature) [11, 12]:

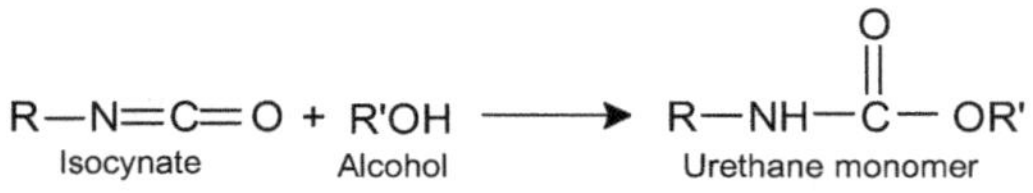

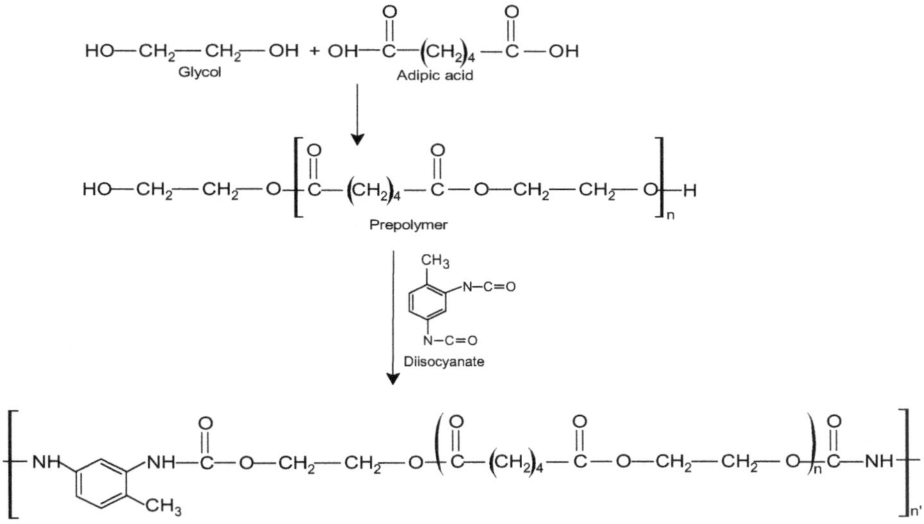

Formation of foamed polyurethane requires the bubbling of carbon dioxide (CO_2) in a reaction, which can be obtained by the addition of water into isocyanate, converting the isocyanate group to an amino group with the liberation of CO_2:

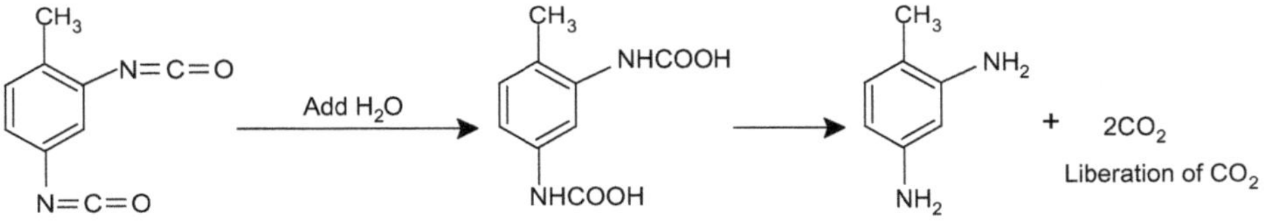

(h) Phenol-formaldehyde resins: Phenol-formaldehyde resins are among the earliest synthetic polymers, prepared by the reaction of phenol with formaldehyde. When an excess of formaldehyde reacts with phenol in the presence of either an acidic or basic catalyst in water, the reaction produces a mixture of low-molecular-weight prepolymers known as hydroxymethylphenols. These prepolymers contain reactive ortho- and para-substituted groups on the aromatic ring, which later undergo further condensation reactions to form higher molecular weight resins [13]:

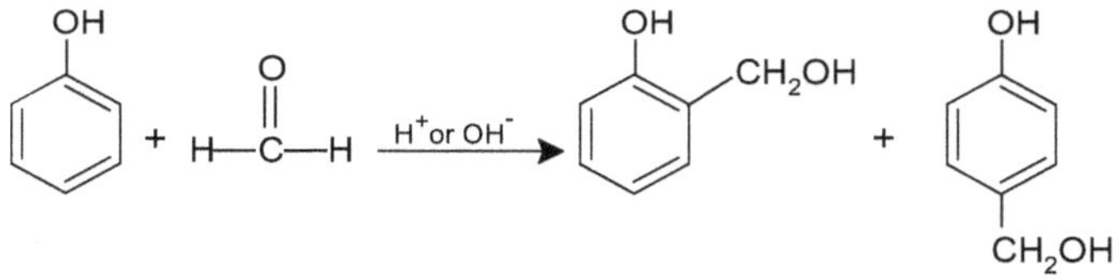

ortho and *para* hydroxymethylphenol

Ortho- and *para*-hydroxymethylphenol are more reactive than phenol toward formaldehyde, which rapidly undergo further substitution to form di and trisubstituted products:

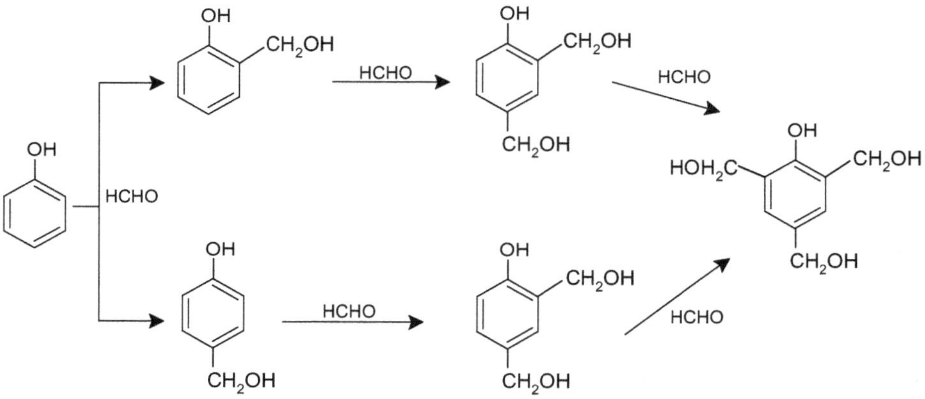

Ortho- and *para*-hydroxymethylphenol form dinuclear and polynuclear phenols linked together by methylene groups through a self-condensation reaction. Two types of resins can be synthesized depending on the ratio of phenol and hydroxymethylphenol, as shown below.

Phenol-formaldehyde resins are excellent wood adhesives for plywood due to their good moisture resistance. Phenolic resins are used in insulating and heat-resistant objects, such as appliance handles and brake linings.

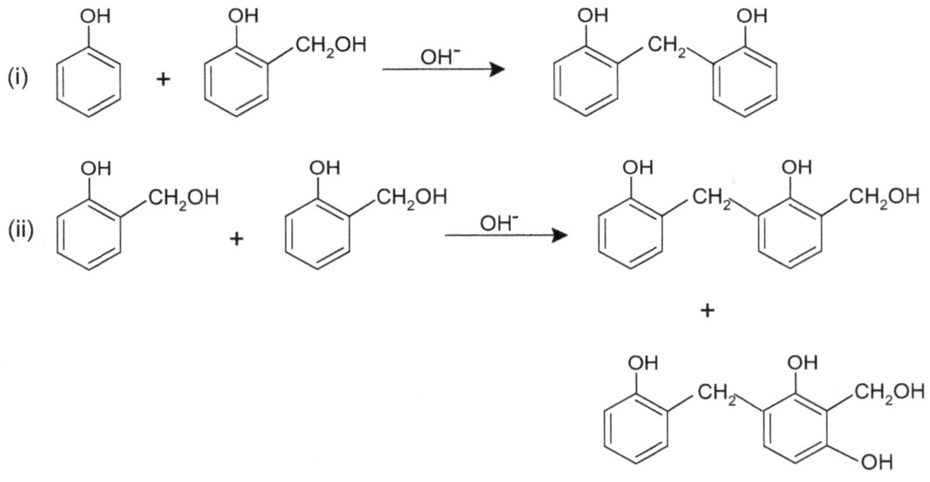

(i) Bakelite: Bakelite, chemically known as polyoxybenzylmethylenglycolanhydride, is a thermosetting phenol-formaldehyde resin, and is recognized as the first synthetic plastic. It is prepared by the condensation reaction of phenol with an excess of formaldehyde in the presence of a basic catalyst such as sodium carbonate (Na_2CO_3). Bakelite is a hard, stable, and chemically resistant material. Owing to its excellent electrical insulating properties and ease of molding, it has found widespread applications [14].

Applications of Bakelite include:

i. Electrical devices such as sockets, switches, insulation wires, and automobile distributor caps.
ii. Household articles like clocks, toys, and kitchenware.
iii. Components requiring durability, thermal resistance, and chemical stability.

Preparation of Bakelite involves several complex stages:

Phenol reacts with formaldehyde in the presence of either an acidic catalyst (such as HCl or $ZnCl_2$) or a basic catalyst (such as NH_3). The initial products are ortho-hydroxybenzyl and para-hydroxybenzyl alcohols, which subsequently undergo further condensation, forming compounds linked through $-CH_2-$ bridges (i) and (ii) (used in phenol-formaldehyde resin section). The condensation product formed is called Novolac [12]:

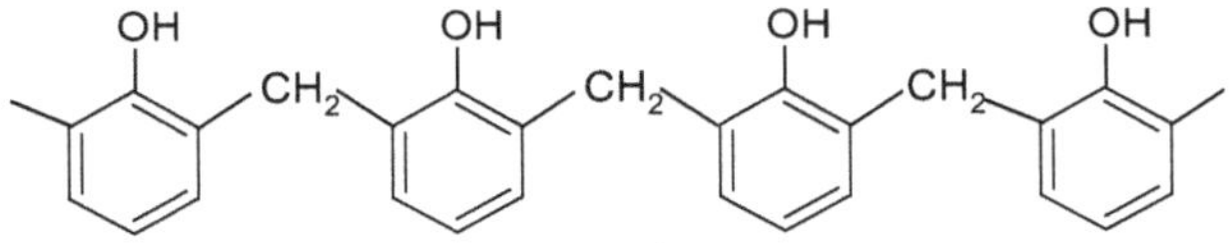

Novolac

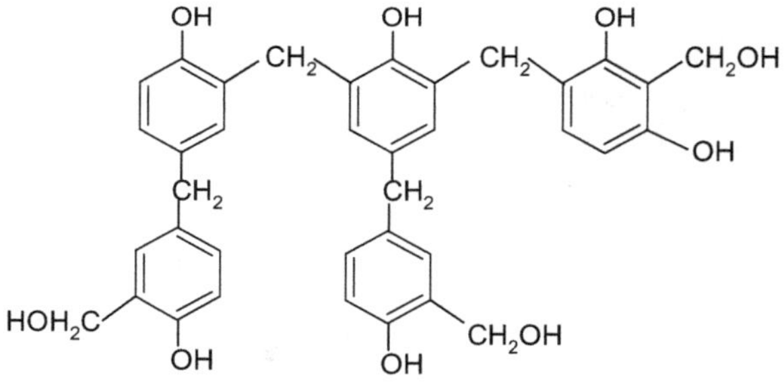

Resole resin

If the molar ratio of formaldehyde to phenol is less than 1 (excess phenol), and an acidic catalyst is used, the product is a novolac resin (as above). If the molar ratio of formaldehyde to phenol is greater than 1 (excess formaldehyde), and a basic catalyst is used, the condensation product is a resole resin. When novolac resin is heated repeatedly with hexamethylenetetramine (hexamine) or with excess formaldehyde, it undergoes extensive cross-linking. The final product is the thermosetting polymer Bakelite, which is hard, chemically resistant, and dimensionally stable [14]:

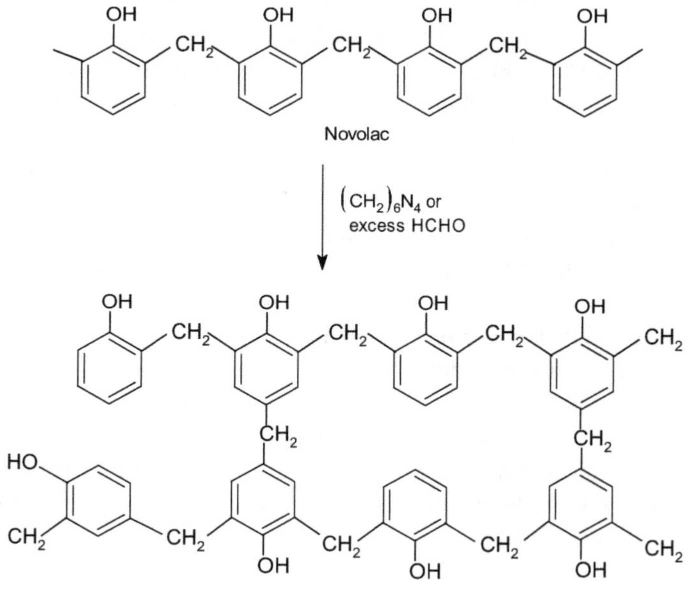

Bakelite

(j) Urea-formaldehyde resins: It is a thermosetting polymer obtained by the combination of urea and formaldehyde in an alkaline catalyst. The first product is monomethylol and dimethylol urea:

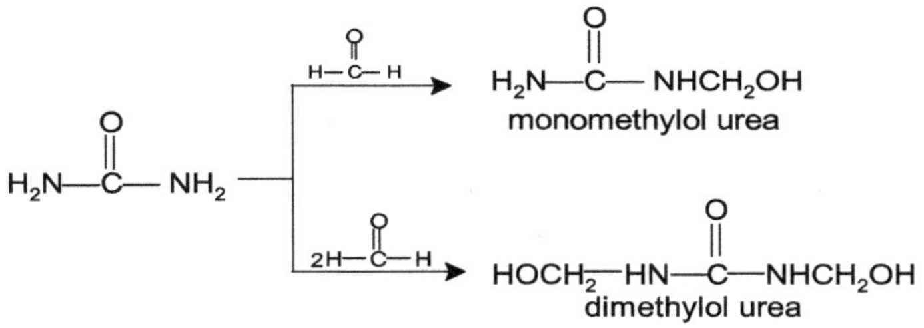

Then, monomethylol and dimethylol ureas react to give a linear polymer, which undergoes reaction with the CH_2OH group to give cross-linking by the methylene group (CH_2):

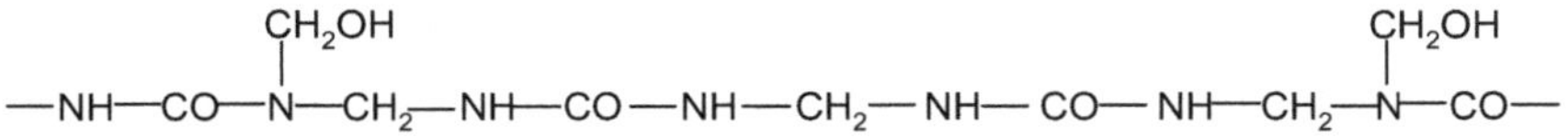

Cross-linking by the CH_2 group gives urea-formaldehyde resins:

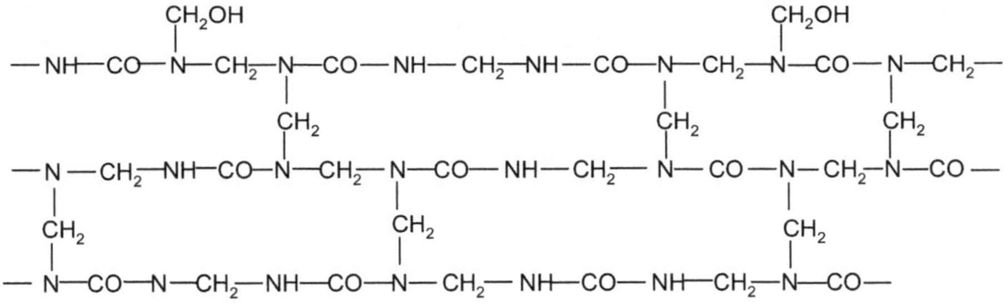

It has high strength, rigidity, excellent thermal and electrical insulation properties, resistance to chemicals and moisture, and is easily of molded and shaped. It acts as a binder, holding wood fibers together. These materials are used in building, furniture, and interior design [15].

(k) Melamine-formaldehyde resins (Melmac): Melamine-formaldehyde resin is produced by the polycondensation reaction of melamine with formaldehyde, resulting in a highly cross-linked, three-dimensional, thermosetting network. Once cured, the resin becomes irreversibly hard, which imparts excellent mechanical strength, thermal resistance, and chemical stability, and is widely used in coatings, laminates, adhesives, and textiles. The synthesis involves the reaction of melamine with formaldehyde. Melamine can be prepared by using cyanamide (i) or by urea (ii):

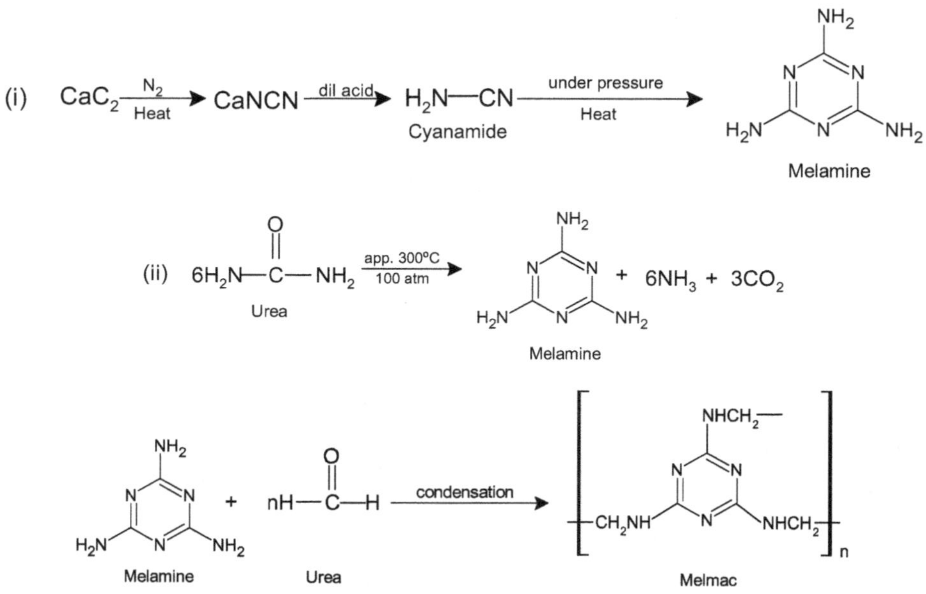

(l) Epoxy resins: Epoxy resins are synthetic polymers obtained from the reaction of epichlorohydrin with compounds containing at least two active hydrogen atoms, such as aminophenols, polyphenolic compounds, diamines, heterocyclic amides/imides, and aliphatic diols. The most widely used epoxy resin monomer is diglycidyl ether of bisphenol A (DGEBA) [16]:

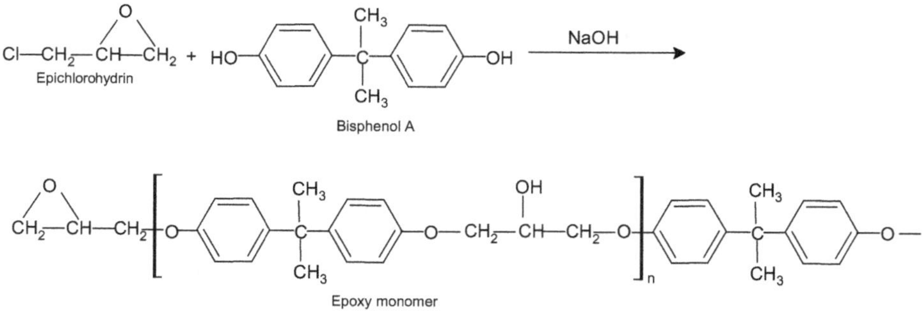

The reactive oxirane (epoxy) group present in epoxy monomers can undergo ring-opening reactions with a wide variety of curing agents, including aliphatic and aromatic amines, amidoamines, phenols, polyamides, anhydrides, thiols, and acids. These curing reactions lead to the formation of a highly cross-linked thermosetting network. The final epoxy products are rigid and brittle in nature, due to the presence of numerous cross-linked chains, but they possess excellent adhesion, chemical resistance, electrical insulation, and mechanical strength, making them extremely valuable in industrial applications.

1.2.6 Classification Based on Mechanical Properties

Polymers can also be grouped on the basis of their mechanical properties, such as elasticity, toughness, and hardness. This classification highlights their ability to withstand stress and deformation, guiding their use in specific structural and functional applications.

They are classified as:

(i) Elastomers

(ii) Fibers

(i) Elastomers: An elastomer has flexibility, stretchability, and is a randomly oriented amorphous polymer containing cross-links (Figure 1.11). In elastomers, the weak van der Waals forces between the polymer chains are not strong enough to keep the chains fixed in place. As a result, the chains can slide past one another when an external force is applied, allowing the material to stretch considerably. Once the force is removed, the chains return to their original random arrangement, enabling the elastomer to regain its initial shape. Unsaturated elastomers containing double bonds (natural rubber) have exceptional elasticity, whereas saturated elastomers, lacking double bonds in their polymer chains, excel in stability and resistance to degradation (silicone rubber). It has exceptional thermal stability, electrical insulation, and biocompatibility. Examples include silicone rubber, neoprene rubber, butyl rubber, and polyurethane [17].

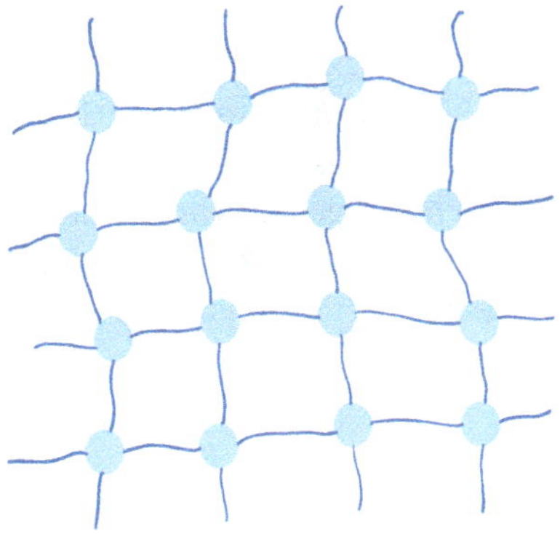

Figure 1.11: Cross-links in polymer chains in an elastomer.

(ii) Fibers: These are thread-like polymers that can be used as fabrics (Figure 1.12). Fabric is a textile material, formed by weaving multiple natural or synthetic fibers together. Fibers have an abundance of hydrogen bonds, which provide high tensile strength and high modulus to fibers. Polymer fibers are traditionally classified as natural and synthetic. Natural fiber is found in nature, such as vegetable and animal fibers. Cotton is the most famous example of a vegetable fiber. It is made up of cellulose or small amounts of hemicellulose and lignin, while animal fiber is produced by animals, for example, wool and silk, from protein. Synthetic fibers are composed of synthetic polymers, which have high strength, high relative expansion, elasticity, and low residual strain. Examples include polyester fiber, polyamide fiber, aromatic polyamide fiber, polyacrylonitrile fiber, polyurethane fiber, polyolefin fiber, and PVC fiber [18].

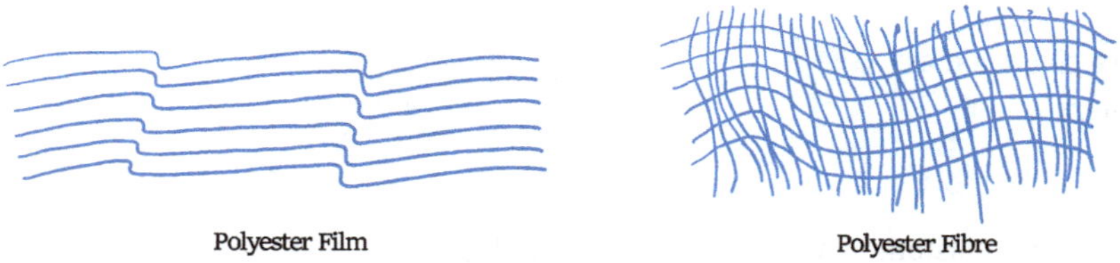

Figure 1.12: Arrangement of polyester threads in polyester film and polyester fiber.

1.3 Ziegler-Natta Catalyst

In 1953, German chemist Karl Ziegler developed a catalytic system capable of polymerizing ethylene into a linear, high-molecular-weight polyethylene, using a combination of a transition metal halide and an alkyl compound of a main group element (e.g., Al with $TiCl_4$). Shortly after, in 1954, Italian chemist Giulio Natta extended this work by polymerizing α-olefins into stereoregular polymers, obtaining either isotactic or syndiotactic structures depending on the catalyst employed. For their groundbreaking contributions, Ziegler and Natta were jointly awarded the Nobel Prize in Chemistry in 1963 [1].

A Ziegler-Natta catalyst typically consists of a transition metal compound (Group IV metals such as Ti, Zr, or Hf) and an organoaluminum cocatalyst. Examples of catalytic systems include $TiCl_4 + Et_3Al$ and $TiCl_3 + AlEt_2Cl$. Both heterogeneous and homogeneous Ziegler-Natta catalysts are used. Today, the polymerization of alkenes using this catalytic system remains one of the most important processes in the polymer industry, enabling large-scale production of polyethylene:

$$(C_2H_5)_3Al + TiCl_4 \longrightarrow (C_2H_5)_2AlCl + C_2H_5TiCl_3$$
$$C_2H_5TiCl_3 \longrightarrow \dot{C}_2H_5 + TiCl_3$$
$$(C_2H_5)_3Al + TiCl_3 \longrightarrow (C_2H_5)_2AlCl + C_2H_5TiCl_2$$
$$C_2H_5TiCl_2 \longrightarrow \dot{C}_2H_5 + TiCl_2.$$
$$(C_2H_5)_3Al + C_2H_5TiCl_3 \longrightarrow (C_2H_5)_2AlCl + (C_2H_5)_2TiCl_2$$
$$(C_2H_5)_2TiCl_2 \longrightarrow \dot{C}_2H_5 + C_2H_5TiCl_2$$

1.3.1 Activation of Ziegler-Natta Catalyst

In the crystal structure, titanium atoms are coordinated with six chlorine atoms on reaction with $Al(C_2H_5)_3$. There is an exchange of the C_2H_5 group between Ti and Al, with the removal of a Cl atom from Ti, which creates an empty orbital on Ti surface. This is called activation of the Ti complex with the cocatalyst. The catalyst is activated by the coordination of $AlEt_3$ to the Ti atom and the reduction of titanium(IV) to titanium(III):

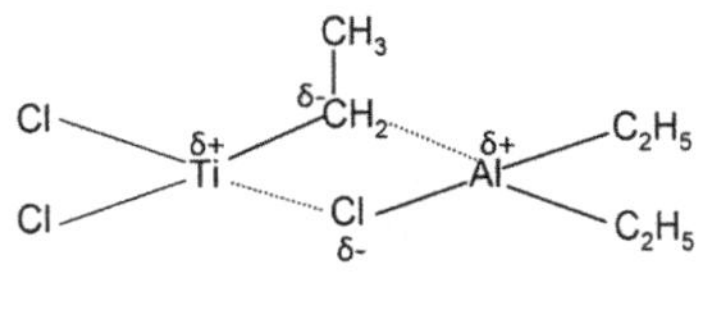

1.3.2 Working of Ziegler-Natta Catalyst

The polymerization process begins with the generation of the active catalytic species when the transition metal halide interacts with the organoaluminum cocatalyst. The next step involves the coordination of the alkene to this active center. Here, the π-electrons of the alkene interact with the electron-deficient metal center (such as Ti^{4+}). Simultaneously, the alkene undergoes insertion into the metal-carbon bond: the negatively charged alkyl group (e.g., an ethyl group) attacks the alkene, leading to the formation of a new C–C bond. This reaction results in the growth of the polymer chain. After the insertion, the active site on the catalyst is regenerated, allowing successive alkene molecules to coordinate and insert. This repeating cycle of coordination-insertion ensures continuous polymer chain propagation, making the Ziegler-Natta system highly efficient for producing long-chain polymers with controlled stereochemistry:

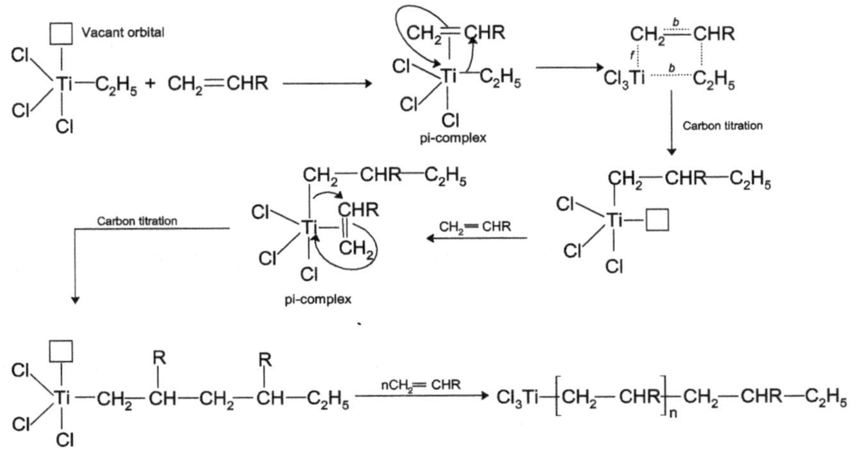

Polymer chain growth can be interrupted by chain transfer, a process in which the growing polymer radical abstracts a hydrogen atom from a chain transfer agent. This event terminates the original chain while simultaneously initiating a new one on the transfer agent.

HDPE (High density polyethylene), a crystalline polymer, exhibits superior strength and heat resistance compared to LDPE (Low density polyethylene). It is produced using the Ziegler-Natta catalytic process and typically contains 4,000–7,000 ethylene units per chain, corresponding to molecular weights in the range of 100,000–200,000 amu. An advanced form, ultra-high-molecular-weight polyethylene (UHMWPE), contains over 100,000 ethylene units per chain, with molecular weights ranging from 3,000,000 to 6,000,000 amu. Owing to its exceptional toughness and wear resistance, UHMWPE is used in bearings, conveyor belts, prosthetic devices, and bulletproof vests [19].

1.4 Synthetic Rubber

1.4.1 Buna-S

Buna-S, commonly known as styrene-butadiene rubber (SBR), is a copolymer of butadiene and styrene, usually combined in a ratio of 3:1. The polymerization is carried out by a free radical emulsion process at around 50 °C for 12–15 hrs, using free-radical initiators [1]. To maintain the strength and quality of the final product, stabilizers are added during synthesis. Compared to natural rubber, SBR offers several advantages, including superior abrasion resistance, resistance to cracking, and improved aging characteristics. Owing to these properties, it is produced on a large scale and is extensively used in the manufacture of automobile and truck tires, as well as in footwear, conveyor belts, and other mechanical goods:

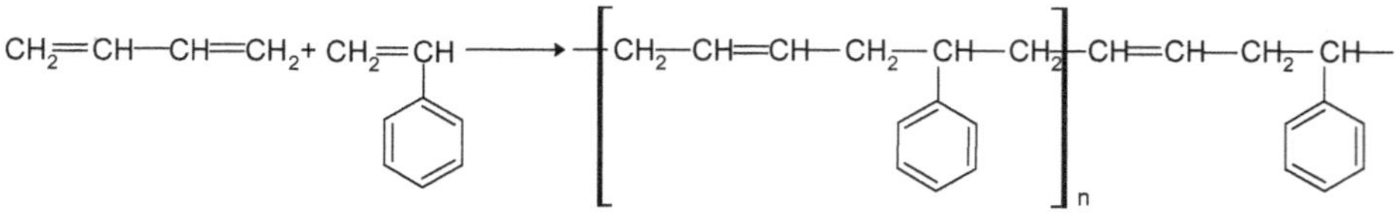

1.4.2 Buna-N or Nitrile Butadiene Rubber, Nitrile Rubber (NBR)

It can be deduced that a nitrile compound is used for the synthesis of Buna-N. The simplest reagent, acrylonitrile, is used in a range of 18–50%. The percentage of acrylonitrile determines the resistance of Buna-N to temperature and other materials. The higher the acrylonitrile, the greater the abrasion and tear resistance and higher the tensile strength. It executes excellent properties in petroleum-based fuels and oils, vegetable oils, silicone oils and grease, dilute acids, and water below 100 °C (212 °F):

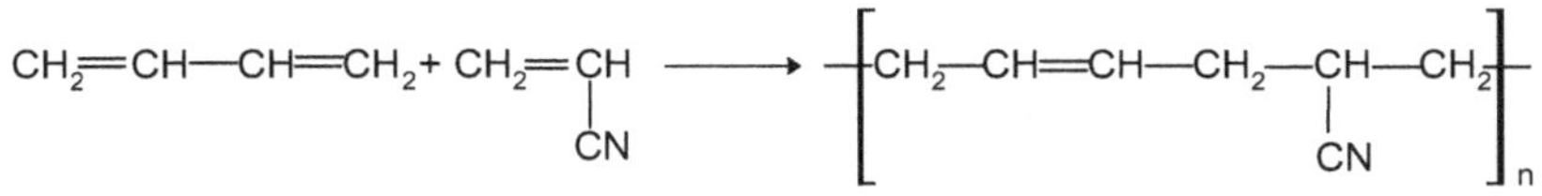

$$CH_2\!\!=\!\!CH\!-\!CH\!\!=\!\!CH_2 + CH_2\!\!=\!\!\underset{\underset{CN}{|}}{CH} \longrightarrow \left[CH_2\!-\!CH\!\!=\!\!CH\!-\!CH_2\!-\!\underset{\underset{CN}{|}}{CH}\!-\!CH_2 \right]_n$$

1.4.3 Butyl Rubber

Butyl rubber, or polyisobutylene, is a vinyl elastomer. A total of 2% butadiene and 98% isobutylene are copolymerized by cationic vinyl polymerization at very low temperatures. The use of butadiene generates double bonds, which, in turn, result in the generation of crosslinks by vulcanization. It shows flexibility, is biocompatible, resists acidic and alkaline chemicals and ozone weathering, and has good ageing properties:

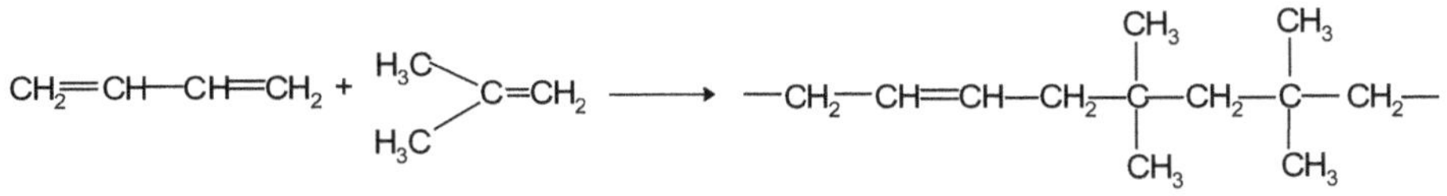

$$CH_2\!\!=\!\!CH\!-\!CH\!\!=\!\!CH_2 + \underset{H_3C}{\overset{H_3C}{>}}C\!\!=\!\!CH_2 \longrightarrow -CH_2\!-\!CH\!\!=\!\!CH\!-\!CH_2\!-\!\underset{\underset{CH_3}{|}}{\overset{\overset{CH_3}{|}}{C}}\!-\!CH_2\!-\!\underset{\underset{CH_3}{|}}{\overset{\overset{CH_3}{|}}{C}}\!-\!CH_2\!-$$

1.4.4 Neoprene

Neoprene is composed of polymerized chloroprene, where these molecules are chemically bonded together to form a durable and flexible material. It exhibits excellent chemical resistance, weather resistance, and outstanding durability. It is used in the automotive sector, for gaskets and seals, ensuring dependable airtight and leak-free seals:

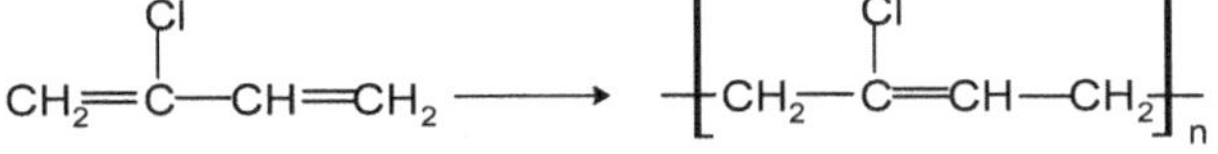

$$CH_2\!\!=\!\!\underset{\underset{Cl}{|}}{C}\!-\!CH\!\!=\!\!CH_2 \longrightarrow \left[CH_2\!-\!\underset{\underset{Cl}{|}}{C}\!\!=\!\!CH\!-\!CH_2 \right]_n$$

1.5 Molecular Weight of Polymers

Molecular weight is a key factor influencing the properties of polymers. Unlike small molecules, polymers consist of long chains made up of repeating units, and their molecular weight can range from a few hundred to several million, depending on the number of repeating units present in the chain. Because a polymer sample generally contains chains of varying lengths, its molecular weight is not expressed as a single value but rather as an average molecular weight [20].

1.5.1 Number Average Molecular Weight (M_n)

The number average molecular weight (M_n) is defined as the total weight of all polymer molecules in a sample divided by the total number of molecules present. It represents the simple arithmetic mean of the molecular weights of all polymer chains. This parameter is especially useful for correlating with certain physical properties such as tensile strength, impact resistance, and hardness. The mathematical expression for M_n is:

$$M_n = \frac{\sum N_i M_i}{\sum N_i} \tag{1.1}$$

Assume that there is a polymer mixture that contains N_1 molecules with molecular weight M_1, N_2 molecules with molecular weight M_2, and so on. If N_i molecules are with molecular weight M_i, then:

$$N = N_1M_1 + N_2M_2 + \cdots + N_iM_i = \sum N_iM_i \tag{1.2}$$

and $N_1 + N_2 + \cdots. + N_i = \sum N_i$

where M_i is the molecular weight of a chain and N_i is the number of chains of that molecular weight. Thus, M_n reflects the distribution of molecular weights across the sample. It can often be predicted from polymerization mechanisms and is experimentally determined using colligative property measurements, such as end-group analysis, osmometry, or cryoscopy.

1.5.2 Weight Average Molecular Weight (M_w)

The weight average molecular weight (M_w) gives greater emphasis to the heavier polymer chains in a sample, making it particularly useful for understanding properties influenced by high-molecular-weight fractions, such as viscosity and mechanical strength. Mathematically, M_w is defined as:

$$M_w = \frac{\sum N_i M_i^2}{\sum N_i M_i} \tag{1.3}$$

Alternatively, if w_i represents the weight of polymer molecules with molecular weight M_i, then the total weight of the polymer sample is:

$$w = \sum w_i = w_1 + w_2 + \cdots\cdots\cdots + w_i \tag{1.4}$$

Since the number of moles $n_i = \frac{w_i}{M_i}$, w_i express as n_iM_i. Substituting this into the equation, M_w can also be written as:

$$M_w = \frac{\sum n_i M_i^2}{\sum n_i M_i} \tag{1.5}$$

M_w is typically measured using light scattering or ultracentrifugation techniques, which are sensitive to the contribution of heavier polymer chains.

1.5.3 Polydispersity Index (PDI)

The PDI is a measure of the distribution of molecular weights within a polymer sample. It is defined as the ratio of the weight average molecular weight to the number average molecular weight:

$$\text{PDI} = \frac{M_w}{M_n} \tag{1.6}$$

If all polymer molecules in a sample have the same molecular weight, the system is called monodisperse, and the PDI equals 1. In practice, most synthetic polymers are polydisperse, meaning they contain chains of different lengths and molecular weights. The greater the deviation of PDI from unity, the broader the molecular weight distribution. Examples include:

Natural biopolymers such as proteins often have PDI = 1, reflecting their uniform chain length.

(i) Well-controlled synthetic polymers produced by modern techniques may have PDI values as low as 1.02–1.10.
(ii) Polymers formed via step-growth polymerization usually exhibit PDI values close to 2.0.
(iii) Polymers formed by chain-growth polymerization can have broader distributions, with PDI values ranging from 1.5 to 20, depending on the reaction conditions.

Thus, PDI provides valuable insight into the quality, uniformity, and performance characteristics of a polymer. Some of the commonly used polymers are given in Table 1.1.

Table 1.1: Commonly used polymers in everyday life.

Polymer used	Monomer unit	Structure of polymer
Polythene	$nCH_2{=}CH_2$	$-[CH_2-CH_2]_n-$
Polypropylene	$nCH_2{=}CH-CH_3$	$-[CH_2-CH(CH_3)]_n-$
Polystyrene	$nCH_2{=}CH(C_6H_5)$	$-[CH_2-CH(C_6H_5)]_n-$
Polyacrylonitrile or acrilan	$nCH_2{=}CH(CN)$	$-[CH_2-CH(CN)]_n-$
Polyvinylacetate	$nCH_2{=}CH(OCOCH_3)$	$-[CH_2-CH(OCOCH_3)]_n-$
Polyvinylidene chloride	$nCH_2{=}CCl_2$	$-[CH_2-CCl_2]_n-$
Polytetrafluoroethylene (Teflon)	$nCF_2{=}CF_2$	$-[CF_2-CF_2]_n-$
Poly(methyl methacrylate) (PMMA)	$nCH_2{=}C(CH_3)(COOCH_3)$	$-[CH_2-C(CH_3)(COOCH_3)]_n-$

1.6 Applications of Polymers in Various Sectors

The introduction of polymers into the industrial sector has revolutionized modern technology due to their versatility, durability, and adaptability [21]. Their unique properties, such as lightweight, chemical resistance, flexibility, and ease of processing, have enabled their use across a wide range of industries. Today, polymers are indispensable in areas including food packaging, electronic components, medical devices, construction, automotive, aerospace, and energy applications. Some of the most significant industrial applications of polymers are discussed below:

1. Healthcare sector: In medicine, polymers play a critical role due to their biocompatibility and biodegradability. For instance, sutures made of polyglycolic acid and polylactic acid are designed to dissolve inside the body without leaving harmful residues, eliminating the need for surgical removal. Similarly, medical devices such as heart valves and catheters are fabricated from polyurethane, a polymer known for its flexibility and compatibility with biological systems.

2. Aerospace industry: In aerospace applications, polymers are valued for their lightweight nature, durability, and thermal resistance. A prime example is polyether ether ketone, a high-performance engineering polymer that withstands extreme temperatures while maintaining mechanical strength. It is widely used in structural aircraft components, engine parts, and insulation materials, thereby reducing overall weight, while enhancing fuel efficiency and performance.

3. Energy sector: Conductive polymers have opened new avenues in energy-related applications. For example, poly(3,4-ethylenedioxythiophene):polystyrene sulfonate (PEDOT:PSS) is a widely used conducting polymer due to its excellent electrical properties and processability. It plays an important role in the fabrication of solar cells, batteries, and flexible electronic devices, where lightweight and flexible materials are essential.

4. Oil and gas sector: In the oil and gas industry, polymers serve as protective and structural materials due to their chemical resistance and mechanical strength. Polymers, such as epoxy resins, PU, and fluoropolymers (polytetrafluoroethylene (PTFE)), are extensively used as protective coatings for pipelines and storage tanks, preventing corrosion under harsh operating conditions. In addition, HDPE, PP, and PVC are employed for the safe transport of hydrocarbons and corrosive liquids, owing to their durability and resistance to aggressive chemicals.

5. Construction sector: Polymers have transformed the construction industry owing to their lightweight, durability, corrosion resistance, and cost-effectiveness. One of the most common examples is HDPE, which is widely used in water supply pipelines because of its excellent mechanical strength, long service life, and low installation cost. Beyond piping systems, HDPE also finds applications in food containers, chemical storage vessels, and protective linings, where its resistance to impact, harsh chemicals, and moisture makes it a reliable material.

1.7 Emerging Applications of Advanced Polymers

1.Biodegradable polymers: Eco-friendly biodegradable polymers are gaining importance in both the packaging industries and the biomedical fields. They are widely used in the manufacture of temporary medical devices, such as prosthetics and implants, with materials like silicones valued for their versatility and ability to mimic the properties of natural tissues.

2. Polymers in drug delivery: Polymers have shown remarkable potential in controlled and targeted drug delivery systems. Materials such as polymer-based hydrogels, micelles, and nanoparticles enable the precise release of therapeutic agents, ensuring that drugs reach specific tissues effectively, and thereby, improving patient outcomes.

3. Advanced polymer composites: Incorporating nanomaterials, such as graphene-based nanoparticles, significantly enhances the electrical and thermal conductivity of polymers. These advanced composites are increasingly used in flexible electronics, high-performance coatings, and precision sensors, where multifunctional properties are essential.

4. Teflon (PTFE): PTFE, commonly known as Teflon, exhibits outstanding thermal stability, chemical resistance, and electrical insulation. These properties make it a valuable material for cables, electronic circuits, nonstick cookware coatings, and certain medical applications, where high performance under extreme conditions is required.

The development of advanced polymers is reshaping modern industries by offering sustainable, multifunctional, and high-performance solutions. From biodegradable polymers that reduce environmental impact to smart drug-delivery systems that revolutionize healthcare, these materials are addressing some of the most pressing global challenges. Advanced polymer composites and specialty materials, such as PTFE, further extend their applications into electronics, aerospace, energy, and medical technology, demonstrating remarkable versatility. As research continues, advanced polymers are expected to play a central role in creating a more innovative, efficient, and sustainable future.

Practice Questions

1. What are natural and synthetic polymers?
2. What do you understand by tacticity? How are polymers classified according to their tacticity?
3. Write down the three differences between thermoplastic and thermosetting polymers with examples.
4. What are elastomers? Give examples.

5. What is the difference between addition and condensation polymerization?

6. What do you understand by an initiator in the polymerization process? How do they assist the process of polymerization?

7. Determine the final product in the reaction given below:

a) 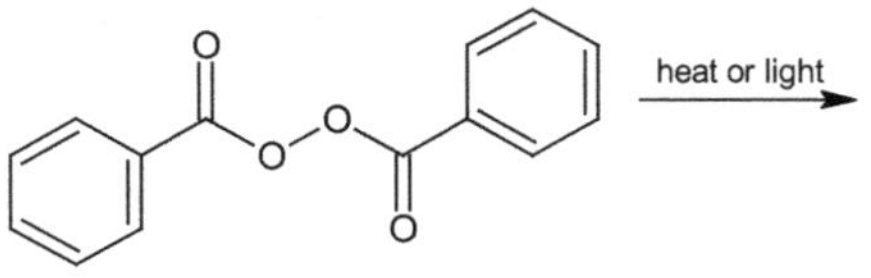

b) 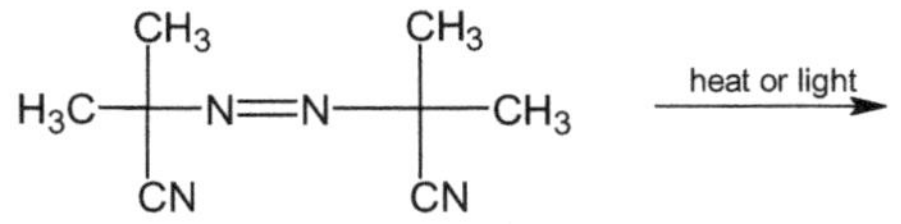

8. What is vulcanization of rubber? How does it affect the strength of rubber?

9. What is polydispersity index? How does it affect the properties of the polymer?

10. Give four applications of polymers in different fields.

11. Branched-chain and cross-linked polymers have different properties. Explain.

12. Give the structure of the following polymers:
(a) Bakelite (b) Teflon (c) Dacron (d) Novolac (e) Melmac

13. Why Ziegler-Natta catalyst holds such importance in the polymerization process?

14. An unsaturated molecule undergoes addition polymerization. Explain.

15. How does the chain transfer reagent affect the length of polymers?

16. What are the uses of BF_3, $TiCl_4$, $AlCl_3$, and $SnCl_4$ in polymerization process? Give a mechanism for condensation polymerization.

17. What are the different types of resins prepared by the polymerization process? Give detailed examples.

18. How does the condensation polymerization differ from the addition polymerization? Explain with examples.

19. Polyamides are an important class of which type of polymers? Give three types of polyamides.

20. Give the polymeric unit and the names of different monomers:

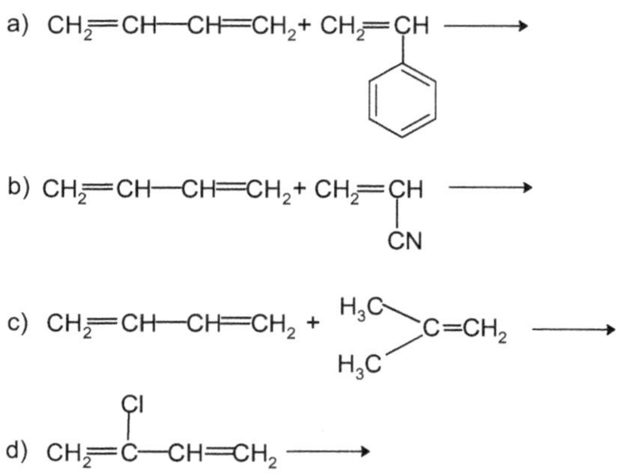

21. Polymer prepared by caprolactam when heated at high temperature in the presence of water is:
(i) Nylon 6 (ii) Nylon 6, 6 (iii) Dacron (iv) Nylon 6,10

22. Which of the following is a synthetic polymer?
(i) Rubber (ii) Silk (iii) Cellulose (iv) Nylon 6

23. Which of the following is a natural polymer?
(i) DNA (ii) Nylon 6 (iii) Rayon (iv) Resin

24. Which of the following polymers is characterized as polyester?
(i) Buna-S
(ii) Poly(ethylene phthalate) alkyd resin
(iii) Melamine
(iv) Novolac

25. Which one of the following polymers is not an example of condensation polymerization?
(i) Nylon 6,6 (ii) Bakelite (iii) Chloroprene (iv) Nylon 6

26. Which of the following natural polymers has structural similarity with synthetic polyamides?
    (i) Polysaccharide (ii) Protein (iii) Cotton (iv) Starch

27. Example of a biodegradable polymer is:
    (i) Epoxy resin (ii) Melamine (iii) Dextron (iv) Terylene

28. The monomers for the synthesis of Nylon 6, 6 are:
    (i) $H_2N\ (CH_2)_6\ NH_2$ and $HOOC(CH_2)_4COOH$
    (ii) $H_2N\ (CH_2)_4\ NH_2$ and $HOOC(CH_2)_4COOH$
    (iii) $H_2N\ (CH_2)_6\ NH_2$ and $HOOC(CH_2)_6COOH$
    (iv) $H_2N\ (CH_2)_4\ NH_2$ and $HOOC(CH_2)_6COOH$

29. Which of the following polymers is synthesized by free radical polymerization?
    (i) Melamine (ii) Teflon (iii) Nylon-6, 6 (iv) Terylene

30. The initiator used in the cationic polymerization is?
    (i) $SnCl_4$ (ii) $AlCl_3$ (iii) BF_3 (iv) All the above

Hints for Practice Questions

21. (i)

22. (iv)

23. (i)

24. (ii)

25. (iii)

26. (ii)

27. (iii)

28. (i)

29. (ii)

30. (iv)

References

[1] Singh J, Yadav LDS. *Organic Chemistry*. Vol. 3. Meerut: Pragati Prakashan; 2013.
[2] Big Chemical Encyclopedia [Internet]. Available from: https://chempedia.info/
[3] Copolymer: Synthesis, Polymerization, Blends [Internet]. Britannica. Available from: https://www.britannica.com/science/copolymer
[4] Thermoplastic: Definition, Properties, Examples & Applications https://www.chemistrylearner.com/thermoplastic
[5] Thermoset [Internet]. Chemistry Learner. Available from: Thermoset Plastic: Definition, Properties, and Examples.
[6] Condensation Polymerization: Characteristics, Polymerization, Addition Polymerization [Internet]. Available from: https://www.example.com/condensation-polymerization (replace with actual URL if available)
[7] Polyester: Synthetic Fibers, Textiles, Clothing [Internet]. Britannica. Available from: https://www.britannica.com/science/polyester
[8] Polyamide: Properties, Types, Advantages, Applications [Internet]. Available from: https://www.example.com/polyamide
[9] Nylon 6 [Internet]. EuroPlas. Available from: https://europlas.com.vn/en-US/blog-1/nylon-6-
[10] Nylon 66 Fiber Application [Internet]. Textile Learner. Available from: https://textilelearner.net/nylon-66-fiber-application
[11] Polyurethane: Foam, Synthesis, Plastics [Internet]. Britannica. Available from: https://www.britannica.com/science/polyurethane
[12] Polyurethane: Learn Meaning, Preparation, Properties and Uses [Internet]. Available from: https://www.example.com/polyurethane
[13] Science Info [Internet]. Available from: https://scienceinfo.com/
[14] Bakelite: Definition, Preparation, and Properties [Internet]. Collegedunia. Available from: https://collegedunia.com/exams/Bakelite-Definition-Preparation-and-Properties
[15] Urea Formaldehyde Resin [Internet]. Chemicals Learning. Available from: https://www.chemicalslearning.com/2024/12/urea-formaldehyde-resin-
[16] Epoxy Resin: Types, Uses, Properties & Chemical Structure [Internet]. Omnexus SpecialChem. Available from: https://omnexus.specialchem.com/Epoxy-Resin-Types-Uses-Properties-Chemical-Structure
[17] Elastomer:Definition,Types,Properties,Examples and Uses https://www.chemistrylearner.com/elastomer
[18] what-is-synthetic-fabric-types-properties-and-uses/https://www.textileindustry.net/
[19] *Stereochemistry of Polymerization: Ziegler–Natta Catalysts* [Internet]. OpenStax. Available from: https://openstax.org/books/organic-chemistry/pages/stereochemistry-of-polymerization-ziegler-natta-catalysts
[20] Agilent Technologies. Agilent 5990-7890EN: Technical Overview [Internet]. Available from: https://www.agilent.com/cs/library/technicaloverviews/Public/5990-7890EN.pdf
[21] Applications of Polymers in Various Industries [Internet]. Polyintec. Available from: https://polyintec.com/applications-of-polymers-in-various-industries

Jaspreet Kaur and Mukesh*

2 Chemistry of Polymerization

Abstract: This chapter provides a comprehensive overview of the chemistry of polymerization, focusing on the fundamental processes through which monomers unite to form polymers with distinct structures and properties. It begins with addition (chain-growth) polymerization, a process in which unsaturated monomers add successively to reactive sites without by-product formation. The mechanisms and kinetics of various chain growth polymerization such as free radical, cationic, and anionic polymerization are described in detail, highlighting the steps of initiation, propagation, and termination. Attention is then given to step-growth (condensation) polymerization, where polymer chains are produced through repeatitive condensation reactions between multifunctional monomers, often accompanied by the elimination of small molecules such as water. The chapter also explores coordination polymerization, emphasizing the role of transition metal catalysts in controlling polymer, tacticity, and microstructure.

Finally, different polymerization techniques, including bulk, solution, suspension, and emulsion polymerization methods, are discussed, with emphasis on their principles, advantages, limitations, and industrial significance. Together, these topics provide a strong foundation for understanding how polymers are synthesized and tailored to meet the requirements of diverse technological and industrial applications.

2.1 Introduction

Polymers are all around us, found in everyday materials such as plastics, fibers, rubbers, paints, and adhesives, as well as in natural substances like proteins, cellulose, and DNA. What makes polymers so important is not only their abundance but also their ability to be designed with a wide range of mechanical, thermal, and chemical properties. The secret behind this versatility lies in the way small building blocks, called monomers, join together in a process known as polymerization. Polymerization is more than just a chemical reaction; it is the foundation of modern materials science. By understanding how polymer chains form, grow, and arrange themselves, chemists and engineers can design materials that are light yet strong, flexible yet durable, or even biocompatible for medical use. This explains why polymerization plays such a critical role in industries ranging from packaging and textiles to electronics, aerospace, and medicine. Basically, primary type of polymerization are:

1. Addition (chain-growth) polymerization
2. Condensation (step-growth) polymerization
3. Coordination polymerization

https://doi.org/10.1515/9783112208236-002

In this chapter, the fundamental types of polymerization are explored and understand how they differ in mechanisms and outcomes [1–5]. Special emphasis will be given to the kinetics of chain growth, the chemistry of step-growth reactions, and the role of catalysts in coordination polymerization. The practical techniques, such as bulk, solution, suspension, and emulsion polymerization, are also examined, that make large-scale production possible. By the end of this chapter, students will appreciate not only the chemical principles of polymerization but also how these principles translate into the design of real-world materials that shape our daily lives.

2.2 Addition Polymerization

Addition polymerization, also known as chain-growth polymerization, is one of the most common and widely used methods for synthesizing polymers. In this process, unsaturated monomers, typically containing a double bond (such as in alkenes, vinyl derivatives, or acrylates), undergo successive addition reactions to form long polymer chanis [1–4]. A key feature of addition polymerization is that the reaction takes place without the elimination of by-products. As a result, the polymer structure is essentially a direct repetition of the monomer unit.

This process proceeds through a chain reaction mechanism consisting of three fundamental stages:

1. Initiation
   - (i) Reactive species such as free radicals, cations, or anions are generated from specific initiators (e.g., peroxides, Lewis acids, or alkali metals).
   - (ii) These reactive intermediates attack the π-bond of the monomer, creating an active site at its end.
2. Propagation
   - (i) The active chain end adds successive monomer molecules one after another.
   - (ii) Each addition regenerates the reactive center/active site, allowing the chain to grow rapidly.
   - (iii) This step is usually much faster than initiation, leading to the formation of high-molecular-weight polymers within a short time.
3. Termination
   - (i) The chain growth stops when the reactive center is destroyed/deactivated.
   - (ii) This can occur by combination (two active chains join together), disproportionation (transfer of a hydrogen atom between chains), or through chain stopping processes such as the action of inhibitors and impurities.

Because of its efficiency, addition polymerization is widely used in industry. It enables the synthesis of a variety of important commercial polymers, including:

(i) Polyethylene (PE): used in packaging, bottles, and films.
(ii) Polypropylene (PP): used in fibers, ropes, and household goods.
(iii) Polystyrene (PS): used in insulation, plastic cutlery, and containers.
(iv) Polyvinyl chloride (PVC): used in pipes, flooring, and electrical insulation.

The ability of addition polymerization to produce high-molecular-weight polymers quickly and economically has made it central to both industrial applications and academic studies in polymer science.

Addition polymerization can be done by different mechanisms. Thus, addition polymerization can be divided, based on its mechanisms, as:
1. Free radical addition polymerization
2. Cationic addition polymerization
3. Anionic addition polymerization

2.2.1 Free Radical Polymerization

Free radical polymerization is one of the most widely used and versatile methods for synthesizing polymers. It relies on the high reactivity of free radicals to convert unsaturated monomers, such as alkenes or styrene derivatives, into long polymer chains with substantial molecular weight. The process follows a chain reaction mechanism consisting of three distinct stages: initiation, propagation, and termination. During these stages, transient radical species drive rapid chain growth, enabling the formation of polymers under relatively mild conditions. The versatility of this method lies in its ability to tolerate a wide variety of monomers and functional groups, making it suitable for producing everything from everyday plastics like PE and PS to specialized resins and coatings. This has established free radical polymerization as a cornerstone of both industrial practice and academic research.

2.2.1.1 Initiators for Free Radical Polymerization

Free radical polymerization requires initiators that can generate reactive radicals. Common initiation methods include:
(i) **Thermal initiation** – Heating the initiator to break a weak bond and produce radicals.
(ii) **Photo-initiation** – Using ultraviolet or visible light to cleave bonds in photosensitive compounds.
(iii) **Redox initiation** – Generating radicals through redox reactions, often used in aqueous or low-temperature systems.

(iv) **Radiation initiation** – Employing high-energy radiation (e.g., gamma rays or electron beams) to form radicals directly from monomers or solvents.
(v) **Electrochemical initiation** – Free radicals are generated at the electrode surface by applying electrical potential.

2.2.1.2 Common Initiators

Although heating a vinyl monomer can break the π-bond to form bifunctional free radical capable of initiation polymerization. Peroxides (e.g., benzoyl peroxide) are widely used due to the relatively weak O–O bond, which breaks easily under heat to form radicals. For example:

$$(C_6H_5CO)_2O_2 \rightarrow 2C_6H_5COO^{\cdot} \rightarrow C_6H_5^{\cdot} + CO_2$$

Each benzoyloxy radical then loses carbon dioxide, generating a phenyl radical. The resulting phenyl radical reacts with a monomer, such as styrene, to initiate chain growth:

$$C_6H_5^{\cdot} + CH_2 = CH - Ph \rightarrow C_6H_5 - CH_2 - \dot{C}H - Ph$$

Azo compounds (e.g., azobisisobutyronitrile [AIBN]) decompose upon heating by breaking the nitrogen-nitrogen bond, releasing nitrogen gas and generating two cyanoisopropyl radicals. These radicals add to monomer double bonds, starting polymerization:

$$(CH_3)_2C(CN) - N = N - C(CN)(CH_3)_2 \rightarrow 2(CH_3)_2C(CN)^{\cdot} + N_2$$

$$2(CH_3)_2C(CN)^{\cdot} + CH_2 = CH - R \rightarrow (CH_3)_2C(CN) - CH_2 - \dot{C}H - R$$

In practice, peroxides and azo compounds are most commonly used. This preference arises from the general trend in bond dissociation energies, C–H > C–C > C–N > O–O. As bonds such as C–H, C–C, and C–N have higher energies compared to the weaker O–O bond in peroxides making them easier to cleave and more efficient for radical generation. Redox initiators involve electron transfer reactions. For example, potassium persulfate reacts with ferrous ions to generate sulfate radical anions, which initiate polymerization:

$$S_2O_8^{2-} + Fe^{2+} \rightarrow SO_4^{\cdot -} + SO_4^{2-} + Fe^{3+}$$

$$SO_4^{\cdot -} + CH_2 = CH - R \rightarrow OSO_3 - CH_2 - \dot{C}H - R$$

Photo-initiators, such as benzoin ethers, absorb UV or visible light, cleave, and generate radicals that attack the monomer double bonds. These radicals add to the monomer double bond, starting polymerization:

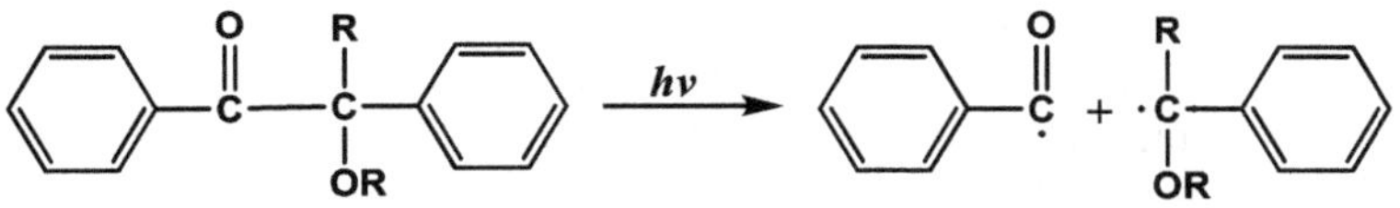

In certain situations, high-energy radiation, including gamma rays and electron beams, can directly break bonds in monomers or solvents, producing free radicals that initiate polymerization.

The initiation step is crucial in free radical polymerization because it generates the active radicals needed to start chain growth. Peroxides, azo compounds, redox systems, photo-initiators, and radiation sources are all commonly employed to create these radicals efficiently. The choice of initiator depends on the reaction conditions, monomer type, and desired polymer properties.

2.2.1.3 Kinetics of Free Radical Chain Polymerization

2.2.1.3.1 Initiation Step

In free radical polymerization, the speed of the reaction described by the following equation is closely linked to the rate at which the initiator decomposes to form radicals:

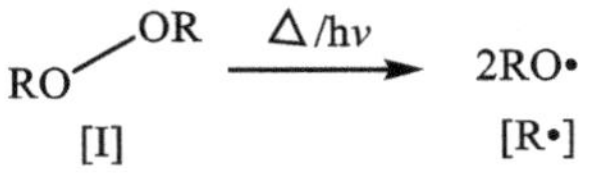

where R is PhCO, initiator (I) is 2RO·

The rate of initiator (decomposition, R_d) can be expressed as:

$$-\frac{d[\mathrm{I}]}{dt} = R_d = k_d[\mathrm{I}] \quad (2.1)$$

where k_d is the decomposition rate constant and [I] is the concentration of the initiator. The decomposition rate constant (k_d) depends on temperature and can be calculated using the Arrhenius equation:

$$k_d = Ae^{-\frac{E_a}{RT}} \quad (2.2)$$

Here, A is the frequency factor, representing the number of effective collisions per unit time. E_a is the activation energy required for bond cleavage. R is the gas constant, and T is the absolute temperature.

Combining these, the rate of decomposition of the initiator becomes:

$$R_\mathrm{d} = Ae^{-\frac{E_\mathrm{a}}{RT}}[\mathrm{I}] \quad (2.3)$$

This equation shows that the rate of initiator decomposition increases with both temperature and initiator concentration, leading to a higher generation of free radicals to start the polymerization process. A free radical chain reaction begins when a reactive radical (RO·) attacks the double bond of a vinyl monomer, forming a new radical at the end of the growing chain. For example, in PP polymerization:

$$\mathrm{R}\cdot + \mathrm{M} \rightarrow \mathrm{RM}\cdot$$

k_1 is the rate constant of the initiation reaction:

$$-\frac{d[\mathrm{M}]}{dt} = R_\mathrm{i} = k_\mathrm{i}[\mathrm{R}^\cdot][\mathrm{M}] \quad (2.4)$$

where R_i represents the rate of initiation and R• represents the free radical from the initiator.

The reactivity of free radicals (R·) is extremely high. As soon as they are formed, they attack the monomer double bond and initiate the polymerization process. Using the steady-state approximation (SSA) (from chemical kinetics), we assume that the rate of production of free radicals is equal to their rate of disappearance:

$$R_\mathrm{i} = R_\mathrm{d} \quad (2.5)$$

$$R_\mathrm{i} = 2k_\mathrm{d}[\mathrm{I}] \quad (2.6)$$

Note on the factor "2": each initiator molecule decomposes to form two radicals. Therefore, a factor of 2 appears in eq. (2.6). In contrast, the earlier rate expression for the initiator decomposition equation (2.1) describes only the decomposition of the initiator molecules, not the number of radicals produced. Similarly, in termination steps, each termination event eliminates two growing radicals, and hence a factor of 2 is also included in rate expressions.

Equation (2.6) assumes that all free radicals generated are effective in initiating chain growth. However, in practice, some radicals are lost due to side reactions (e.g., recombination). To account for this, an efficiency factor (f) is introduced [1, 4] $\mathrm{R_i} = 2fk_\mathrm{d}[\mathrm{I}]$. Here:

(i) f represents the fraction of free radicals that successfully initiate chain growth.

(ii) The value of f usually lies between 0.6 and 1.0. For example, if $f = 0.6$, it implies that 60% of generated radicals effectively initiate polymerization, while the rest are lost in side reactions.

2.2.1.3.2 Propagation Step

Once initiation occurs, the polymer chain begins to grow via the propagation reaction. This is a bimolecular process in which a newly formed radical (RM·) reacts with another monomer molecule (M):

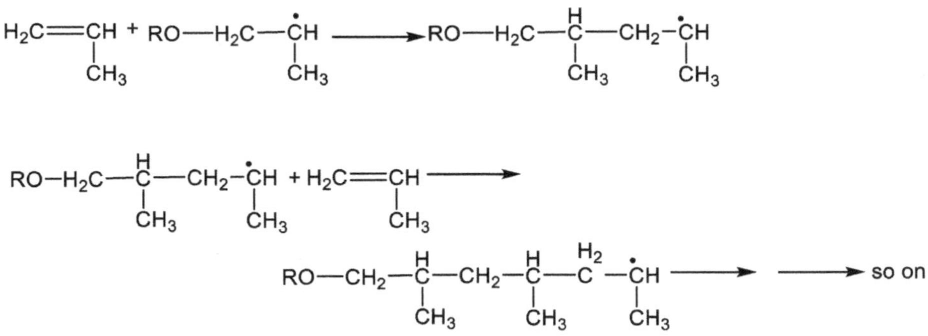

The radical site shifts to the chain end, allowing successive additions of monomer units. This step repeats many times, leading to rapid chain growth and the formation of high-molecular-weight polymers.

In vinyl monomers with the general structure $CH_2 = CH–R$, the two carbons of the double bond are often referred to by distinct names:

(i) The head carbon is the one attached to the substituent group (–R).
(ii) The tail carbon is the methylene carbon ($–CH_2$), which is bonded to more hydrogen atoms.

During the propagation step of chain-growth polymerization, the incoming monomer unit can attach to the growing polymer radical in different ways, leading to four possible arrangements:

(i) Head-to-tail
(ii) Head-to-head
(iii) Tail-to-tail
(iv) Tail-to-head

Among these, the head-to-tail arrangement is by far the most common and energetically favored. This preference arises because:

(i) It minimizes steric hindrance between substituent groups (–R), since bulky side groups are positioned farther apart.
(ii) It provides greater electronic stability to the growing radical, as the radical center is more stable when it forms at the carbon adjacent to the substituent group.

As a result, most vinyl polymers (such as PE, PP, PS, and PVC) predominantly exhibit a head-to-tail configuration along their backbone. The other arrangements (head-to-head or tail-to-tail) are possible but occur only to a limited extent [4].

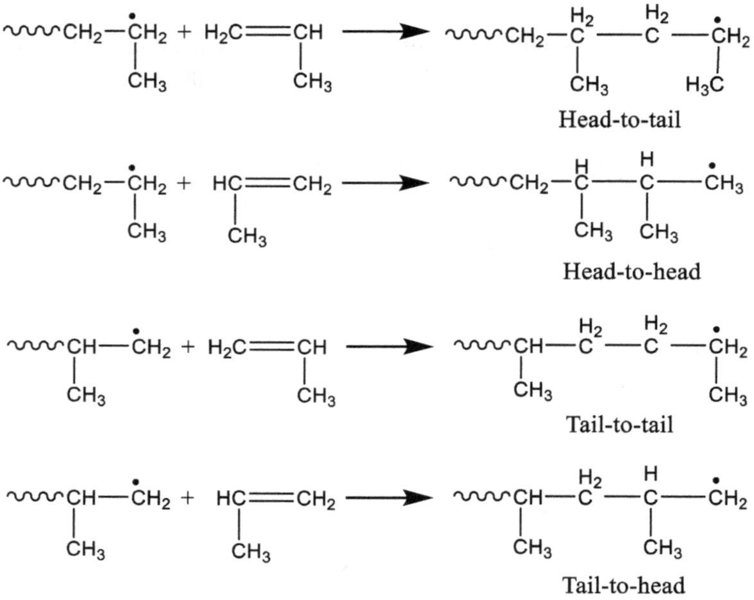

The generalized propagation steps can be represented as:

$$RM\bullet + M \rightarrow RMM\bullet$$

$$RMM\bullet + M \rightarrow RMMM\bullet$$

$RMMM\bullet + M \rightarrow RMMM\bullet$......chain keeps growing until it gets terminated

This process continues until the active radical is deactivated in a termination step.

The rate of monomer consumption during propagation can be expressed as:

$$-\frac{d[M]}{dt} = R_p = k_p[M^\bullet][M] \tag{2.7}$$

where k_p is the rate constant of propagation, R_p is the rate of propagation, [M•] is the concentration of active growing chains, and [M] is the concentration of monomer.

Experimental evidence indicates that the specific propagation rate constant k_p is essentially independent of chain length. Therefore, a single value of k_p can be used for all propagation steps.

2.2.1.3.3 Termination Step

The growth of a polymer chain eventually ceases when the active radical is destroyed. Two major termination pathways exist:

1. Combination (coupling): Two growing macroradicals combine to form a single dead polymer chain:

$$RM_{n1}M\cdot + \cdot MM_{n2}R \rightarrow RM_{n1}M - MM_{n2}R$$

Here, two radicals are consumed and one high-molecular-weight polymer is formed:

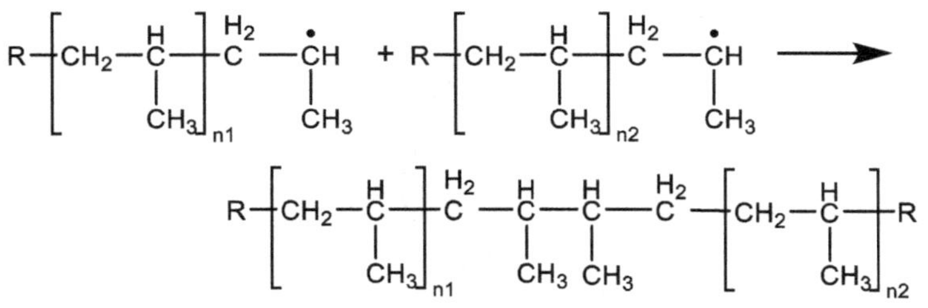

The rate of radical consumption due to termination is:

$$-\frac{d[M^{\cdot}]}{dt} = R_t = 2k_t[M^{\cdot}]^2 \qquad (2.8)$$

where k_t is the rate constant of termination.

2. Disproportionation: In this process, a hydrogen atom is transferred from the end of one growing chain to the radical site of another. The result is two nonradical chains:
(i) One saturated
(ii) One with an unsaturated end group

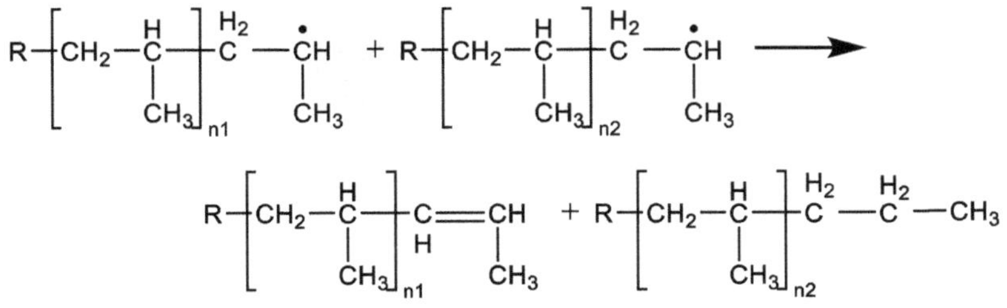

In free radical polymerization, it is generally assumed that the concentration of radicals quickly reaches a steady state. Thus, the rate of radical generation equals the rate of radical consumption:

$$R_i = R_t \qquad (2.9)$$

Equating initiation and termination:

$$2fk_d[I] = 2k_t[M^{\cdot}]^2 \qquad (2.10)$$

$$[M^{\cdot}]^2 = \frac{fk_d[I]}{k_t} \tag{2.11}$$

$$[M^{\cdot}] = \left(\frac{fk_d[I]}{k_t}\right)^{\frac{1}{2}} \tag{2.12}$$

Substituting radical concentration into the propagation rate law (eq. (2.7)):

$$R_p = k_p[M^{\cdot}][M] \tag{2.13}$$

$$R_p = k_p\left(\frac{fk_d[I]}{k_t}\right)^{1/2}[M] \tag{2.14}$$

$$R_p = k_p\frac{k_d{}^{1/2}}{k_t{}^{1/2}}f^{1/2}[I]^{1/2}[M] \tag{2.15}$$

From this result (eq. (2.15)):
- $R_p \propto$ [M] (first order in monomer), f is independent of [M] in case initiator efficiency is high; with low efficiency may depend on monomer concentration
- $R_p \propto [I]^{1/2}$ (half order in initiator)

2.2.1.2 Kinetic Chain Length (ν)

Kinetic chain length is defined as the average number of monomer molecules consumed per effective radical generated. It reflects how many monomers are added to a growing chain before termination occurs:

$$\nu = \frac{\text{rate of propagation}}{\text{rate of initiation}} = \frac{\text{rate of propagation}}{\text{rate of termination}}$$

$$\nu = \frac{R_p}{R_i} = \frac{R_p}{R_t} \tag{2.16}$$

Substituting expressions for R_p and R_i:

$$\nu = \frac{R_p}{R_i} = \frac{k_p\frac{k_d{}^{1/2}}{k_t{}^{1/2}}(f[I])^{1/2}[M]}{2fk_d[I]} \tag{2.17}$$

$$\nu = \frac{R_p}{R_i} = \frac{k_p[M]}{2(k_t k_d f[I])^{1/2}} \tag{2.18}$$

Thus, kinetic chain length depends:
(i) Directly on monomer concentration [M] and k_p
(ii) Inversely on the square root of initiator concentration $[I]^{1/2}$, k_t, k_d, f.

2.2.1.3 Degree of Polymerization (D_p)

The degree of polymerization ($\overline{D_p}$) describes the average number of monomer units incorporated into a polymer chain before chain termination occurs. It is directly linked to the kinetic chain length (ν), which expresses how many monomer molecules are consumed per active radical generated [4].

1. Degree of polymerization and termination pathways: Depending on how termination occurs, the relationship between $\overline{D_p}$ and ν changes:
(i) Termination by coupling: Two growing radicals combine, so one dead polymer chain contains two radical-initiated segments:

$$\overline{D_p} = 2\nu \tag{2.19}$$

(ii) Termination by disproportionation: A hydrogen atom is transferred, producing two polymer chains per termination event:

$$\overline{D_p} = \nu \tag{2.20}$$

A general expression may be written as:

$$\overline{D_p} = \frac{\nu}{N} \tag{2.21}$$

where $N = 0.5$ for termination by coupling, and $N = 1$ for termination by disproportionation.

Thus, termination mode strongly influences chain length.

2. Chain transfer reactions: In addition to termination by coupling or disproportionation, polymer growth may be interrupted by chain transfer reactions. In a chain transfer process, the active radical site shifts from one growing chain to another molecule, halting the growth of the first chain but creating a new radical that can initiate further polymerization.

General form of chain transfer:

$$R{-}CH_2{-}CHX\bullet + ZX \rightarrow R{-}CH_2{-}CH_2X + Z\bullet$$

where $R{-}CH_2{-}CHX\bullet$: growing polymer chain radical, ZX: chain transfer agent, $R{-}CH_2{-}CH_2X$: terminated polymer chain (growth stops here), and $Z\bullet$: a new radical formed from ZX, which may start a new polymer chain.

Key effect: Chain transfer reduces the molecular weight of the polymer because each transfer creates a shorter chain and initiates a new one.

Modified kinetic chain length for chain transfer reactions: The expression for kinetic chain length, previously defined as:

$$\nu = \frac{R_\mathrm{p}}{R_\mathrm{t}} \tag{2.22}$$

applies when chain termination occurs via radical coupling or disproportionation. However, in the presence of chain transfer, the effective kinetic chain length is reduced, since growth is interrupted more frequently. The modified relation is:

$$D_\mathrm{p} = \nu = \frac{R_\mathrm{p}}{R_\mathrm{t}} = \frac{R_\mathrm{p}}{R_\mathrm{tr}} \tag{2.23}$$

where R_tr is the rate of chain transfer. Substituting:

$$\nu = \frac{k_\mathrm{p}[\mathrm{M}][\mathrm{M}^\cdot]}{k_\mathrm{tr}[\mathrm{ZX}][\mathrm{M}^\cdot]} \tag{2.24}$$

$$\nu = \frac{k_\mathrm{p}[\mathrm{M}]}{k_\mathrm{tr}[\mathrm{ZX}]} \tag{2.25}$$

The ratio $\frac{k_\mathrm{tr}}{k_\mathrm{p}}$ is defined as the chain transfer constant C_ZX of the transfer agent:

$$\frac{k_\mathrm{tr}}{k_\mathrm{p}} = C_\mathrm{ZX} \tag{2.26}$$

where C_ZX is the transfer constant of compound ZX.

2.2.1.4 Ceiling Temperature

Polymerization reactions, like most chemical processes, are reversible. While the forward process involves propagation (addition of monomer units to the growing chain), the reverse process is called depropagation, where the active radical eliminates a monomer unit and shortens the chain:

$$\mathrm{RM}^\cdot + \mathrm{M} \rightleftharpoons \mathrm{RMM}^\cdot$$

At ordinary polymerization temperatures, the forward propagation rate greatly exceeds the rate of depropagation, allowing high-molecular-weight polymers to form. However, as the temperature increases:

(i) Both propagation and depropagation rates increase.
(ii) But the rate of depropagation rises more sharply, because depolymerization is entropically favored.

At a certain critical temperature, the two rates become equal. Beyond this point, polymerization cannot proceed because the polymer chains continually revert back to monomer. This critical point is known as the ceiling temperature (T_c) [4]. At T_c, the system reaches equilibrium, and polymer and monomer coexist without net polymer growth.

Thermodynamic Basis

The condition for equilibrium is:

$$\Delta G_p = \Delta H_p - T\Delta S_p = 0 \tag{2.27}$$

where ΔG_p is the Gibbs free energy change of polymerization, ΔH_p is the enthalpy change of polymerization, and ΔS_p is the entropy change of polymerization.

Rearranging gives:

$$T_c = \frac{\Delta H_p}{\Delta S_p} \tag{2.28}$$

Thus:

(i) Since ΔH_p (change in enthalpy of polymerization) is usually negative (exothermic), and ΔS_p (change in entropy during polymerization) is also negative (ordering of chains), T_c is a positive finite value.

(ii) Above T_c, $\Delta G_p > 0$, making polymerization thermodynamically unfavorable, and depropagation dominates.

2.2.2 Cationic Polymerization

Cationic polymerization is a chain-growth polymerization process in which the active centers are positively charged carbocations. This type of polymerization is particularly suitable for electron-rich monomers, such as isobutylene, styrene derivatives, and vinyl ethers, which can readily stabilize a carbocationic center [1,5]. The process is strongly influenced by carbocation stability, solvent polarity, temperature, and the nature of the counterion (gegenion) associated with the cation. This technique enables the design of specialty elastomers, resins, and advancedfunctional materials.

2.2.2.1 Mechanism of Cationic Polymerization

1 Initiation Step

The initiation step typically requires a Lewis acid catalyst (e.g., BF_3, $AlCl_3$, or $TiCl_4$) and a small amount of a proton-donating cocatalyst (such as water or alcohol). The Lewis acid alone cannot generate active carbocation; it needs the cocatalyst to provide a proton.

For example, with $BF_3 + H_2O$

a. Proton generation: Cations are always paired with an oppositely charged ion, commonly referred to as a counterion or gegenion:

$$\underset{\text{Lewis acid}}{BF_3} + \underset{\text{Lewis base (cocatalyst)}}{H_2O} \rightleftharpoons \underset{\text{Catalyst-cocatalyst complex}}{H^+[BF_3OH^-]}$$

The stability of the carbocations formed during polymerization is crucial for both the initiation and propagation phases. For olefins, the reactivity sequence is $(CH_3)_2C{=}CH_2$ being the most active, followed by CH3($CH{=}CH_2$), and then $CH_2{=}CH_2$. In the case of para–substituted styrenes, the substituent activity order is OCH_3, then CH_3, followed by H, and finally Cl. Along with solvent, the process also depends on gegenion's nucleophilicity and the monomer's electrophilicity. Ionic polymerizations may result in rearrangements.

b. Carbocation formation (using isobutylene as monomer): The proton adds to the double bond of isobutylene, creating a stable tertiary carbocation, paired with the counterion $[BF_3OH]^-$. This carbocation acts as the active species that initiates chain growth:

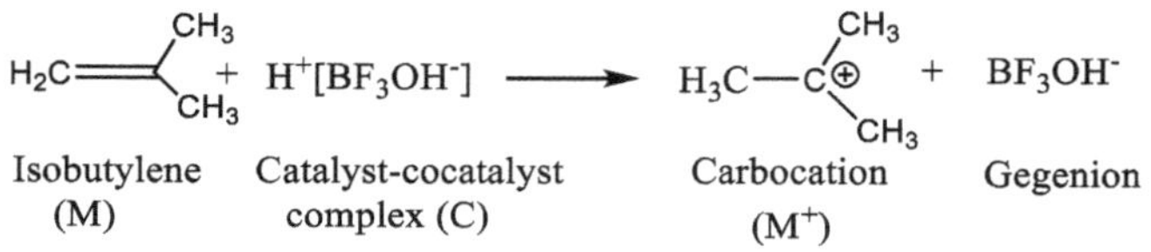

The rate of initiation is directly proportional to both the monomer concentration [M] and the concentration of the catalyst-cocatalyst complex [C]:

$$R_i = k_i[M][C] \tag{2.29}$$

2 Propagation Step

Once formed, the carbocation adds sequentially to new monomer units in a head-to-tail fashion, preserving the positively charged active center at the chain end. The propagation rate is given as:

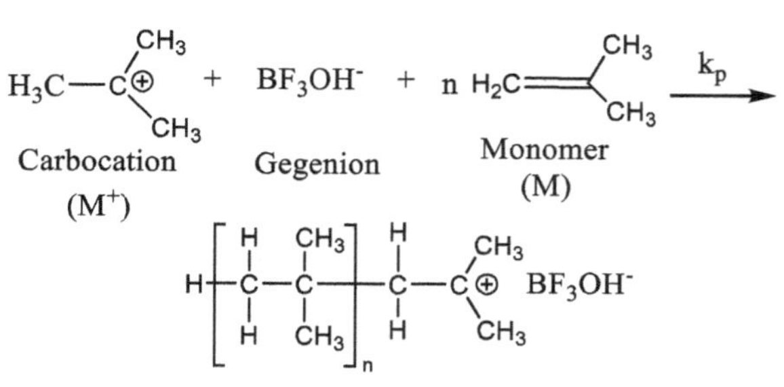

$$R_p = k_p[M^+][M] \tag{2.30}$$

where k_p is propagation rate constant, $[M^+]$ is the concentration of active carbocationic centers, and [M] is the monomer concentration. The rate constant for propagation (k p) remains essentially constant throughout the reaction for each propagation step.The process is significantly affected by the solvent's dielectric constant. In solvents that have a high dielectric constant, the separation between the carbocation and its counterion (gegenion) is promoted, resulting in a faster propagation rate. The solvent strongly influences propagation. High dielectric constant solvents promote ion pair separation (cation–counterion dissociation), leading to faster propagation.

Factors Influencing Reactivity

(i) Carbocation stability: More substituted carbocations are more stable → higher reactivity.

$$(CH_3)C = CH_2 > CH_3CH = CH_2 > CH_2 = CH_2$$

(ii) Substituent effects: Electron-donating groups (EDGs) ($-OCH_3$, $-CH_3$) increase reactivity, while electron-withdrawing groups (EWGs) (–Cl) reduce it. In case of para substituted styrene, the substituent activity order is OCH_3 then CH_3 followed by H and finally Cl.

(iii) Solvent polarity: Polar solvents stabilize carbocations and counterions, promoting propagation.

(iv) Counterion effects: Weakly nucleophilic counterions (like gegenion $[BF_3OH]^-$) favor longer chain growth.

3 Termination Step

Termination generally occurs through first-order decomposition of the macrocarbocation–counterion complex:

$$R_t = k_t[M^+] \tag{2.31}$$

Here, the active ionic species decomposes, regenerating BF_3 and water while producing a neutral polymer chain. Alternatively, chain transfer reactions may dominate. In such cases, a proton is transferred from the growing polymer chain to another monomer molecule, creating a new carbocation that can initiate further growth:

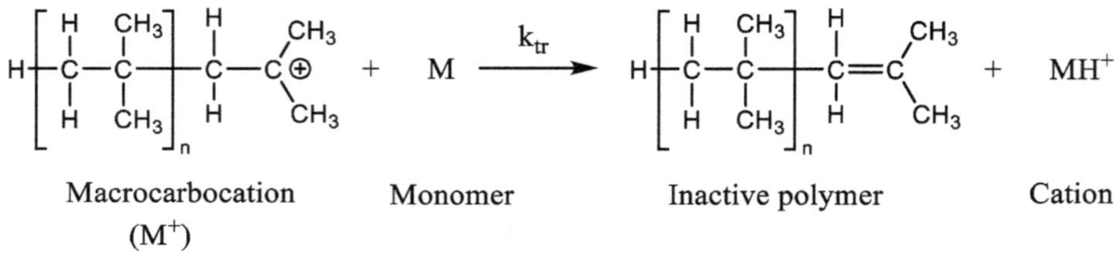

$$R_{tr} = k_{tr}[M^+][M] \tag{2.32}$$

This process reduces the chain length of individual polymers but maintains overall chain growth. In such cases the degree of polymerization (D_p) is eqaul to kinetic chain length *(ν)*.

Steady-state approximation (SSC): Because it is experimentally difficult to measure concentrations of reactive intermediates like [M^+], the SSA is applied:

$$R_i = R_t \tag{2.33}$$

$$k_i[M][C] = k_t[M^+] \tag{2.34}$$

$$[M^+] = \frac{k_i}{k_t}[M][C] \tag{2.35}$$

Substituting this value of $[M^+]$ into the expression for R_p:

$$R_p = k_p[M^+][M] = \frac{k_p k_i}{k_t}[C][M]^2 = k'[C][M]^2 \tag{2.36}$$

Thus, the polymerization rate depends quadratically on monomer concentration and linearly on catalyst concentration.

2.2.2.1.2 Degree of Polymerization (D_p)

The average degree of polymerization ($\overline{D_p}$) depends on the dominant termination pathway:

– When termination is by macrocabocation-gegenion dissociation:

$$\overline{D_p} = \frac{R_p}{R_t} = \frac{k_p[M][M^+]}{k_t[M^+]} = \frac{k_p}{k_t} = k''[M] \tag{2.37}$$

– When chain transfer dominates:

$$\overline{D_p} = \frac{R_p}{R_{tr}} = \frac{k_p[M][M^+]}{k_{tr}[M][M^+]} = \frac{k_p}{k_{tr}} = k'' \tag{2.38}$$

In either case, the degree of polymerization is independent of initiator concentration, unlike in free radical polymerization.

2.2.2.1.3 Key Factors Affecting the Rate and Molecular Weight

- Solvent polarity – polar solvents enhance propagation by promoting ion-pair separation.
- Carbocation stability – more substituted carbocations propagate more effectively.
- Counterion stability – weakly nucleophilic counterions (e.g., $[BF_3OH]^-$) favor longer chain growth.
- Initiator strength – initiators that generate protons easily (high electropositivity) accelerate the process.

2.2.2.1.4 Activation Energy in Cationic Polymerization

In cationic polymerizations, the activation energies of the elementary steps [1] generally follow the order:

$$E_{\mathrm{tr}} > E_{\mathrm{i}} > E_{\mathrm{p}}$$

where E_{tr} is the activation energy for chain transfer, E_{i} for initiation, and E_{p} for propagation.

According to the Arrhenius equation:

$$k = Ae^{\frac{-E_{\mathrm{a}}}{RT}} \tag{2.39}$$

the rate constant (k) depends exponentially on the activation energy (E_{a}), where A is the frequency factor, R is the universal gas constant, and T is the absolute temperature. The overall activation energy of polymerization can be expressed as:

$$E_{\mathrm{a}} = E_{\mathrm{tr}} + E_{\mathrm{i}} + E_{\mathrm{p}} \tag{2.40}$$

Interestingly, in many cationic polymerizations, the overall activation energy is negative. As a result, the polymerization rate (R_{p}) decreases with increasing temperature. This behavior is attributed to the fact that chain transfer has a higher activation energy than propagation. Therefore, as the temperature rises, chain transfer becomes more competitive relative to propagation. Consequently, the average degree of polymerization ($\overline{D_{\mathrm{p}}}$) decreases at elevated temperatures, leading to polymers of lower molecular weight.

2.2.3 Anionic Polymerization

Anionic polymerization is a chain-growth polymerization that proceeds through the formation of carbanions as the active centers. It is typically initiated by strong nucleophiles such as alkali metals, metal amides, or organometallic compounds. These initiators generate carbanions that readily attack electron-deficient monomers containing electron-withdrawing substituents. Representative examples include styrene, acrylonitrile, and conjugated dienes.

A key feature of anionic polymerization is its ability to proceed under "living" conditions, where the active sites remain stable in the absence of impurities or intentional termination. This property allows precise control over chain length and molecular architecture, enabling the synthesis of block copolymers, well-defined elastomers, and advanced functional polymers. Historically, anionic polymerization played a pivotal role in the development of synthetic rubbers. In the early 1900s, butadiene was polymerized using alkali metals dissolved in liquid ammonia. By the 1940s, this approach was replaced with metal alkyl initiators, such as *n*-butyllithium, which offered greater stability and ease of handling.

Reactivity of monomers: The ability of monomers to undergo anionic polymerization is strongly governed by the electronic effects of substituents [1] on the double bond:

(i) Electron withdrawing groups (EWGs): Enhance reactivity by stabilizing the carbanion (e.g., –CN in acrylonitrile and $-COOCH_3$ in methyl methacrylate).
Electron donating groups (EDGs): Reduce susceptibility to anionic attack, favoring cationic mechanisms.
Example reactivity order (with amide ion as initiator): acrylonitrile > methyl methacrylate > styrene > butadiene

(ii) Substituents at the β-carbon further modulate reactivity. A methyl substituent tends to decrease reactivity by donating electron density, whereas a chlorine substituent increases reactivity by withdrawing electron density.

Despite the difference in the nature of the reactive intermediates, the overall process including initiation, propagation, and chain termination—follows a mechanistic pattern analogous to that observed in cationic polymerization. The initiation step involves the generation of a carbanion, which subsequently propagates through successive monomer additions. Termination may occur via chain transfer or other deactivating processes, similar in concept to those found in cationic systems.

2.2.3.1 Mechanism and Kinetics Anionic Polymerization

1 Initiation Step

Initiation involves the nucleophilic attack of the initiator on the monomer to generate a carbanion. For instance, potassium amide (KNH_2) can initiate acrylonitrile polymerization:

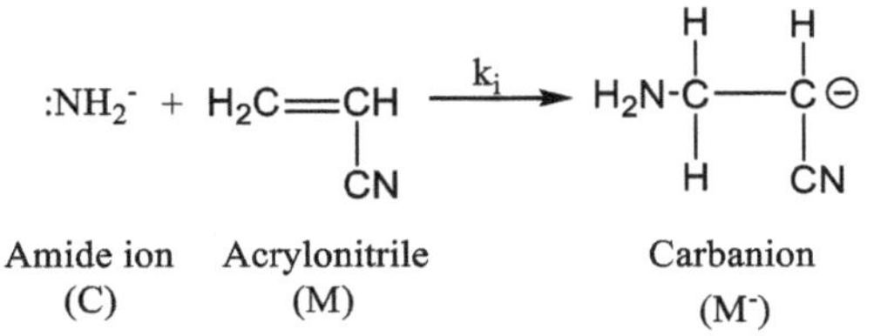

where M^- is the carbanionic active center paired with a counterion. The rate of initiation is expressed as:

$$R_i = k_i[C][M] \tag{2.41}$$

where [C] is the initiator concentration, [M] is the monomer concentration, and k_i is the rate constant for initiation.

2 Propagation Step

Once formed, the carbanion adds successively to monomer molecules in a head-to-tail fashion, preserving the negative charge at the chain end:

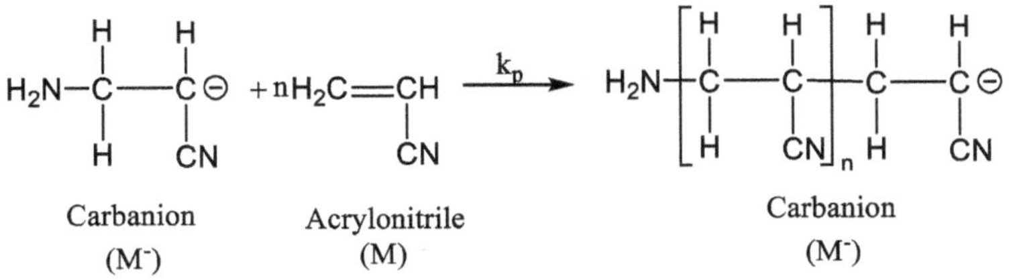

The rate of propagation is given by:

$$R_p = k_p[M][M^-] \tag{2.42}$$

where $[M^-]$ is the concentration of carbanionic species.

3 Termination and Chain Transfer Step

Unlike cationic polymerization, termination is not spontaneous in anionic polymerization. The active carbanions are relatively stable in the absence of impurities, and polymerization may continue indefinitely unless terminated intentionally by proton donors (e.g., water, alcohol, and ammonia) or electrophiles.

(i) Termination (via proton donor):

$$R_t = k_t[M^-] \tag{2.43}$$

(ii) Chain transfer (to ammonia):

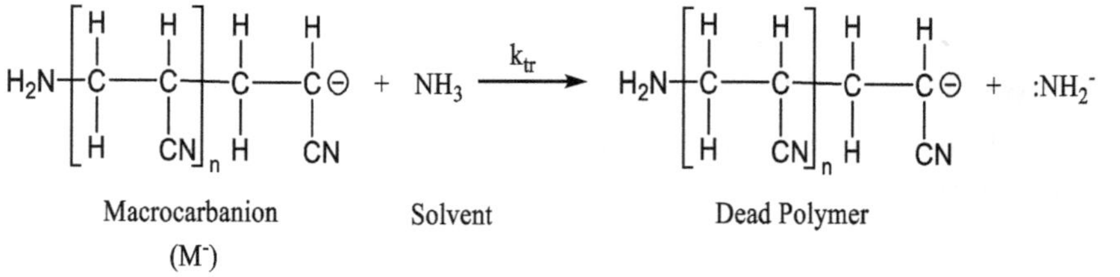

$$R_{tr} = k_{tr}[M^-][NH_3] \tag{2.44}$$

This transfer reaction produces a neutral polymer chain and simultaneously generates a new carbanion, allowing further propagation. In systems without termination or chain transfer, living polymerization occurs which is characterized by persistant active sites.

Because measuring $[M^-]$ directly is difficult, the steady-state approximation is used. Assuming:

$$R_i = R_{tr} \tag{2.45}$$

$$k_i[C][M] = k_{tr}[M^-][NH_3] \tag{2.46}$$

$$[M^-] = \frac{k_i[C][M]}{k_{tr}NH_3} \tag{2.47}$$

Substituting this value of $[M^-]$ into the expression for R_p equation (2.42):

$$R_p = k_p \frac{k_i[C][M]}{k_{tr}NH_3}[M] \tag{2.48}$$

$$R_p = \frac{k_p k_i}{k_{tr}} \frac{[C][M]^2}{NH_3} \tag{2.49}$$

$$R_p = k' \frac{[C][M]^2}{NH_3} \tag{2.50}$$

Thus, the rate of polymerization is directly proportional to the initiator and square of monomer concentration, and inversely proportional to ammonia concentration.

2.2.3.3.2 Degree of Polymerization

The average degree of polymerization ($\overline{D_p}$) depends on whether termination or chain transfer dominates. For chain transfer to ammonia:

$$\overline{D_p} = \frac{R_p}{R_{tr}} = \frac{k_p[M^-][M]}{k_{tr}[M^-]NH_3} \tag{2.51}$$

$$\overline{D_p} = k' \frac{[M]}{NH_3} = \frac{R_p}{R_{tr}} \tag{2.52}$$

This shows that molecular weight decreases as ammonia concentration increases.

2.2.3.4 Activation Energy and Temperature Dependence

In anionic systems, the activation energy for chain transfer is generally higher than that for propagation. The overall activation energy is about 38 kcal mol^{-1}. As a result:

(i) Increasing temperature raises the polymerization rate (R_p).
(ii) However, the average molecular weight decreases with temperature, due to enhanced chain transfer reactions.

2.2.3.5 Factors Influencing Polymerization Rate

Several parameters critically influence anionic polymerization:

(i) Solvent dielectric constant: polar solvents promote ion separation, increasing propagation rates.
(ii) Nature of initiator: initiators with higher electronegativity or polarity stabilize the carbanion better.
(iii) Resonance stabilization of carbanion: conjugated or aromatic substituents stabilize the active center.
(iv) Counterion solvation: effective solvation reduces ion pairing and enhances reactivity.
(v) Initiator polarity: nonpolar initiators (e.g., Grignard reagents) are effective with strongly electron-withdrawing monomers, while polar initiators (e.g., *n*-butyllithium) are needed for less activated monomers.

2.3 Copolymerization

The copolymerization process fundamentally resembles homopolymerization, which involves just one type of monomer. However, when multiple monomers are involved, differences in their reactivity can result in polymers with distinct structures and compositions. Depending on the type of monomers and the method used, copolymers can be formed via either step-growth or chain-growth polymerization.

A copolymer is not a mere physical combination of the individual homopolymers $[M_1]_n$ and $[M_2]_n$ [2]. Instead, it consists of monomer units $[M_1]$ and $[M_2]$ covalently bonded within the same polymer chain. The sequence of these units is governed by the relative reactivity ratios of the monomers. This molecular-level integration gives copolymers unique properties that are often distinct from those of the individual homopolymers or their simple blends.

2.3.1 Natural and Biological Copolymers

Although most naturally occurring polymers (like cellulose, starch, or natural rubber) are homopolymers, many essential biological macromolecules are copolymers:
(i) Proteins → copolymers of various amino acids.
(ii) Nucleic acids (DNA and RNA) → copolymers of nucleotide units.

These examples highlight the critical role of copolymerization in life processes.

2.3.2 Synthetic Copolymers

In industrial context, copolymerization provides a powerful way to tailor polymer properties for specific applications. Some important examples include:
(i) Styrene-butadiene rubber (SBR): Copolymer of styrene (S) and butadiene (B), the most common synthetic rubber used in automobile tires.
(ii) Acrylonitrile-butadiene-styrene (ABS): A versatile engineering plastic made from acrylonitrile, butadiene, and styrene monomers.
(iii) Spandex (Elastane): A block copolymer containing alternating rigid polyurethane blocks and flexible polyester blocks, giving it both strength and elasticity.

2.3.3 Significance of Copolymerization

By combining different monomeric units in a controlled manner, wide range of properties can be achieved:
(i) Enhanced mechanical strength and elasticity.

(ii) Improved chemical and thermal resistance.
(iii) Unique physical properties (e.g., transparency, toughness, and flexibility).
(iv) Materials tailored for advanced applications such as fibers, elastomers, plastics, coatings, and biomedical polymers etc.

2.3.4 Types of Copolymers

Copolymers are classified according to the arrangement of different monomer units along the polymer chain [2]. The major types are:

1. **Alternating copolymers**
   (i) Structure: Monomers are incorporated in a strictly alternating sequence: $(M_1M_2)_n$.
   (ii) Characteristics: These copolymers have a highly regular structure, resembling homopolymers in uniformity but combining properties of both monomers.
   (iii) Example: Maleic anhydride-styrene copolymer, used in adhesives and coatings.
2. **Random copolymers**
   (i) Structure: Monomer units are distributed along the chain in a statistical manner without a fixed order: $M_1M_1M_2M_1M_2M_2$, etc.
   (ii) Characteristics: The arrangement depends on the reactivity ratios of the monomers. These polymers often display averaged properties of both monomers.
   (iii) Example: Styrene-acrylonitrile (SAN), used in automotive and household applications.
3. **Block copolymers**
   (i) Structure: Large segments (blocks) of one type of monomer are covalently bonded to blocks of another monomer: $(M_1)_n(M_2)_n$.
   (ii) Characteristics: The distinct block arrangement gives rise to phase-separated domains, imparting unique mechanical and thermal properties.
   (iii) Example: Spandex (polyurethane-polyester), and styrene-butadiene-styrene (SBS) triblock copolymer used in thermoplastic elastomers.
4. **Graft copolymers**
   (i) Structure: Chains of one monomer are grafted as side branches onto the backbone of another polymer: Backbone: M_1, Branches: M_2, depicted in Figure 2.1.
   (ii) Characteristics: The combination of a backbone and grafted side chains results in materials with combined properties of both polymers.
   (iii) Example: High-impact PS (HIPS), where PS is grafted onto a polybutadiene backbone.

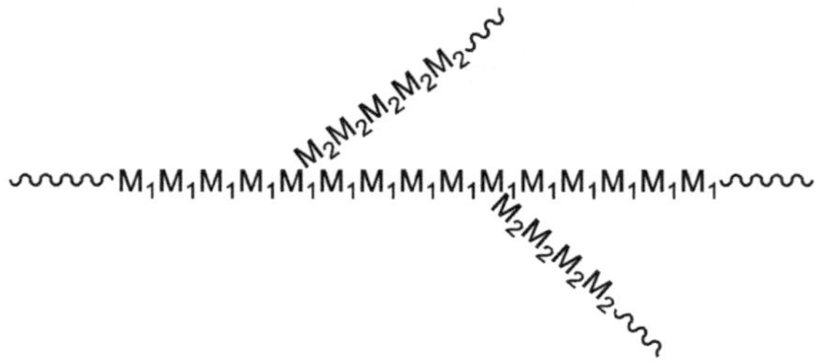

Figure 2.1: Graft copolymer.

2.3.5 Kinetics of Copolymerization

When two different monomers participate in copolymerization, the reaction produces two distinct propagating chain ends. One chain end is capped with monomer M_1 and the other with monomer M_2. These are denoted as M_1^* and M_2^*, where the asterisk (*) represents the active propagating site. Depending on the polymerization mechanism, this active site may be a free radical, a carbocation, or a carbanion.

If the reactivity of the growing chain depends only on the terminal monomer unit, then four possible propagation steps can take place. In each case, either monomer (M_1 or M_2) may add to a chain whose terminal unit is M_1^* or M_2^*. Out of the four possibilities, two steps (reactions 1 and 3) correspond to homopolymerization (self-propagation) and the other two (reactions 2 and 4) correspond to cross-propagation (heteropolymerization). The propagation reactions and their rate expressions are summarized in Table 2.1.

Table 2.1: The propagation reactions and their rate expressions.

Serial No.	Rection Scheme	Rate expression
1	$M_1^* + M_1 \rightarrow M_1M_1^*$	$R_{11} = k_{11}[M_1^*][M_1]$
2	$M_1^* + M_2 \rightarrow M_1M_2^*$	$R_{12} = k_{12}[M_1^*][M_2]$
3	$M_2^* + M_2 \rightarrow M_2M_2^*$	$R_{22} = k_{22}[M_2^*][M_2]$
4	$M_2^* + M_1 \rightarrow M_2M_1^*$	$R_{21} = k_{21}[M_2^*][M_1]$

Here: k_{11} is the rate constant for the addition of M_1 to a chain ending in M_1,
k_{12} is the rate constant for the addition of M_2 to a chain ending in M_1,
k_{22} is the rate constant for the addition of M_2 to a chain ending in M_2,
k_{21} is the rate constant for the addition of M_1 to a chain ending in M_2.

To simplify the description of copolymerization behavior, monomer reactivity ratios are introduced:

$$r_1 = \frac{k_{11}}{k_{12}}; r_2 = \frac{k_{22}}{k_{21}} \tag{2.53}$$

These ratios express the relative tendency of a propagating radical (or ion) to add its own type of monomer compared to the other type. Experimental evidence shows that the values of these propagation constants are essentially independent of the length of the growing polymer chain. Instead, they depend mainly on two factors:

(i) The structure of the incoming monomer
(ii) The nature of the active chain end

Thus, the kinetics of a binary copolymerization system can be fully described by the four fundamental propagation steps outlined above.

At the beginning of copolymerization, the rate of consumption of the two monomers from M_1 and M_2 from the reaction mixture can be expressed as follows:

Rate of depletion of M_1 is

$$-\frac{d[M_1]}{dt} = k_{11}[M_1^*][M_1] + k_{21}[M_2^*][M_1] \tag{2.54}$$

Rate of depletion of M_2 is

$$-\frac{d[M_2]}{dt} = k_{22}[M_2^*][M_2] + k_{12}[M_1^*][M_2] \tag{2.55}$$

By dividing eq. (2.54) by eq. (2.55), the ratio of rates at which monomers M_1 and M_2 are incorporated into the polymer can be written as:

$$\frac{d[M_1]}{d[M_2]} = \frac{k_{11}[M_1^*][M_1] + k_{21}[M_2^*][M_1]}{k_{22}[M_2^*][M_2] + k_{12}[M_1^*][M_2]} \tag{2.56}$$

This expression represents the copolymer composition. Rearrangement yields:

$$\frac{d[M_1]}{d[M_2]} = \frac{[M_1](k_{11}[M_1^*] + k_{21}[M_2^*])}{[M_2](k_{22}[M_2^*] + k_{12}[M_1^*])} \tag{2.57}$$

To eliminate the radical concentrations $[M_1^*]$ and $[M_2^*]$ in eq. (2.57), a steady-state approximation is applied:

$$\frac{d[M_1^*]}{dt} = \frac{d[M_2^*]}{dt} = 0 \tag{2.58}$$

This leads to the condition:

$$R_{12} = R_{21} \tag{2.59}$$

$$k_{21}[M_2^*][M_1] = k_{12}[M_1^*][M_2] \quad (2.60)$$

Rearranging for $[M_1^*]$:

$$M_1^* = \frac{k_{21}[M_2^*][M_1]}{k_{12}[M_2]} \quad (2.61)$$

Substituting $[M_1^*]$ in eq. (2.57):

$$\frac{d[M_1]}{d[M_2]} = \frac{[M_1]\left(k_{11}\frac{k_{21}[M_2^*][M_1]}{k_{12}[M_2]} + k_{21}[M_2^*]\right)}{[M_2](k_{22}[M_2^*] + k_{12}\frac{k_{21}[M_2^*][M_1]}{k_{12}[M_2]})} \quad (2.62)$$

Dividing eq. (2.62) with k_{21} and introducing the monomer reactivity ratios, and canceling terms in $[M_2^*]$, we arrive at:

$$\frac{d[M_1]}{d[M_2]} = \frac{[M_1]\left(\frac{k_{11}[M_2^*][M_1]}{k_{12}[M_2]} + [M_2^*]\right)}{[M_2]\left(\frac{k_{22}[M_2^*]}{k_{21}} + \frac{[M_2^*][M_1]}{[M_2]}\right)} \quad (2.63)$$

$$\frac{d[M_1]}{d[M_2]} = \frac{[M_1]\left(\frac{r_1[M_1]}{[M_2]} + 1\right)}{[M_2]\left(r_2 + \frac{[M_1]}{[M_2]}\right)} \quad (2.64)$$

which can be rewritten as:

$$\frac{d[M_1]}{d[M_2]} = \frac{[M_1]}{[M_2]}\left(\frac{\frac{r_1[M_1]}{[M_2]} + 1}{\frac{[M_1]}{[M_2]} + r_2}\right) \quad (2.65)$$

Multiplication by eq. (2.65) with $[M_2]$ yields what we called the "copolymerization equations" (eq. (2.66)), often referred to as the Mayo-Lewis equation, which provides the copolymer composition without requiring the knowledge of the concentration of free radicals:

$$n = \frac{d[M_1]}{d[M_2]} = \frac{[M_1]}{[M_2]}\left(\frac{r_1[M_1] + [M_2]}{[M_1] + r_2[M_2]}\right) \quad (2.66)$$

$$n = \frac{d[M_1]}{d[M_2]} = \left(\frac{[M_1]r_1[M_1] + [M_1][M_2]}{[M_2][M_1] + [M_2]r_2[M_2]}\right) \quad (2.67)$$

By further manipulation, dividing numerator and denominator of Eq. (2.67) by $[M_1][M_2]$:

$$n = \frac{d[M_1]}{d[M_2]} = \frac{r_1[M_1]/[M_2] + 1}{r_2[M_2]/[M_1] + 1} \tag{2.68}$$

If the monomer feed composition is represented as:

$$x = \frac{[M_1]}{[M_2]} \tag{2.69}$$

then:

$$n = \frac{r_1 x + 1}{r_2/x + 1} \tag{2.70}$$

This form clearly shows the dependence of copolymer composition (n) on the feed composition (x) and the reactivity ratios (r_1, r_2).

2.3.6.1 Significance of Reactivity Ratios

(i) $r_1 > 1$: Chain end M_1^* strongly prefers to add M_1 → tendency toward homopolymerization or block formation.
(ii) $r_1 < 1$: Chain end M_1^* favors cross-propagation with M_2.
(iii) $r_1 = 0$: Monomer M_1 cannot undergo homopolymerization.
(iv) $r_1 \approx r_2 \approx 1$: Random copolymer formation (not alternating).
(v) $r_1 \approx r_2 \approx 0$: Strong tendency toward alternating copolymerization.
(vi) $r_1 r_2 \approx 1$: Near-ideal random copolymerization (almost perfect random distribution of monomer along polymer chain).
(vii) $r_1 > 1$, $r_2 > 1$: Block copolymer, strong preference for adding same monomer.

In conclusion, the copolymerization equation provides a quantitative framework for predicting the composition of a copolymer from the feed composition and the monomer reactivity ratios. It also highlights that the copolymer composition is generally different from the feed composition, except under special conditions, such as azeotropic copolymerization [1,4], (copolymer composition is constant and matches monomer feed composition throughout polymerization).

2.4 Step Polymerization or Condensation Polymerization

Condensation polymerization, also known as step-growth polymerization, is the process by which polymers are formed through the gradual reaction of bifunctional or polyfunctional monomers. Unlike chain-growth polymerization, where polymer chains grow rapidly from active centers, step polymerization proceeds through a series of reactions in which monomers, dimers, trimers, and longer oligomers successively combine to form High-molecular-weight polymers.

This type of polymerization is especially important in nature. Many naturally occurring polymers, such as proteins, cellulose, and polyesters, are produced by condensation processes. Historically, some of the earliest synthetic polymers, for example, Bakelite, were also obtained by this method. Step-growth reactions often resemble typical organic condensation reactions, such as esterification or amide formation, where small by-products (commonly water, HCl, or methanol) are eliminated during bond formation.

Key characteristics of step polymerization include:

1. Slow polymer build-up: High-molecular-weight polymers are obtained only at very high conversions of monomer. At lower conversions, the mixture contains mostly low-molecular-weight species.
2. High activation energy: The process generally proceeds more slowly than chain-growth polymerization, requiring higher temperatures or catalysts.
3. Chain length independence: The rate of reaction does not depend on the size of the reacting molecules; a dimer can react with a trimer as readily as two monomers can react with each other.
4. Gradual growth: Polymerization progresses step by step, unlike chain-growth processes, where polymer chains elongate rapidly once initiated.

In contrast, chain-growth polymerizations (such as those of vinyl monomers) are characterized by rapid propagation due to a reactive chain end and lower activation energy barriers.

2.4.1 Kinetics of Step-Growth Polymerization

In step-growth (condensation) polymerization, monomers containing two reactive functional groups react in a stepwise manner, gradually building longer polymer chains [1, 5]. Each step typically involves the elimination of a small by-product, such as water, methanol, or HCl. A common example is the formation of polyesters, produced by the condensation of a dicarboxylic acid with a diol.

Example: Polyester formation:

$$\mathrm{HOOC-C_6H_4-COOH+HO-CH_2CH_2-OH \longrightarrow [-OC-C_6H_4-COOCH_2CH_2O-]_{\mathit{n}}+2\mathit{n}H_2O}$$

In this reaction, terephthalic acid (a diacid) reacts with ethylene glycol (a diol) to form polyester linkages, releasing water as a by-product.

Experimental studies have shown that certain diacids can catalyze their own reactions, enhancing polymerization rates even in the absence of an added catalyst. In such systems, the rate of polymerization depends directly on the concentrations of the functional groups – carboxylic acid (–COOH) and hydroxyl (–OH):

$$\text{Rate} \propto [\text{COOH}][\text{OH}] \tag{2.71}$$

If [*A*] is the concentration of diacid, and [*D*] is the concentration of diol, then the polymerization rate may be written as:

$$-\frac{d[\text{A}]}{dt} = k[\text{A}][\text{D}][\text{A}] \text{ (when acid catalysis its own reaction)} \tag{2.72}$$

$$-\frac{d[\text{A}]}{dt} = k[\text{A}]^2[\text{D}] \tag{2.73}$$

For an equimolar mixture of diacid and diol, this simplifies to:

$$-\frac{d[\text{A}]}{dt} = k[\text{A}]^3 \tag{2.74}$$

On integrating eq. (2.74) between limits $[\text{A}] = [\text{A}]_0$ at $t = 0$ and $[\text{A}] = [\text{A}]_t$ at time t:

$$2kt = \frac{1}{[\text{A}_t]^2} - \frac{1}{[\text{A}_0]^2} \tag{2.75}$$

The extent of reaction (*p*) is defined as the fraction of functional groups that have reacted consequently (1-p) corresponds to fraction of groups that are still unreacted:

$$p = \frac{[\text{A}]_0 - [\text{A}]_t}{[\text{A}]_0} \tag{2.76}$$

$$[\text{A}]_t = [\text{A}]_0(1-p) \tag{2.77}$$

Substituting $[A]_t$ into eq. (2.75) gives:

$$\frac{1}{(1-p)^2} = 1 + 2[\text{A}]_0^2 kt \tag{2.78}$$

Equation (2.78) is observed to fit experimental data well, especially after about 80% conversion. Thus, a plot of $\frac{1}{(1-p)^2}$ versus time should yield a straight line with slope $2[\text{A}]_0^2 k$. The value of k can then be extracted, and the activation energy is determined from its temperature dependence.

2.4.1.1 Degree of Polymerization and the Carothers Equation

The number average degree of polymerization ($\overline{D_\text{p}}$) is defined as:

$$\overline{D_\text{p}} = \frac{\text{Number of original molecules}}{\text{Number of molecules at a specific time } t} = \frac{N_0}{N} \tag{2.79}$$

Since

$$\frac{N_0}{N} = \frac{A_0}{A_t} \tag{2.80}$$

$$\overline{D_\text{p}} = \frac{A_0}{A_t} = \frac{A_0}{A_0(1-p)} = \frac{1}{(1-p)} \tag{2.81}$$

Equation (2.81) is known as the Carothers equation, derived by Wallace Carothers during his pioneering work on polyamide (nylon) synthesis.

At very high conversion ($p \to 0.9999$):

$$\overline{D_\text{p}} = \frac{1}{(1-p)} = \frac{1}{(1-0.9999)} = \frac{1}{0.0001} = 10,000$$

The degree of polymerization of 10,000, as calculated for nylon, is sufficiently high to yield fibers with excellent mechanical strength.

2.4.2 Polymerization with the Addition of an Acid Catalyst

When an external acid catalyst is introduced into the system, the rate of polyesterification increases significantly. The catalyst provides additional protons (H^+), which facilitate the esterification reaction between the carboxyl (–COOH) and hydroxyl (–OH) groups.

The general rate expression in the presence of an acid catalyst can be written as:

$$-\frac{d[\text{A}]}{dt} = k'[\text{A}][\text{D}][\text{H}^+] = k[\text{A}][\text{D}] \tag{2.82}$$

where [A] is the concentration of diacid, [D] is the concentration of diol, $[\text{H}^+]$ is proton concentration from the acid catalyst, k' is the rate constant, and k is $k'[\text{H}^+]$ = effective rate constant.

If the diacid and diol are present in equimolar quantities, the expression simplifies to:

$$-\frac{d[\text{A}]}{dt} = k[\text{A}]^2 \tag{2.83}$$

This represents a second-order rate law.

On integrating eq. (2.83) between limits $[\text{A}] = [\text{A}]_0$ at $t = 0$ and $[\text{A}] = [\text{A}]_t$ at time t:

$$kt = \frac{1}{[\text{A}]_t} - \frac{1}{[\text{A}]_0} \tag{2.84}$$

Using the extent of reaction, p, as in eq. (2.77), eq. (2.84) becomes:

$$\frac{1}{(1-p)} = 1 + [A]_0 kt \tag{2.85}$$

This relation shows that a plot of $\frac{1}{(1-p)}$ versus time should yield a straight line, with slope equal to $[A]_0 k$. From the slope, the rate constant can be determined, and its temperature dependence can further be used to calculate the activation energy of the reaction.

2.5 Coordination Polymerization

Before 1950, the only ethylene polymer available commercially was produced through high-pressure radical polymerization of ethylene through free radical polymerization process. This material, had several drawbacks:
i) Excessive chain branching
ii) Low crystallinity
iii) Poor mechanical strength

Although some early reports of coordination polymerization appeared in 1947 [6], the real breakthrough came in the late 1940s and early 1950s. Independent contributions from Marvel, Hogan, Banks, and Karl Ziegler opened the door to the production of linear PE.

Karl Ziegler, who later won the Nobel Prize, revolutionized polymer chemistry by demonstrating that ethylene could be polymerized at low pressures and near-room temperature using a novel catalyst system. This system consisted of triethylaluminum ($AlEt_3$) and titanium tetrachloride ($TiCl_4$). The product obtained, high-density PE (HDPE), was far more crystalline and mechanically stronger than the earlier branched PE. Building on Ziegler's discovery, Giulio Natta, another Nobel laureate, used the same catalytic approach to polymerize propylene. Remarkably, this led to the synthesis of highly crystalline and stereoregular PP. The catalysts developed by Ziegler and Natta, now widely known as Ziegler-Natta catalysts (ZNCs), have since transformed both academic research and industrial polymer production.

For those interested in the historical development of this field, the Nobel lectures of Ziegler (1964) and Natta (1965) remain highly recommended resources. Before exploring the details of coordination polymerization, it is important to understand the concept of stereoregularity in polymers, which plays a key role in determining the properties of the resulting materials.

2.5.1 Stereoregularity of Polymers and Their Types

In polymer chemistry, stereoregularity refers to the systematic spatial arrangement of substituent groups attached to the polymer backbone. This concept is especially important for polymers formed from monomers containing asymmetric (chiral) carbon atoms such as vinyl monomers because the relative orientation of the substituent groups along the chain strongly influences the material's properties. The degree of stereoregularity determines whether a polymer can pack efficiently into a crystalline structure or remains largely amorphous. Crystalline polymers generally possess higher strength, rigidity, and melting points compared to their amorphous counterparts.

2.5.1.1 Types of Stereoregularity

Polymers containing chiral centers may exhibit different stereochemical arrangements, which are classified into three main types:
1. Isotactic polymers

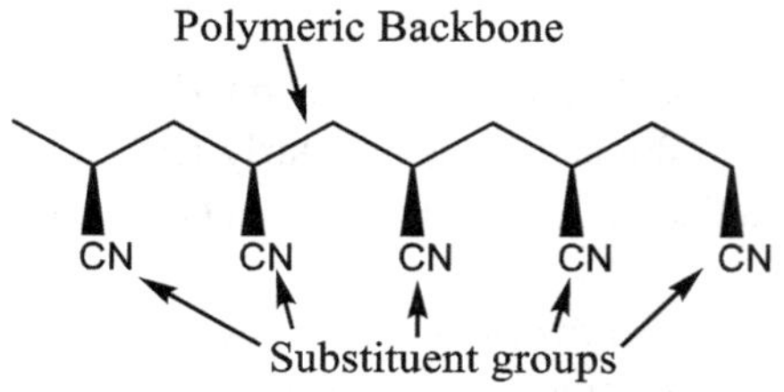

Isotactic polymer: Substituent groups aligned on the same side of the asymmetric carbon.

Note: Hydrogen atoms are not displayed to enhance structural clarity:

(i) In this arrangement, all substituent groups are oriented in the same direction relative to the polymer backbone.
(ii) Isotactic polymers pack efficiently into crystalline structures, which enhance their mechanical strength, stiffness, and melting temperature.
(iii) Example: Isotactic PP, which is widely used for packaging and fibers.

2. Syndiotactic polymers

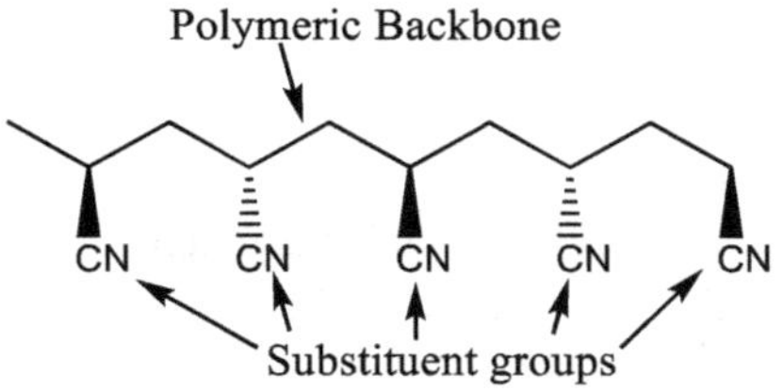

Syndiotactic polymer: Substituent groups on alternating sides of the asymmetric carbon:

(i) Here, substituent groups alternate regularly on opposite sides of the polymer chain.
(ii) This alternating arrangement also promotes crystallinity, though typically to a lesser extent than isotactic forms.
(iii) Syndiotactic polymers often show unique thermal behavior and balanced mechanical properties, making them valuable in specialty applications.

3. Atactic polymers

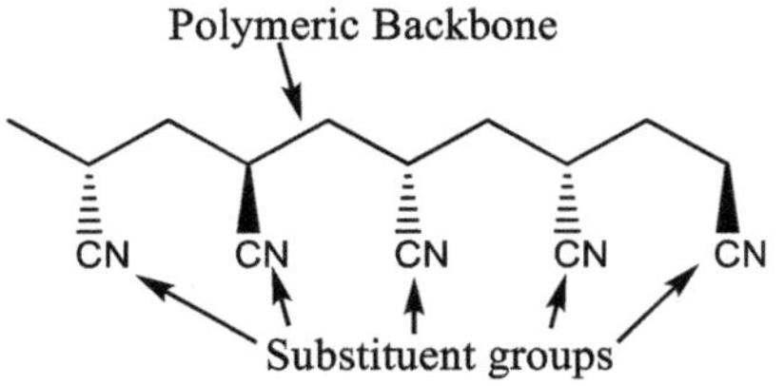

Atactic polymer: Substituent groups distributed randomly on either side of the asymmetric carbon:

(i) In this case, substituent groups are distributed randomly along the chain, with no regular order.
(ii) The irregularity prevents close packing, making these polymers largely amorphous, with low crystallinity, softness, and reduced mechanical strength (Melting point is exhibited by polymers with crystalline structure).
(iii) Example: Atactic PP, which is soft, sticky, and often used as an adhesive component.

Origin of stereoregularity: The stereoregularity of a polymer is primarily determined by the polymerization method and the catalyst employed. Transition-metal catalysts, such as Ziegler-Natta and metallocenes, are able to direct the spatial arrangement of monomer units during chain growth, resulting in isotactic or syndiotactic polymers. In contrast, free-radical polymerization does not offer such control and therefore generally produces atactic polymers with random substituent orientation.

2.5.2 Importance and Application of Stereoregularity

The stereochemical order in a polymer chain has a profound effect on its structure–property relationship.

Key influences include:

1. Crystallinity: Regular stereoregular chains (isotactic or syndiotactic) can pack efficiently, producing higher crystallinity. This leads to enhanced tensile strength, modulus, and thermal resistance.
2. Thermal properties: Both the glass transition temperature (T_g) and melting temperature (T_m) are strongly influenced by stereoregularity. Isotactic and syndiotactic polymers exhibit higher T_g and T_m values than atactic forms.
3. Solubility and processability: Atactic polymers, being amorphous, are generally more soluble and easier to process, but they lack the mechanical integrity for certain applications.
4. Optical activity: In some stereoregular polymers, the ordered arrangement of chiral centers can impart optical activity, useful in advanced applications such as chiral separation membranes and optoelectronic devices.

Applications based on stereoregularity: Control over stereoregularity has enabled the design of polymers with tailored properties for specific applications:

1. Isotactic PP: A highly commercial material valued for its strength, fatigue resistance, and high melting point, making it suitable for automotive parts, packaging, and textiles.
2. Syndiotactic PS: Exhibits superior chemical resistance and dimensional stability compared to atactic PS, making it useful in electronics and engineering plastics.
3. Atactic polymers: Although less crystalline, their softness, flexibility, and tackiness make them ideal for adhesives, sealants, and coatings.

2.5.3 Coordination Polymerization

A coordination polymer is a coordination compound in which repeating coordination units extend in one, two, or three dimensions. In the context of polymer chemistry, coordination polymerization refers to a chain-growth mechanism, where transition-metal catalysts coordinate monomers to a metal center and insert them sequentially into a growing polymer chain. This process offers precise control over molecular weight, microstructure, and stereoregularity [1].

Coordination polymerization using complex transition-metal catalysts is complex with respect to its mechanism, kinetics and applications. The most important catalysts are typically solid, heterogeneous systems (e.g., Ziegler-Natta catalysts), although soluble catalysts are also employed. The complexity increases because the same catalyst can promote different mechanisms – cationic, anionic, or even free-radical depending on the reaction conditions.

In industrial practice, coordination polymerization is often conducted:

(i) using catalysts suspended as fine solid particles in an inert medium (as in fluidized-bed processes), or
(ii) with catalysts immobilized on solid supports for enhanced stability and control.

This approach underlies the large-scale production of important commercial polymers such as HDPE, isotactic PP, and stereoregular PS, which would not be feasible through radical processes.

2.5.4 Ziegler-Natta Catalysts (ZNCs)

ZNCs are the most influential class of catalysts used in coordination polymerization:

2.5.4.1 Composition of ZNCs

A typical ZNC system consists of two essential components:

1. Transition metal compound (catalyst):
   (i) Generally derived from Groups IV B (4) to VIII B (10) of the periodic table.
   (ii) Common examples include halides or alkoxides of Ti, V, Cr, Mo, Co, and Ni.
   (iii) The polymerization reaction occurs at the transition metal center.
2. Organometallic compound (cocatalyst):
   (i) Usually derived from lighter elements of Groups I A (1) to III A (13), such as Li, Mg, or Al.
   (ii) Aluminum alkyls (e.g., triethylaluminum and $Al(C_2H_5)_3$) are most widely used.
   (iii) Their primary role is to activate the transition metal compound, often by reducing it to a lower oxidation state, and generating the active sites for monomer coordination and insertion.

Although both are crucial, the term "catalyst" is usually applied to the transition metal compound since it is the actual site of monomer polymerization, while the organometallic serves as a supporting role. Some widely used combinations of transition metal compounds and organoaluminum cocatalysts that serve as Ziegler-Natta catalysts are summarized in Table 2.2.

Table 2.2: Common Ziegler-Natta catalysts.

Component type	Examples	Role in catalysis
Transition metal compounds (Groups IV–VIII)	$TiCl_4$, $TiCl_3$, VCl_4, $VOCl_3$, $CrCl_3$, $MoCl_5$, $CoCl_2$, $NiCl_2$	Active sites for monomer coordination and insertion
Alkylaluminum compounds (cocatalysts)	$Al(C_2H_5)_3$ (TEA), $Al(i\text{-}C_4H_9)_3$ (TIBA), $Al(C_2H_5)_2Cl$, $Al(C_2H_5)Cl_2$	Alkylating agents, activators, and regulators of polymerization
Other organometallic compounds (Groups I–III)	*n*-BuLi, Et_2Mg, $Et_3Al_2Cl_3$	Initiators or modifiers for catalytic activity
Supported catalysts	$TiCl_4/MgCl_2$, $TiCl_3/Al_2O_3$	Improved stereoregularity and activity (used in modern industrial processes)

2.5.4.2 Mechanism of Polymerization with ZNCs

To better illustrate and understand the coordination polymerization process, let us consider titanium-based Ziegler-Natta catalysts. In these systems, exchange reactions occur between the transition metal compound and the organoaluminum cocatalyst. During these interactions, titanium in the +4 oxidation state (Ti(IV)) is partially reduced to Ti(III), generating the active sites required for polymerization.

In both the monometallic and bimetallic mechanisms, the first step involves the formation of a π-complex between the olefin and the vacant d-orbital of titanium. Natta favored a bimetallic pathway, in which the process proceeds through a cyclic, electron-deficient transition complex.

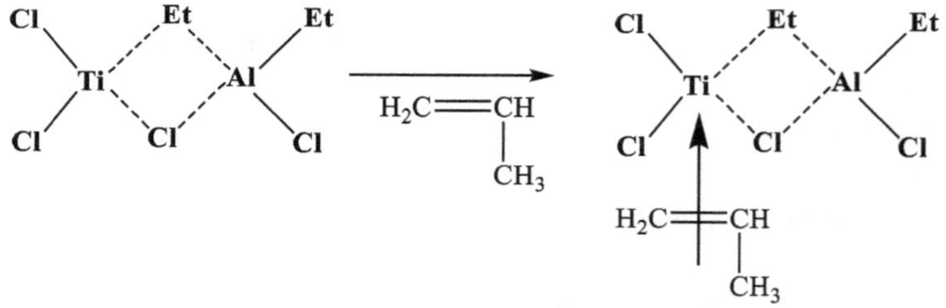

However, the widely accepted monometallic mechanism can be summarized as follows:

1. Formation of the active center: The cocatalyst (typically triethylaluminum) reacts with titanium tetrachloride to produce ethyltitanium chloride, which serves as the active species.

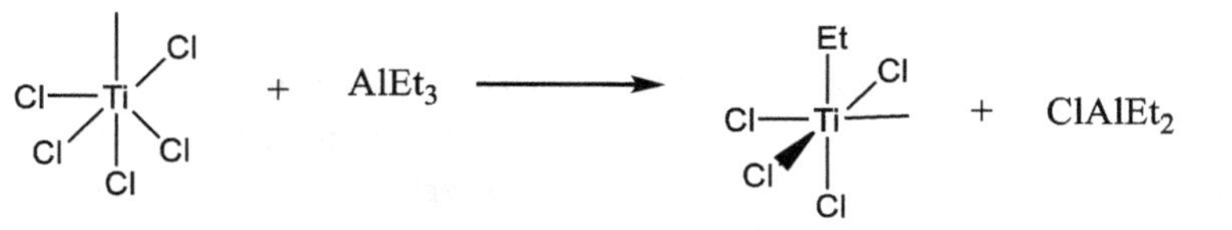

Titanium Chloride | Triethyl aluminium | Ethyltitanium chloride (active center) | Diethyl aluminium chloride

2. π-Complex formation: The incoming monomer (e.g., propylene) coordinates with the titanium center through its double bond.

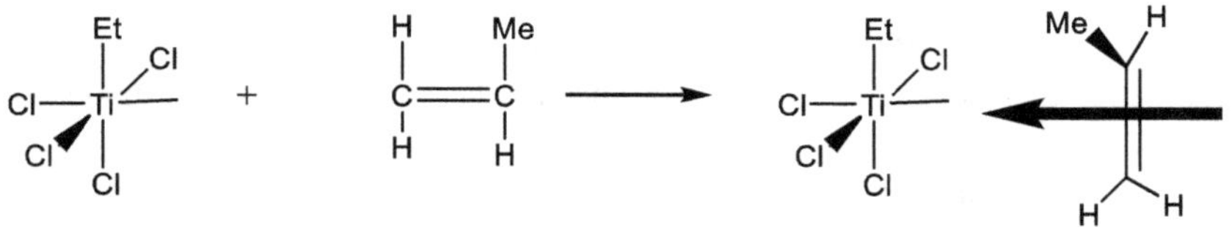

Ethyltitanium chloride | Propylene | Complex

3. Insertion step (chain propagation): The coordinated monomer inserts into the metal-carbon bond at the growing polymer chain end via a transition state complex. This regenerates the active site and placing the ethyl group at the terminus of a newly added propylene unit.

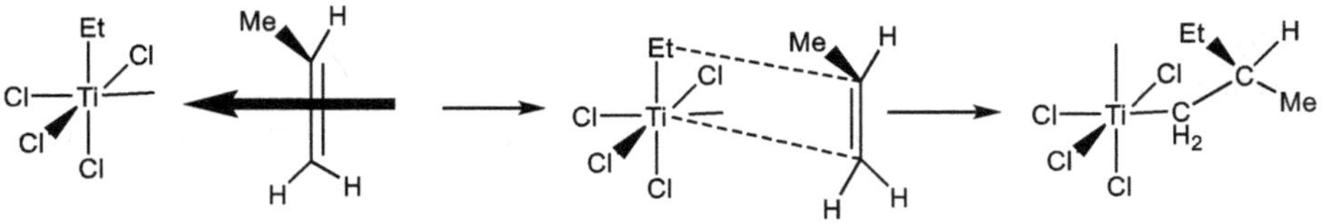

Complex | Transition state | Active center

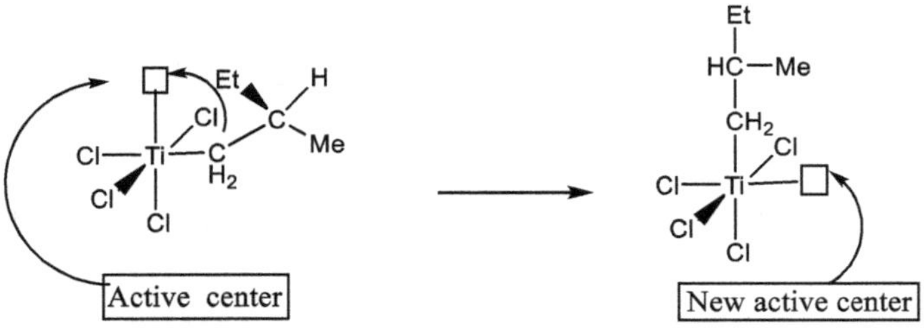

4. Stereoregulation: During propylene polymerization, the orientation of the incoming monomer is influenced by the surface geometry of the titanium salt – particularly structural irregularities such as edges, corners, and defect sites. This ensures that growth occurs in a controlled, linear fashion.

The cocatalyst not only initiates the process by forming ethyltitanium chloride but also maintains the activity of the catalytic site throughout polymerization. The degree and type of stereoregulation – isotactic, syndiotactic, or atactic growth – can be fine-tuned by varying the catalyst formulation, the choice of transition metal/co–catalyst pair, and reaction conditions.

This mechanism underpins many industrial-scale polyolefin syntheses, making Ziegler-Natta catalysts one of the most important classes of catalysts in modern polymer chemistry.

Titanium compounds, such as titanium tetrachloride, are typically used in the solid state, and the precise mechanism of polymerization may vary depending on the catalyst formulation and operating conditions.

During propylene polymerization, the incoming monomer positions itself between the titanium center and the terminal carbon of the propagating chain. This insertion step occurs at the surface of the titanium salt, most likely at structural irregularities such as corners, edges, or defects that act as active sites. Monomer addition consistently takes place at the chain end, ensuring that the polymer grows in a linear manner. The co-catalyst, triethylaluminum, plays a vital role in initiating the process. It reacts with the titanium-containing species to produce ethyltitanium chloride, which serves as the active polymerization site. This process forms the foundation of many industrial-scale polyolefin syntheses and remains a cornerstone of modern polymer chemistry.

As illustrated in the figures below, the same sequence of monomer coordination and insertion that occurs during initiation is repeated throughout propagation, ensuring the preservation of the stereoregular structure of the growing polymer chain.

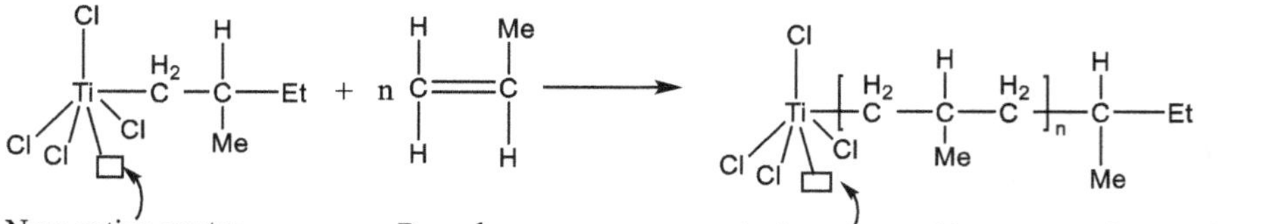

New active center — Propylene — Active center of isotactuc polypropylene

2.6 Polymerization Techniques

The method used to carry out polymerization [1,4] significantly influences the molecular weight, structure, properties, and applications of the resulting polymer. For students, it is important not only to understand the basic polymerization mechanisms but also the practical techniques used to carry out these reactions. The major polymerization techniques include:

1. Bulk polymerization
2. Solution polymerization
3. Suspension polymerization
4. Emulsion polymerization

These techniques are designed to:

(i) Control the rate of reaction.
(ii) Regulate molecular weight and molecular weight distribution.
(iii) Influence the microstructure of polymers (linear, branched, cross-linked, or stereoregular).
(iv) Ensure safety and feasibility for large-scale industrial production.

While bulk and solution polymerizations are commonly used in laboratory-scale synthesis, suspension, emulsion, and gas-phase methods dominate industrial processes. Understanding these techniques connects polymer chemistry with practical material design and processing.

2.6.1 Bulk Polymerization

Bulk polymerization involves polymerizing a pure monomer without the use of any solvent. A small amount of initiator is added to start the reaction, which is carried out either at constant temperature or under controlled heating. The process can be batch, semibatch, or continuous. Polymers commonly produced using this technique include PMMA (Polymethyl Methacrylate), PS, and PVC.

Advantages:

(i) Produces high-purity polymers with no solvent contamination.
(ii) High polymerization rate.
(iii) Simple process with minimal equipment.
(iv) No need for polymer-solvent separation.

Disadvantages:

(i) Poor heat dissipation, which may cause autoacceleration (Trommsdorff effect).
(ii) High viscosity as the polymer forms, making mixing difficult.
(iii) Not suitable for very large-scale production unless careful cooling is applied.

2.6.2 Solution Polymerization

In solution polymerization, the monomer and initiator are dissolved in a solvent, and the reaction occurs in this solution. The polymer forms within the solution and is later recovered by precipitation or solvent evaporation. This technique is commonly used for polymers such as polyacrylonitrile (PAN), polyvinyl acetate, and styrene-butadiene copolymers.

Advantages:
(i) Good heat control, as the solvent acts as a heat sink.
(ii) Easier management of highly exothermic reactions.
(iii) Lower viscosity facilitates better mixing.
(iv) Allows production of polymers with controlled molecular weight.

Disadvantages:
(i) Solvent removal adds extra cost.
(ii) Potential environmental pollution due to solvents.
(iii) Slightly lower polymer purity due to residual solvent.
(iv) Polymerization is generally slower than in bulk processes.

A comparison of the key features, advantages, and disadvantages of bulk and solution polymerization is summarized in Table 2.3.

Table 2.3: Comparison between solution and bulk polymerization.

Feature	Bulk polymerization	Solution polymerization
Reaction medium	Pure monomer (no solvent)	Monomer dissolved in a solvent
Heat control	Poor; heat buildup can occur	Good; solvent absorbs heat
Viscosity during reaction	High, becomes very viscous	Low to moderate; easier mixing
Purity of polymer	Very high	Slightly lower due to residual solvent
Polymerization rate	Fast	Slower than bulk
Recovery of polymer	Simple; no solvent separation needed	Requires solvent removal (precipitation and evaporation)
Equipment complexity	Simple	More complex; solvent handling and recovery needed
Control of molecular weight	Limited	Better control possible
Safety	Risk of overheating and autoacceleration	Safer due to heat dissipation

2.6.3 Suspension Polymerization

In suspension polymerization, the monomer, typically a liquid, is dispersed as small droplets in a nonsolvent, usually water, with the help of stabilizers or suspending agents (as shown in Figure 2.2). An initiator, soluble in the monomer, is added to start the polymerization within each droplet, and the mixture is stirred continuously to maintain a uniform suspension and ensure effective heat distribution. After the polymerization is complete, the resulting bead-like polymer particles are separated from the water, washed, and dried. This technique is widely used for polymers such as PVC, PS, and PAN. It is particularly suitable for producing polymers in pellet or bead form, which facilitates later processing and molding.

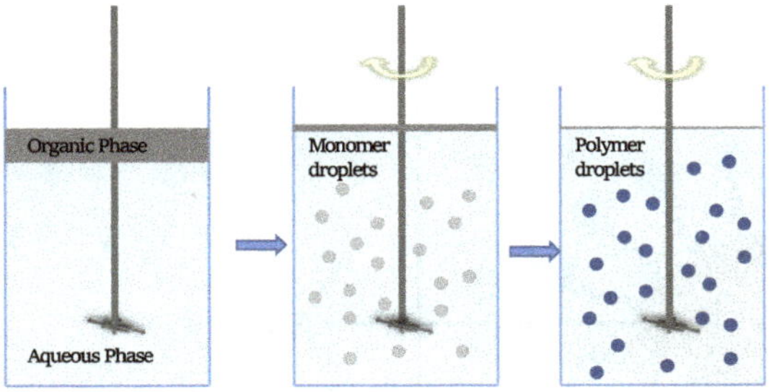

Figure 2.2: Schematic representation of the suspension polymerization process.

Advantages:
(i) Good heat dissipation due to the water medium.
(ii) Easier control of particle size and shape.
(iii) Produces free-flowing polymer beads that are easy to handle.
(iv) Reduced viscosity compared to bulk polymerization, allowing better mixing.

Disadvantages:
(i) Requires stabilizers or suspending agents, which may need removal. Since emulsion stability is dependent on surfactant concentration and the surfactant may need to be removed for some polymer applications.
(ii) More complex equipment than bulk or solution polymerization.
(iii) Not suitable for monomers that are highly soluble in water.
(iv) Applicable only to monomers that are not fully water–soluble.

2.6.4 Emulsion Polymerization

Emulsion polymerization is a heterogeneous polymerization technique in which monomer droplets are finely dispersed in water, with the aid of surfactants or emulsifying agents. Unlike suspension polymerization, the reaction primarily occurs in the micelles rather than in the bulk monomer droplets, producing high-molecular-weight polymers as stable latex particles.

Key components:

1. Monomer: Water-insoluble monomers such as styrene, vinyl acetate, and butadiene.
2. Water: Acts as the continuous phase.
3. Surfactant/emulsifier: Agents like sodium dodecyl sulfate (SDS) help form micelles and stabilize polymer particles.

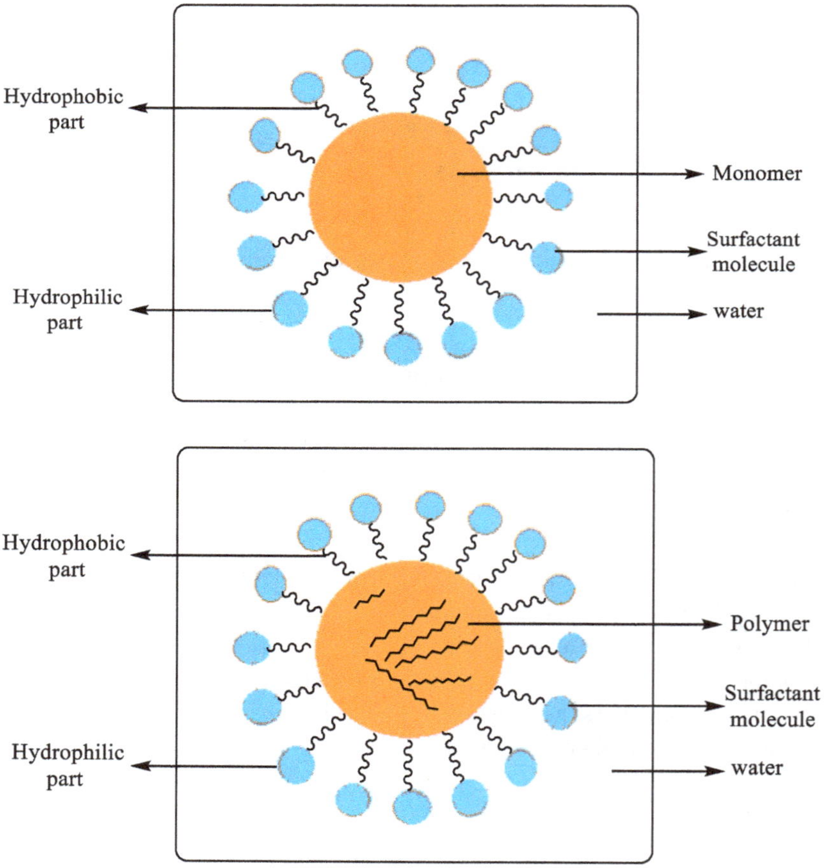

Figure 2.3: Representation of (a) Micelle formation (b) Polymer formation for Emulsion polymerization process.

4. Initiator: Water-soluble initiators, for example, potassium persulfate ($K_2S_2O_8$), generate free radicals in the aqueous phase, initiating polymerization.

Mechanism/steps:
1. Formation of micelles: Surfactant molecules aggregate in water, forming micelles that solubilize monomer molecules.
2. Initiation: A water-soluble initiator decomposes to generate free radicals, which react with monomer molecules within the micelles.
3. Propagation: Radical-monomer reactions continue inside the micelle, growing polymer chains.
4. Polymer particle growth: Growing polymer chains aggregate to form colloidal polymer particles (latex).
5. Termination: Free radicals are terminated by combination or disproportionation.

The process described above, from micelle formation to polymer particle growth, is illustrated in Figure 2.2.

Characteristics:
(i) Produces high-molecular-weight polymers rapidly.
(ii) Polymerization occurs mostly in micelles, not in bulk monomer droplets.
(iii) Results in stable aqueous polymer latexes.

Advantages:
(i) Produces high-molecular-weight polymers rapidly.
(ii) Good heat control due to the aqueous medium.
(iii) Generates stable latexes, which are easy to handle and process.

Disadvantages:
(i) Requires careful control of surfactant concentration. Since emulsion stability is dependent on surfactant concentration andthe surfactant may need to be removed for some polymer applications.
(ii) Postpolymerization latex recovery may involve additional steps.
(iii) Some residual surfactant may remain in the polymer, affecting properties.
(iv) Applicable only to monomers that are not fully water-soluble.

Table 2.4 provides a comparative overview of suspension and emulsion polymerization, detailing the major distinctions in aspects such as definition, form of monomer, type of dispersing agent and initiator, particle size of the polymer, polymerization rate, heat transfer efficiency, and final product characteristics. It further illustrates the underlying mechanisms, benefits, and representative polymers obtained by each technique, offering insight into their unique features and practical significance in industrial applications.

Table 2.4: Summary of the key differences between suspension and emulsion polymerization, highlighting their mechanisms, reaction media, advantages, and typical applications.

Feature	Suspension polymerization	Emulsion polymerization
Definition	Polymerization of monomer droplets suspended in a continuous phase (usually water) with the help of a stabilizer	Polymerization of monomer emulsified in water using surfactants, initiated by water-soluble initiators
Monomer form	Liquid monomer; droplets are relatively large (1–1,000 μm)	Insoluble monomer; droplets are very small (0.1–1 μm)
Dispersing agent	Protective colloids or stabilizers (e.g., gelatin and polyvinyl alcohol)	Surfactants (e.g., SDS and CTAB) forming micelles
Initiator	Water-insoluble initiator (e.g., benzoyl peroxide)	Water-soluble initiator (e.g., potassium persulfate)
Polymer particle size	Large, bead-like particles (micron-sized)	Very small, colloidal particles (nanometer-sized)
Rate of polymerization	Relatively slower; depends on droplet size	Faster; polymerization mainly in micelles
Heat transfer	Easy, due to water as the continuous phase	Very efficient, small particles, and water as continuous phase
Examples of polymers	PVC, polystyrene (beads), and PMMA	SBR (styrene-butadiene rubber), polyvinyl acetate latex, and acrylic latex
End product	Solid beads or granules	Colloidal latex (dispersion in water)
Mechanism	Polymerization occurs inside monomer droplets	Polymerization occurs inside micelles; monomer diffuses from droplets to micelles

This chapter has explored the fundamental mechanisms, kinetics, and practical techniques of polymerization, emphasizing how different methods influence polymer structure, molecular weight, and properties. By understanding the principles behind chain-growth, step-growth, and coordination polymerizations, as well as the role of catalysts and stereoregularity, students can appreciate the connection between molecular-level processes and macroscopic polymer behavior. The discussion of bulk, solution, suspension, and emulsion polymerization highlights how reaction conditions and medium selection impact polymerization efficiency and product form. Overall, mastering these concepts provides a solid foundation for the rational design, synthesis, and application of polymers in both academic research and industrial practice.

Practice Questions

1. Give examples for the initiators for cationic, anionic, and radical polymerization (two for each type).

2. What is the role of an inhibitor in free radical polymerization? Explain.

3. Describe the mechanism and kinetics by applying the steady state approximation to free radical polymerization.

4. Explain Mayo and Lewis to find monomer reactivity ratios.

5. Give the advantages of coordination polymerization over free radical polymerization.

6. What technique would you choose for producing a polymer of acrylonitrile and polyisobutylene: (i) cationic or (ii) anionic?

7. Define: Ceiling temperature of a polymer, telomerization, chain transfer, living polymerization, T_m of a polymer (thermal analysis), Flory solvent, and Flory temperature.

8. What is kinetic chain length? How is it related to the degree of polymerization?

9. Consider the following reactivity ratios for the copolymerization of two monomer pairs:
   $r_1r_2 = 1$, $r_1r_2 > 1$, $r_1r_2 < 1$

10. Specify the type of copolymer (random/alternate/block).

11. Explain the mechanism and kinetics of free radical polymerization.

12. Compare suspension and emulsion polymerization techniques by giving examples.

13. Describe the cationic polymerization of monomers along with its significance.

14. Discuss the coordination polymerization mechanism with representative examples.

15. Explain the chain transfer agent. Give one example.

16. Name the steps involved in the free radical polymerization of vinyl monomer. Explain one example.

17. Derive the kinetics of chain length in radical chain polymerization using the steady state approximation.

18. Polymer-like polyacrylonitrile, polyacrylates are often prepared by anionic polymerization. Explain.

19. Differentiate between:
    (i) Cationic and anionic polymerization
    (ii) Solution and bulk polymerization

20. Show the steps of polymerization of styrene initiated by $BF_3 + H_2O$, and discuss the kinetics of polymerization.

21. What is the role of an inhibitor in free radical polymerization?

22. Write down the structure of the polymer obtained from the reaction of the following monomers, and also name the preferred mechanism of the reaction:
    (i) $n\mathrm{HO(CH_2)_4OH} + n\mathrm{OCN(CH_2)_3NCO}$
    (ii) $n\mathrm{ClOC(CH_2)_5COCl} + n\mathrm{NH_2\ (CH_2)_6\ NH_2}$
    (iii) $n\mathrm{CH_2 = CH\text{-}COOCH_3}$

23. Compare the bulk and solution polymerization techniques by giving examples.

24. Explain, with the aid of chemical reactions, the differences between PE synthesized using a Ziegler-Natta catalyst and that PE prepared through free radical initiators.

25. What is a living polymer? Write down the structure of the polymer obtained from the reaction of the following monomers, and also name the preferred mechanism for the reaction:
    (i) $n\mathrm{NH_2{-}(CH_2)_5{-}\ NH_2} + n\mathrm{ClOC{-}(CH_2)_4{-}COCl}$
    (ii) $n\mathrm{HO{-}(CH_2)_4{-}OH} + n\mathrm{OCN{-}CH_2{-}CH_2{-}CH_2{-}NCO}$
    (iii) $n\mathrm{CH_2 = CH{-}CN}$

26. Write the structure and repeat unit of the following polymers:
    (i) Polyacrylonitrile
    (ii) Polyethylene terephthalate

27. Differentiate between random and copolymers with an appropriate example.

28. Give one example of a cationic, anionic, or radical polymerization initiator.

29. Identify the monomer(s) and write the name of the following polymers:

(i)

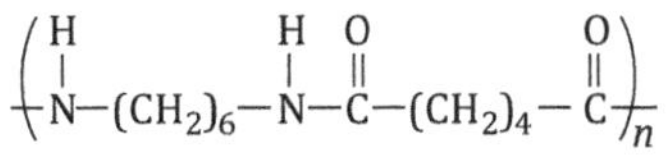

(ii)

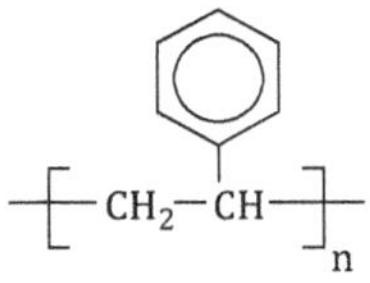

(iii)

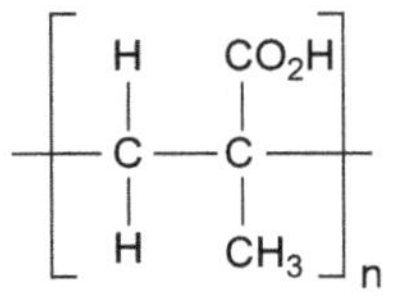

30. Write the structure of repeating units in:
   (i) Polyvinyl alcohol
   (ii) Polyethylene oxide
   (iii) Polyisoprene

Further Reading

[1] Carraher CE Jr. *Seymour's polymer chemistry*. 6th ed. Boca Raton (FL): CRC Press; 2013.
[2] Odian G. *Principles of polymerization*. 4th ed. Hoboken (NJ): John Wiley & Sons; 2004.
[3] Billmeyer FW. *Textbook of polymer science*. 3rd ed. New York: John Wiley & Sons; 1984.
[4] Gowariker VR, Viswanathan NV, Sreedhar J. *Polymer science*. New Delhi: New Age International Publishers; 2010.
[5] Sperling LH. *Introduction to physical polymer science*. 4th ed. Hoboken (NJ): John Wiley & Sons; 2005.
[6] Schildknecht CE, Zoss AO, McKinley C Vinyl Alkyl Ethers. *Ind Eng Chem*. 1947;39(2):180–186.
[7] Natta G. Macromolecular Chemistry: From the Stereospecific Polymerization to the Asymmetric Autocatalytic Synthesis of Macromolecules. Science. 1965 Jan 15;147(3655):261–72.
[8] Ziegler K. Consequences and development of an invention. Rubber Chemistry and Technology. 1965 Mar 1;38(1):23–36.

Ravi Kumar Thankur, Madhuri Chaurasia, and Mohan Kumar*

3 Polymer Solutions

Abstract: Polymer solutions play a pivotal role in science, which makes them an important aid in engineering and living systems. It is found in various things that we use in our daily lives, including clothes, food, and medicines. The fundamental properties of polymer solutions and the parameters that play a crucial role in their behavior are discussed in detail in this chapter. Also, their interactions with solvents and the conditions that regulate their behavior in solution are well explained. The thermodynamic behavior of polymer solutions is reviewed in detail, with a focus on polymer-solvent interactions and thermodynamic theories in terms of their miscibility and phase behavior. A robust conceptual background is built by reviewing seminal theoretical models related to polymer solutions, including Flory-Huggins theory and its variants.

In addition, this chapter covers the importance of polymer solutions in both industrial and biological applications. Polymer solutions are used in plastics, textiles, and coatings, thereby demonstrating their relevance in everyday materials. Furthermore, polymer solutions in biomedical applications, including drug delivery and tissue engineering, will show the potential for future innovations in the healthcare sector. This chapter also discussed conducting materials and smart polymers, which can be utilized in such a broad range of applications, including sensors, electronics, and so on. The chapter would be a very good connection between the knowledge and real-world applications to researchers, students, and professionals.

3.1 Introduction

Polymers are of much more importance to society, and they are extensively used materials in our day-to-day life. These are macromolecules that are composed of repeated monomer units linked together by a covalent bond. They are generally found in three forms: solid, liquid, and gel form, and they are widely used in multiple fields such as medicine, biology, and engineering of materials. Polymer solutions, essentially a uniform mixture that is composed of a polymer dispersed in any suitable solvent, are very important for understanding the polymers behavior in different conditions [1].

Polymer solutions significantly contribute to various industrial sectors such as coatings and paints, glue, drug delivery processes, and environmental remediation [2]. The study of their thermodynamic properties can ensure the engineering of innovative polymeric products in various pharmaceutical, food, and textile industries. The study will also help researchers and scientists to innovate such high-quality materials.

https://doi.org/10.1515/9783112208236-003

Polymer solutions, due to having macromolecular properties, exhibit detailed behavior compared to any molecular solutions, which helps in studying their unique thermodynamic and rheological properties [3].

3.1.1 Historical Background and Key Contributions

The study of polymer solutions and their behavior in different solvents was started by Staudinger and Flory in the 1920s [4, 5]. Staudinger explained the macromolecular theory, which reshaped the research and knowledge of polymer structure, while Flory explained the thermodynamic behavior of polymer solutions, highlighting the interaction theory of polymers and solvents. The Flory-Huggins theory was later established in the 1940s, which proved to be helpful in studying the behavior of polymer solutions and a quantitative idea of their solubility, osmotic, and phase behavior [6–8]. Years of investigations led to the innovation of various experimental techniques such as light scattering, viscometry, and rheology, which helped the researchers to acquire more accurate behavior of polymer solutions. These techniques further helped in the accurate identification of molecular weight, interaction parameters, and conformational changes, encouraging a deeper knowledge of polymer-solvent systems. In the recent past, modern computational techniques such as molecular dynamics simulations and self-consistent field theory have made considerable progress in the study of polymer solution behavior. The modern methodologies and developments in computational modeling and related techniques have expedited the innovative work in polymer science.

3.1.2 Significance of Polymer Solutions

The polymers, being macromolecular and having an adjustable shape and structure, bestow the polymer solutions with distinct behavior against the simple molecular solutions. The innovation in polymer solutions is of huge significance in the scientific as well as industrial growth, and their importance can be seen for several reasons:

1. **Physical aspects of polymers:** The essential characteristics of polymers, such as conformations of the polymer chain, solvent and polymer interactions, and coil-globule conversion, are significantly crucial for knowing the thermodynamic properties of polymer solutions. These studies enable scientists to get a deep understanding of the polymer solutions and work in a positive direction.
2. **Industrial applications:** Polymer solutions have wide applications in various sectors, such as in the manufacturing and processing of spinning fibers, coating substrates, and hydrogels used for medical applications.
3. **Pharmaceutical and biomedical uses:** Polyethylene glycol (PEG) and polyvinyl alcohol (PVA) are those polymers that are dissolved and shaped in such a manner

that they encapsulate drug molecules. Their behavior is distinct in solution, which affects bioavailability and diffusion rate.

4. **Material designing:** The characteristics of polymer solutions and their knowledge can be utilized for designing materials having certain optical, mechanical, and thermal aspects.
5. **Usage in environmental and chemical engineering**: Polymer solutions are widely used for the treatment of wastewater, recovery of oil, and membrane engineering. The water-soluble polymers, for example, perform as flocculants.
6. **Modern and progressive technologies:** Nanotechnology, 3D printing, and designing exclusive materials are the fields where polymer solutions have made a significant contribution. Some polymers, which respond to stimulus, when subjected to a change in pH, temperature, or light, also change their unique behavior also change accordingly. Therefore, these polymers, when present in solution, are able to create resilient and effective materials.
7. **Managing the behavior of polymer solution:** The properties like volatility and viscosity can be controlled to maintain the good quality of the material. Spinning of nylon fibers is an example of the textile sector, where nylon pellets are dissolved in any suitable solvent so as to make the solution viscous, which is then squeezed out to get the fibers.

3.1.3 Distinct Characteristics of Polymer Solutions

Polymer solutions show various distinct properties, which are attributed to their large size against the smaller-sized molecular solutions. The elongated structure of polymers is responsible for their entanglements, viscoelastic properties, and a significant dependency on molecular weight and concentration. In contrast to low-molecular-weight substances that dissolve uniformly, polymers exhibit phase transitions, coil-to-globule transformations, and conformational changes depending on the type of solvent [9]. The major difference between polymer solutions and small molecule solutions is the size difference between the solute and the solvent. Polymers are larger in size, and thus they interact with solvent molecules over a large surface area, resulting in nonideal mixing. Polymer chains also interact intramolecularly either by hydrogen bonding or van der Waals forces, leading to a change in their solution behavior. The viscosity of polymer solutions is mostly concentration dependent. For example, they behave as unconnected coils when present at lower concentrations. On the contrary, they overlap at higher concentrations, resulting in entanglement of chains and thereby altering the solution's viscosity and elasticity. These significant changes are important and utilized in industrial sectors related to coatings, drug formulations, and thickening of food items.

3.2 Properties and Criteria of Polymer Solutions

The distinct behavior of polymers in solution form is mainly dependent on the type of polymer, solvent, and molecular structure of the polymer. Understanding their behavior is most important for researchers to explore various applications of polymers. The longer chain molecules of polymer basically differ in their way of branching, cross-linking, and molecular weight, which control the important parameters like solubility, viscosity, and polymer-solvent interactions. Polymer solutions can readily transform into different conformations depending on their nature. Polymer chains usually enlarge in a preferred or good solvent due to positive interactions with the solvent molecules, while they tend to become smaller in an adverse solvent condition. Polymer solutions show some unique properties in contrast to simple solute-solvent mixtures, which are discussed below.

3.2.1 Properties of Polymers Due to Molecular Weight

1. **Molecular weight:** Polymers having high molecular weight show higher viscosity and elasticity in solution, attributed to their processing and end-use properties.
2. **Polymer chain structure:** Linear, branched, and cross-linked polymers correspondingly exhibit distinct solubility and rheological behaviors. For example, branched polymers show lower viscosity compared to linear ones having the same molecular weight, attributed to decreased entanglement. The shape of a polymer chain in solution mainly depends on the quality of the solvent. A good solvent favors polymer-solvent interactions, leading to chain expansion. In a theta solvent, the polymer behaves as an ideal chain. Poor solvent promotes unfavorable interactions, causing the chain to collapse into a compact structure [10, 11].
3. **Polydispersity:** The distribution of molecular weights (polydispersity index, PDI) affects the viscosity and stability of polymer solutions. A low PDI is basically preferred for applications demanding consistent material characteristics. The development of low-PDI polystyrene required for high-resolution lithography highlights the significance of regulating molecular weight distribution to get the desired performance.

3.2.2 Solvent Properties in Polymer Solution

1. **Polymer-solvent interaction:** The interaction between polymer and solvent molecules plays a crucial role in deciding the solubility and phase behavior of polymer solutions. This interaction is measured by a dimension called the Flory-Huggins interaction parameter (ξ), which denotes the energy involved in the mix-

ing of polymer and solvent. A value of ξ near zero indicates good solubility, and the values more than zero denote poor solubility [6–11].

2. **Nature of the solvent:** Good solvents usually lead to total dissolution of the polymer, forming a homogeneous solution, and poor solvents separate the different phases or form a gel.
3. **The polar nature of the solvent:** If the solvent is polar, it tends to solvate the polar polymers, and vice versa. A polar polymer like PVA is completely dissolved by dimethyl sulfoxide (DMSO), a polar solvent. The selection of solvent when preparing nanoparticles like poly(lactic-co-glycolic acid) (PLGA) for drug delivery affects both the distribution of particle sizes and the drug release characteristics of the nanoparticles.

3.2.3 Effects of Concentration on Polymer Solutions

The characteristics of polymer solutions are much influenced by the polymer concentration, with notable transitions in behavior taking place from dilute to semi-dilute states. The production of hydrogels from poly(*N*-isopropylacrylamide) (PNIPAM) solutions mainly relies on concentration, wherein the lower concentrations result in clear transparent solutions and higher concentrations yield opaque gels due to aggregation. Polymer solutions are also distinguished into three regimes depending on their concentration, as shown in Table 3.1 [12, 13].

Table 3.1: Variation of characteristics of polymer solutions with their concentration.

Dilute system	Semi-dilute system	Concentrated system
The chains of polymer are separated	Polymer chains overlap	Entanglements of polymer chains predominate
No overlapping of individual polymer coils	Intermolecular interactions start to dominate	Viscoelastic behavior starts to appear
Suitable for studying intrinsic properties like diffusion and molecular weight	Viscosity and osmotic pressure increase distinctly	Behavior similar to polymer melts

3.2.4 Rheological Investigation on Polymer Solutions

Rheology plays a pivotal role in the processing and manufacturing of polymers, which involves spinning of fibers, printing, paints and coatings, and film casting [14]. The rheological behavior of a polymer solution is enhanced by raising the concentration of the polymer. The flow properties, like shear thinning or thickening, are highly influential in manufacturing and processing techniques [15]. For instance, the shear-thinning property is good for inkjet printing, which has a smooth flow of ink. There are various important characteristics associated with the rheological behavior of polymer solutions:

3.2.4.1 Viscosity

Polymer solutions are more viscous than pure solute-solvent systems due to the larger size of polymer chains entangling in the solution. The intermolecular interactions become stronger with the increase in concentration of polymers, leading to a nonlinear rise in viscosity. This particular characteristic has a crucial role in techniques like extrusion and injection molding.

3.2.4.1.1 Relation Between Molecular Mass and Concentration on the Viscosity of Concentrated Polymer Solutions

The viscosity of concentrated polymer solutions is related to molecular mass and concentration, which is expressed by the power-law equation as shown in eq. (3.1). This was first proposed by Johnson et al., and it was applied by Onogi et al. [12, 15, 16]:

$$\eta = K\ c^{\alpha} M^{\beta} \tag{3.1}$$

where K denotes a constant term that depends on the nature of the system, M represents molar mass, and c is the concentration of polymer solution. For a polymer melt, the value of β is 3.4, and α ranges between 4.0 and 5.6.

3.2.4.1.2 Role of Temperature on the Viscosity

The effectiveness of temperature on the solvent viscosity (η_s) can be understood by Andrade's equation [17], which is given by the following equation:

$$\eta_s = B\,\exp^{\frac{E_a}{RT}} \tag{3.2}$$

where B is a constant term and E_a is the energy of activation, which usually ranges from 7 to 14 kJ mol^{-1} for solvents.

3.2.4.1.3 Interrelation of Viscosity and Rheology

Polymer solutions generally have a flow behavior of a non-Newtonian type on increasing the velocity gradient of the fluid, which is inversely related to their viscosity. This viscosity and velocity gradient relation is significantly important in coating industries. The polymer-solvent interactions present in any polymer solution are identified by the parameter called the intrinsic viscosity (η). The equation given by Mark-Houwink relates the intrinsic viscosity and molecular weight, which facilitates the characterization of a polymer [18].

Viscosity is directly related to the molecular weight of the polymer or the concentration of the polymer solution. For a dilute solution, viscosity is expressed as follows:

$$[\eta] = KM^{a} \tag{3.3}$$

wheere η signifies the intrinsic viscosity (associated with the solute), M denotes molecular weight, and K and a are constants for the given polymer. At higher concentrations, a non-Newtonian type of behavior is exhibited by the polymer solutions.

3.2.4.2 Phase Behavior

Polymer solutions show various types of phase behaviors such as gel formation, phase transitions, and crystallization. This behavior is mainly due to the nature of the solvent and the macromolecular properties of polymers. These phase behaviors take place at a certain fixed concentration, which depends on the temperature and nature of the solvent.

3.2.4.3 Viscoelasticity

Viscoelastic characteristics of polymer solutions are of practical significance when they flow through any apparatus. Concentrated polymer solutions exhibit higher analogy in the viscoelastic properties as compared to the dilute solutions ($c < c_{cr}$) [19]. For instance, thermoplastic elastomers are developed by a combination of thermoplastics with rubber elasticity by using the viscoelastic properties of the polymer solution made up of linear arrangements of blocks.

3.2.4.3.1 Viscoelastic Properties of Dilute Solutions and Concentrated Solutions

There are two important models related to the viscoelastic nature of polymer solutions, which are discussed below:

(i) **Rouse model**: It is simply the bead-spring model discovered in 1953, which is useful for treating the chain dynamic [20]. It is assumed here that the beads do not possess any excluded volume (considered here as a point), and they do not have any hydrodynamic interactions. In general, Rouse's model applies to moderately concentrated polymer solutions.

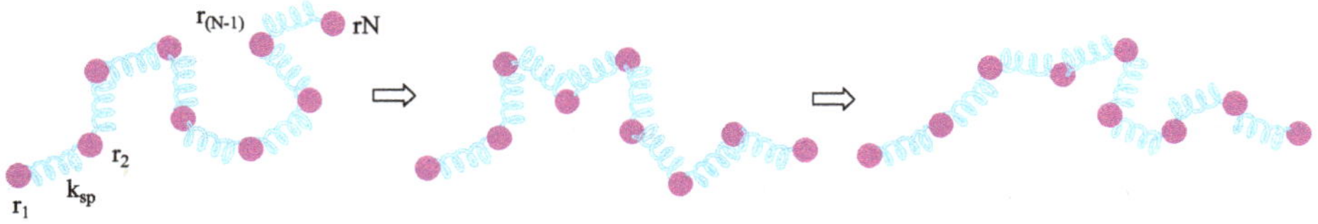

Figure 3.1: Bead-spring model of polymer chain exhibiting different conformational changes with time.

The Rouse model does not give the exact equation for the relaxation time related to different conformational changes. It provides information about the elastic forces exerted on the nth bead (n = 2, 3, . . ., N–1) due to the two adjacent springs connected to the respective beads (Figure 3.1) [19]. The force between (n–1)th and nth beads connected with a spring is given by $k\ (r_{n-1} - r_n)$. Similarly, a force is also exerted by another spring between the adjacent (n + 1)th and the nth bead as $k\ (r_{n+1} - r_n)$. Apart from this, some random force (f_n) is also experienced by the nth bead from the nearest molecules of solvent, which changes with time t. Thus, we have the expression for the motion associated with the nth bead, r_n, given below:

$$\mu(dr_n/dt) = k_{sp}\ (r_{m-1} + r_{m+1} - 2r_n) + f_n(t);\ (n = 1,\ 2,\ 3, \ldots,\ N) \tag{3.4}$$

(ii) **Zimm model:** B. Zimm worked on the improvement of the Rouse model by taking into consideration the hydrodynamic interactions among the beads [21]. After careful investigations, he derived the general expressions for the diffusion coefficient and the relaxation times based on some experiments and results. In general, Zimm's model applies to very dilute polymer solutions.

When there are no hydrodynamic interactions, the other beads are not affected by the movement of the mth bead; it only affects the spring force. However, when hydrodynamic interactions occur then it affects the velocity of the other beads (Figure 3.2) [19]. Thus, the general expression for the motion of the nth bead, r_n, is given as follows:

$$\left(\frac{dr_n}{dt}\right) = \sum_{m=1}^{N} H_{nm}\left[k_{sp}(r_{m-1} + r_{m+1} - 2r_m) + f_m(t)\right];\ (n = 1,\ 2,\ 3, \ldots,\ N) \tag{3.5}$$

where H_{nm} is a tensor of second rank that signifies how the nth bead's velocity is affected by the velocity of the mth bead through the solvent flow between them.

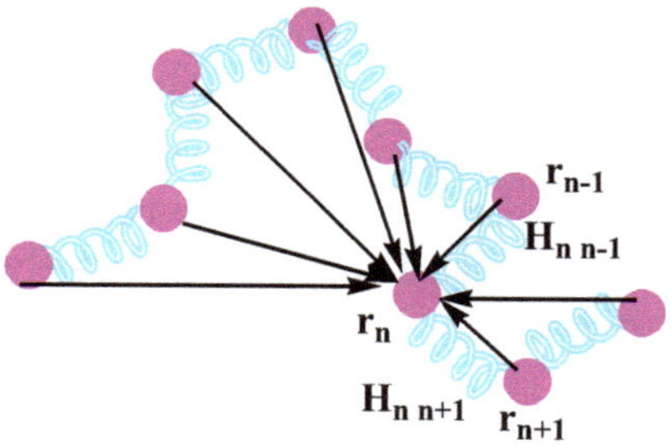

Figure 3.2: Hydrodynamic interaction affecting the motion of the nth bead by the motion of all the other beads.

The Rouse model is generally employed for the linear viscoelastic behavior of both entangled and unentangled semi-dilute solutions, while the Zimm model is utilized for the linear viscoelastic properties of both dilute and semi-dilute solutions. In the 1950s, both Rouse and Zimm presented their classical models, which signify the viscoelastic properties of dilute solutions, and both have bead-spring models. The Zimm model is also called a non-draining assumption, as there is consideration of hydrodynamic interaction between polymer and solvent, whereas, in the Rouse model, it is ignored and hence, called a free-draining assumption. For example, the polystyrene (III) solution containing a high molecular weight has a Rouse free-draining type of behavior. However, the same concentration of polystyrene solution with having low molecular weight behaves as Zimm non-draining.

The stress relaxation modulus is the same for both models, particularly with respect to the sum of N exponential relaxation modes as given below:

$$G(t) = \frac{cRT}{M} \sum_{p=1}^{N} \exp^{\frac{-t}{\tau p}} \tag{3.6}$$

where R represents the gas constant, c is the mass concentration of the polymer, p is the mode index, and τ_p is the respective relaxation time.

3.2.5 Transport and Diffusion Properties of Polymer Solutions

When the molecules are dissolved in a solvent and tend to move, such motion is called solution dynamics. The rate of diffusion of polymers in solution is lower than that of small molecules due to the difference in their hydrodynamic radius, and the mode of their movement is center-of-mass diffusion. The various methods to study polymer diffusion coefficients are dynamic light scattering (DLS) (Figure 3.3) [19] and pulsed-field gradient nuclear magnetic resonance (PFG-NMR). The principle involved in DLS has been used for years, as they are used in a few commercial particle-sizing systems. The analog signal coming out from the photodetector, $I(t)$, is modified into a digital signal by using the pulse-amplifier discriminator. Subsequently, the digital signal gets converted to the corresponding autocorrelation function of the signal with the help of an instrument called an autocorrelator (Figure 3.3).

The transport properties of polymer solutions have great significance in various fields, such as membrane science, controlled drug release, and enhanced oil recovery, where the fluid flow properties are modified by utilizing polymer solutions.

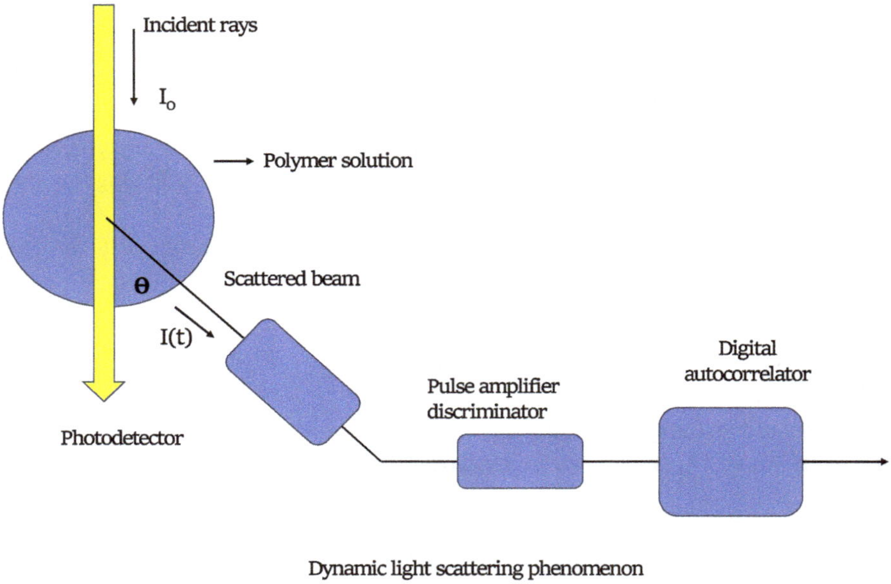

Figure 3.3: Dynamic light scattering (DLS) assembly.

3.2.5.1 The Stokes-Einstein Relation Applicable for Dilute Solutions

When a particle in a suspension of viscous solution, having a friction coefficient (μ), is pulled with a constant velocity v, a force of magnitude $F = \mu v$ needs to be applied constantly to the particle. Einstein demonstrated that the diffusion coefficient (D) of the given particle in a motionless solution at the given temperature (T) is related to the friction coefficient μ by the Nernst-Einstein equation as follows:

$$D = \frac{k_b T}{\mu} \tag{3.7}$$

It was later proved by the scientist Stokes that in the case of a sphere having radius R_s, the coefficient of friction μ is given by the following equation in a solvent of viscosity η_s.

$$\mu = 6\pi\eta_s R_s \tag{3.8}$$

Combining the above two equations, i.e., eqs. (3.7) and (3.8) give the Stokes-Einstein equation [19, 22]:

$$D = \frac{k_b \times T}{6 \times \pi \times n_s \times R_s} \tag{3.9}$$

where D represents the diffusion coefficient of a sphere and R_s stands for the Stokes' radius. The diffusion speed is higher at elevated temperatures, in a lower-viscosity solvent, and for smaller-sized particles. Now consider the Stokes' radius concept for nonspherical molecules. Once the center-of-mass diffusion coefficient (D) has been measured for the suspension or the molecule, the concept of the hydrodynamic radius R_H can be introduced, which is given by the following equation:

$$R_s = \frac{k_b T}{6\pi n_s D} \tag{3.10}$$

In the case of a spherical suspension or molecule, $R_H = R_S$.

3.2.5.2 Osmotic Pressure for Semi-Dilute Solutions

In the case of dilute solutions of neutral polymers, osmotic pressure helps to find the number average molar mass, which is measured as kT per molecule of solute (van't Hoff Law; van't Hoff, 1887) [23]. In the case of semi-dilute solutions containing neutral polymers, osmotic pressure determines the number density of correlation blobs. The characteristics and behavior of semi-dilute solutions are entirely distinct in contrast to dilute solutions. On increasing the concentration ten times, the osmotic pressure increases by a factor of several hundred times. Osmotic pressure is proportional to c in the case of an ideal solution. Moreover, the motion of the chain is slow in the case of semi-dilute solutions because of their entangled nature. Semi-dilute solutions containing a polymer of high molecular weight can hardly flow due to the highly viscous nature of the solutions, and can also behave similarly to elastic rubber. In the case of polymer solutions, osmotic pressure (π) is different, and it deviates from the ideal solution:

$$\Pi = \frac{RT_c}{M_n}\left(1 + B_2 c + B_3 c^2 + \cdots\right) \tag{3.11}$$

where B_2 and B_3 are virial coefficients showing polymer-polymer interactions.

3.2.6 Scattering of Light

The classical light scattering theory was first given by the scientist Rayleigh, which applies to smaller particles whose size lies below the wavelength (λ) of the scattered light:

$$s_{\text{rms}} < \frac{\lambda}{20} \tag{3.12}$$

where s is the radius of gyration and s_{rms} is the root mean square of s. There are some local fluctuations in the refractive index inside the medium present in the solutions, which result in light scattering. The general equation to determine the light scattering from solutions [24, 25] is given as follows:

$$R_\theta = \frac{KRT_c}{d\pi/dc} \tag{3.13}$$

where R_θ is the Rayleigh ratio or refractive index increment and c is the polymer concentration (kg m^{-3}).

Static light scattering (SLS) and DLS are the two methods for the characterization of polymer solutions. SLS gives information about absolute molecular weights, radius of gyration, and second virial coefficients. DLS gives an idea of time-dependent fluctuations and diffusion coefficients. The intensity of scattering depends mainly on the contrast in refractive index and concentration of polymer.

3.2.7 Criteria for Understanding the Properties of Polymer Solutions

Several factors can be considered for a comprehensive knowledge of the behavior of polymer solutions:

1. **Molecular weight of the polymer:** The properties such as viscosity, elasticity, and other bulk properties related to polymer solutions are directly proportional to the size of the polymer molecules. Macromolecular polymers with high molecular mass are more viscous, and those having a lower molecular mass readily dissolve.
2. **Concentration of polymer solution:** The polymer solutions exhibit ideal behavior at lower concentrations, and at higher concentrations, they show nonideal properties like chain entanglement and phase separation.
3. **Nature and quality of the solvent:** The nature of the interaction between the polymer and solvent predicts whether the polymer will swell or collapse. A high-quality solvent develops a swollen polymer configuration, i.e., extended chains, while a low-quality solvent forms a coiled or aggregated polymer.

3.3 Thermodynamics of Polymer Solutions

Thermodynamics provides a detailed knowledge about the stability, miscibility, and phase behavior of polymer solutions, giving a complete framework for their equilibrium behavior [26, 27].

3.3.1 Important Thermodynamic Terms

1. **Gibbs free energy (*G*):** It talks about the spontaneity of mixing of a polymer and solvent. The mixing will be spontaneous if the change in Gibbs free energy is negative ($\Delta G < 0$).

2. **Enthalpy and entropy**: It is the heat absorbed or released during mixing of polymer and solvent, and mixing is exothermic if polymer-solvent interactions are strong. Entropy is the measure of disorder taking place from polymer-solvent mixing, corresponding to the number of possible configurations. The change in enthalpy (ΔH) and the change in entropy (ΔS) when combined give the total free energy change upon mixing. These two factors are responsible for the miscibility of the polymer and solvent.
3. **Relation between free energy and phase behavior:** When polymer and solvent are mixed, the change in free energy is given by the following equation:

$$\Delta G = \Delta H - T\Delta S \tag{3.14}$$

The contributions made by enthalpy are mainly due to polymer-solvent interactions, while the freedom of polymer chains and solvent molecules to undergo any configurational change contributes to the entropy change. These enthalpic and entropic changes are accountable for the phase behavior of polymer solutions, such as phase separation or gel formation. For instance, a poor solvent will give a positive enthalpy of mixing, resulting in phase separation, while the contribution due to entropy dominates in a good solvent, resulting in the enhancement of mixing.

3.3.2 Solubility and Miscibility

The nature of the solvent plays a crucial role in the polymer solutions exhibiting their thermodynamic behavior. A good solvent strongly favors the solvent-polymer interac-

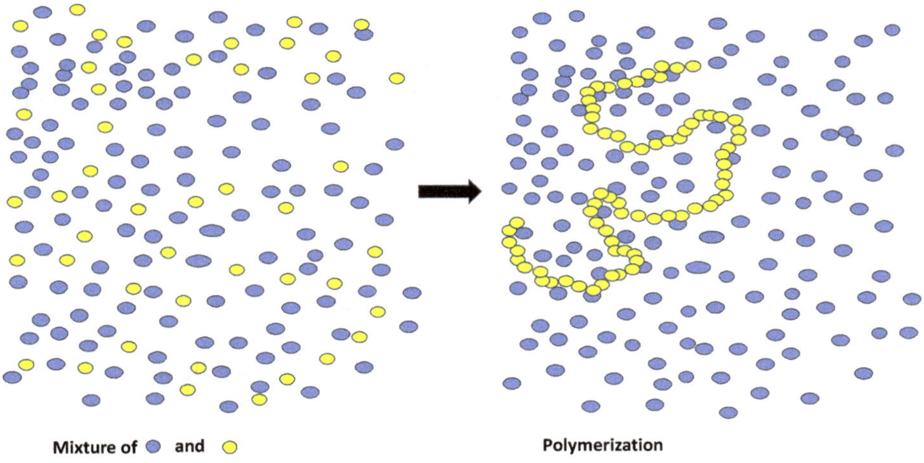

Figure 3.4: The effect of polymerization in a "good solvent."

tion. The characteristics of the polymer solution are controlled by some of the factors, like the solvent and polymer interaction and the degree of polymerization. Figure 3.4 shows the effect of polymerization, which takes place specifically in the "good solvent." The opposite effect is observed in the case of a poor solvent, leading to the accumulation of polymers.

The solubility of polymers in solvents is significantly dependent on two factors: entropy and enthalpy. A negative change in free energy during mixing will result in a miscible system. The theory proposed by Flory-Huggins provides a clear explanation of the theory behind the solubility of polymers in solvents, clearly describing the role of their interaction parameter and the molecular weight. The mean-field theory and the role of interaction parameters in studying the characteristics of polymer solutions will be discussed in the next section.

3.3.3 The Flory-Huggins Mean-Field Theory and Interaction Parameter

3.3.3.1 The Mean-Field Theory

Dissolution decreases the enthalpy, so when the polymer dissolute into a solvent, it also decreases the free energy of the system. It can also be observed from eq. (3.14), when the quantity enthalpy of mixing is smaller than the overall product of the temperature as well as the entropy of mixing, it reduces the free energy. On increasing the ratio of solvent molecules to the polymer, the entropy of the system does not increase, but it increases in the system containing solutes of low molecular mass, and thus, the polymer-solvent systems are immiscible.

The mean-field theory deals with the proper explanation of the miscibility of the polymer-solvent system. It presents an extended form of lattice fluid theory, which was earlier established to account for the miscibility of two low-molecular-weight liquids. Flory successfully introduced the application of the mean-field lattice fluid theory for polymer solutions.

3.3.3.2 Flory-Huggins Mean-Field Theory

It is a simpler version of the so-called lattice chain theory [6–8]. It is one of the most widely used lattice-based models to explain the thermodynamics of polymer solutions involving both the entropy and enthalpy of mixing, considering the most important parameter called the Flory-Huggins interaction parameter (χ), which quantifies the interaction taking place among the molecules of polymer and solvent. If the value of χ is positive, the interactions are unfavorable, which results in phase separation, and a negative value of χ leads to favorable mixing. Generally, a lower value of the chi (χ)

parameter promotes stronger interactions and better solubility. This theory assists in analyzing and comparing the free energy change before and after mixing the polymer and solvent molecules. It is specifically helpful in explaining the phase transition in solutions of a polymer suitable for phase separation.

The arrangement of polymer chains and solvent molecules is done by Flory-Huggins theory by using the lattice model. A simplified version of the lattice model is shown in Figure 3.5 [19, 28], where the given system contains n number of sites. In each of the sites, there is an occupancy of either a molecule of solvent or a monomer of the respective polymer. None of the sites can be occupied twice or left vacant. Every linear chain of polymer is occupied by N sites, which are connected by a series of N–1 bonds. There are N monomers present in the polymer chain, which are arranged into the empty sites in such a way as to get a total n_P number of chains. The vacant sites are subsequently covered by the molecules of solvent.

The volume fraction Φ of the polymer will be calculated by the following equation:

$$n_p = \frac{n_{\text{site}}\Phi}{N} \tag{3.15}$$

If the total number of solvent molecules is n_S and the volume fraction of the solvent is (1–Φ), then:

$$n_s = n_{\text{site}}(1-\Phi) \tag{3.16}$$

Before mixing polymer and solvent:

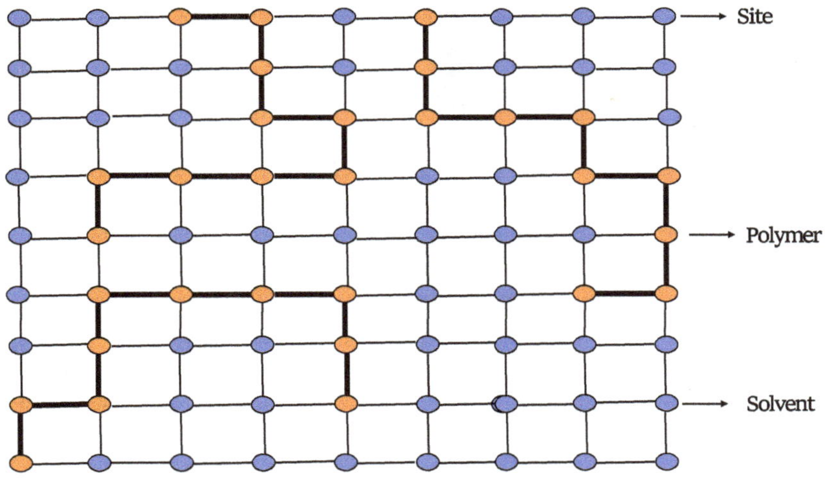

Figure 3.5: Lattice chain model in polymer solution; the blue site is occupied by solvent and the brown site by polymer molecules.

$$\text{Volume utilized by the polymer} = n_p\ N\ v_{\text{site}} \tag{3.17}$$

$$\text{Volume utilized by the solvent} = n_s\ v_{\text{site}} \tag{3.18}$$

where v_{site} is the volume per site. When the polymer and solvent are mixed, there is no change in the total volume ($n_{\text{site}}\ v_{\text{site}}$). Therefore, after mixing:

$$\text{Enthalpy of mixing: } (\Delta H_{\text{mix}})_p = (\Delta U_{\text{mix}})_p \tag{3.19}$$

$$\text{Gibbs free energy change: } \Delta G_{\text{mix}} = \Delta A_{\text{mix}} \tag{3.20}$$

where U is internal energy and A is Helmholtz free energy

3.3.3.3 Role of Entropy and Enthalpy

(i) **Entropy of mixing**: In the case of solutions having small molecules, there will be a larger number of molecules, and hence the entropy of mixing is also large. However, in the solutions of polymer, the entropy of mixing is significantly smaller due to their high molecular weight and low molar concentration. Therefore, Flory and Huggins have accounted for the number of lattice sites occupied by the polymer chain, and they computed the number of alignments of n_P chains on n-site sites and then compared it to the possible number of alignments on n_PN sites before mixing. Thus, they got the expression for the entropy of mixing ΔS_{mix} per site as shown below [6–8]:

$$\text{Flory-Huggins} \frac{-\Delta S_{\text{mix}}}{k_b n_{\text{site}}} = \frac{\Phi}{N} \ln \Phi + (1-\Phi) \ln (1-\Phi) \tag{3.21}$$

The change in entropy of mixing, $\Delta S_{\text{mix,}}$ given by eq. (3.17), is usually higher than the entropy of mixing obtained when an ideal solution containing n_P solute molecules is mixed with the n_S molecules of solvent. The observed deviation is mainly due to a large number of conformations shown by a polymer chain.

(ii) **Enthalpy of mixing and Flory-Huggins parameter (χ)**: The enthalpy of mixing for polymer solutions is due to the interactions taking place between polymer and solvent molecules, and it is measured by the Flory-Huggins interaction parameter (χ) [29]. The entropy of mixing is smaller for a system containing polymer and solvent molecules at low concentrations. Therefore, miscibility is governed by the enthalpy of mixing, which results from the change in the interactions upon mixing.

In the lattice fluid model, interactions take place between the connected neighbors, and these interactions exist in close proximity. Let us denote ε_{PP}, ε_{SS}, and ε_{PS} as the interactions for a polymer-polymer (P-P), a solvent-solvent (S-S), and a polymer-solvent (P-S) contact, respectively (Figure 3.6) [19, 30]. When polymer and solvent mix

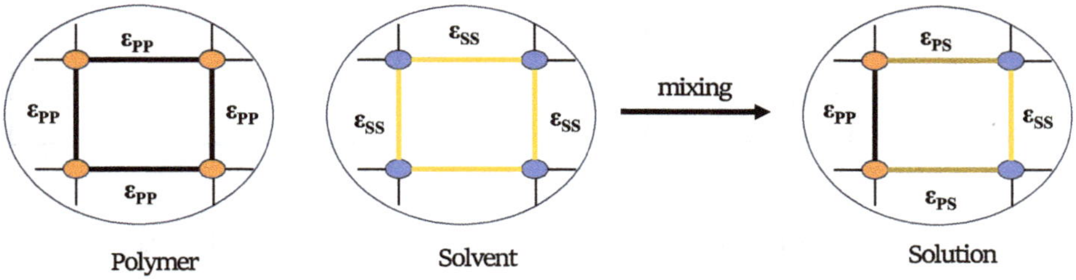

Figure 3.6: Interaction between the nearest neighbors and change in contacts on mixing of polymer and solvent molecules.

with each other, rearrangement of contacts takes place, and there is a change in overall interaction energy. Figure 3.6 shows the interaction between the nearest neighbors and the respective change in contacts.

The **Flory-Huggins parameter (χ)** is mathematically defined as the product of the lattice coordinate (Z) and the change in energy divided by K_BT:

$$\chi = \frac{Z\left[\varepsilon_{\mathrm{ps}} - \left(\varepsilon_{\mathrm{pp}} + \varepsilon_{\mathrm{ss}}\right)/2\right]}{k_B T} \tag{3.22}$$

The symbol χ is an interaction parameter that represents the scale of affinity between the molecules of polymer and solvent. A negative χ value favors the P-S contacts, while a positive χ value promotes P-P and S-S contacts. If the value of χ is low, it leads to better solvation of the polymer and good solubility, whereas a higher value of χ leads to poor solubility, resulting in phase separation.

3.3.4 Phase Separation and Critical Solution Temperatures

The solvent feature can be modified in two ways: firstly, with the change in temperature for a given polymer solution, and secondly modifying the solvent composition. When the temperature is changed, the coexistence curve will be drawn on a temperature-composition plane. Considering temperature (T) as the ordinate in place of χ and solvent composition (Φ) as another ordinate, we can plot two phase diagrams: (a) upper critical solution temperature (UCST) type and (b) lower critical solution temper-

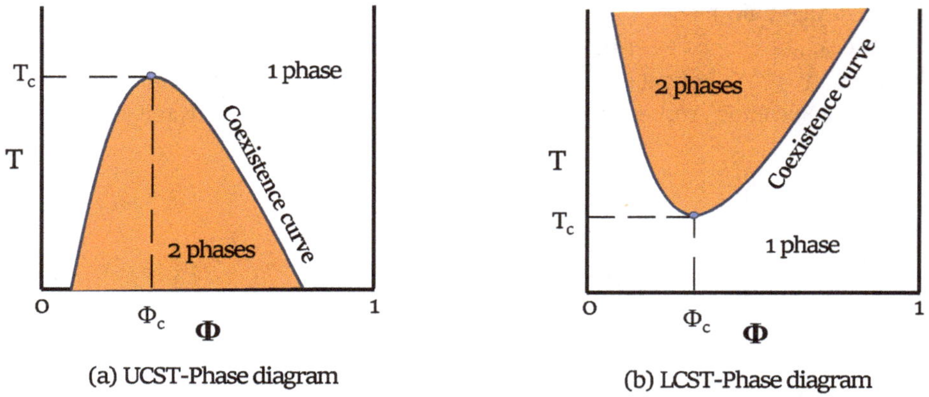

Figure 3.7: A polymer solution phase diagram on temperature-composition ordinates: (a) UCST type and (b) LCST type.

ature (LCST) type (Figure 3.7) [31]. The peaks appearing in the coexistence curve are commonly called critical points. Therefore, at the critical point, Φ_c represents critical composition, and likewise, T_c represents the critical temperature (T_c).

Polymer solutions basically show two types of phase separation, as mentioned below. They are highly significant in drug delivery systems, in which the solubility of the polymer is controlled for releasing the target.

1. **Upper critical solution temperature (UCST):** When the critical temperature (T_c) is present at the apex of the coexistence curve, it is called UCST. The solution changes to a miscible state above this critical temperature, while there occurs a phase separation occurs below this critical point. For example, cooling polymers like polystyrene in cyclohexane shows phase separation.
2. **Lower critical solution temperature (LCST):** When the critical temperature (T_c) is present at the lowest point on the coexistence curve, it is called LCST. The solution becomes miscible at the lower side of this critical point, while the solution is immiscible above this point, leading to phase separation. For example, when polymers like PNIPAM (Poly(N–isopropylacrylamide) are heated, they go through a phase separation.

3.4 Theories Associated with Polymer Solutions

The distinct properties of polymer solutions are due to the macromolecular size of the polymer. Theoretical models illustrate a better understanding of the behavior of polymer solutions; several theories have been developed so far to account for such distinct characteristics. Some of the commonly used theories are discussed here.

3.4.1 Flory-Huggins Theory

The Flory-Huggins theory is the most prevalent model for a deeper understanding of polymer solutions and their characteristics, which describes the entropic and enthalpic interactions between solvent molecules and polymer.

3.4.2 Kirkwood-Buff Theory

The Kirkwood-Buff (KB) theory, proposed in 1951 by the scientists J.G. Kirkwood and F.P. Buff [32], explains the molecular-level studies helpful for understanding the thermodynamics of polymer solutions. It provides a correlation between polymer segments in solution and macroscopic thermodynamic properties like viscosity and diffusion. The limitation of the KB theory lies in assuming equilibrium conditions and not accounting for polymer connectivity or chain conformational entropy. This issue was solved by using intramolecular distribution functions along with Flory-Huggins theory.

3.4.2.1 Basic Principles and Kirkwood-Buff Integrals (KBIs)

The KB theory has established a clear relationship by using the molecular distributions between various components present in solution to the respective thermodynamic properties of the mixture. The theory relies on Kirkwood-Buff integrals (KBIs), and the integral expressions are derived from the radial distribution functions. It extends the relationship between the chemical potential and the spatial pair correlation functions, G_{ij}, also termed KBIs. It provides an expression that links the changes in chemical potential of solution components to the respective concentrations, at constant temperature and pressure. The KBI theory applies to different types of stable mixtures, which may contain multiple or any number of components. The KBI between the ith and jth components' ij is expressed as the spatial integral over the pair correlation function (eq. (3.23)):

$$G_{ij} = \int \left[g_{ij}(R) - 1\right] \, dR \tag{3.23}$$

In the case of spherical symmetry, the KBI expression is given by:

$$G_{ij} = \int_0^\infty \left(g_{ij}(r) - 1\right) \, 4\,\pi\, r^2 dr \tag{3.24}$$

where G_{ij} is the KBI for ith and jth species, $g_{ij}(r)$ is the radial distribution function (RDF), and r is the intermolecular separation.

3.4.2.2 Application of Kirkwood-Buff Theory to Polymer Solutions

Polymer solution explores various applications of KB theory, which may include the understanding of favorable solvation techniques, different conformations of polymer, and thermodynamic behavior of solution [33]. When a polymer is dissolved in an adequate solvent, the local as well as the major part of the composition differs everywhere in the polymer segments. These differences are quantified by the KB theory, which is connected to the quantifiable properties like chemical potentials and activity coefficients. KB theory has been extensively utilized in the case of co-solvency and co-nonsolvency phenomenon, the hydrophobic breakdown of polymers, and protein denaturation by compatible solutes, specifically in polymer systems.

In a dilute polymer solution, let us assume that the polymer is component 1 and the solvent is component 2. The solvent molecules bind to the polymer, and the binding energy will be calculated by the difference as follows:

$$\Delta G = G_{12} - G_{22} \tag{3.25}$$

When the value of ΔG is higher than zero, solvent molecules surround the polymer chain, enhancing the solubility.

3.4.3 Scaling Theory

The scaling theory of polymer solutions established by Pierre-Gilles de Gennes in the 1970s [34] has made a remarkable impact on the investigation of the behavior of polymer solutions. The theory is well applicable to high-molecular-weight polymer solutions and dilute or semi-dilute concentrations. It illustrates how different macromolecular properties of polymer solutions, such as viscosity, the radius of gyration, and diffusion coefficient, scale with molecular weight and related parameters like polymer chain length, concentration, and nature of solvent. The theory also illustrates the behavior of polymer solutions in both the semi-dilute and concentrated regimes, where the chains overlap, creating an interconnected network-type structure. It also talks about the behavior of polymer solutions dependent on concentration, which is the basic reason for the variations in properties such as viscosity and diffusion coefficient. Scaling theory accounts for the interactions due to excluded volume as well as chain conformational statistics, with more accuracy specifically in good solvents and dilute to semi-dilute regimes.

3.4.3.1 Scaling Theory on Polymer Systems with Different Concentration Regimes

Polymer solutions show similar properties at various length scales and, therefore, the behavior of long polymer chains can be explained by power-law relationships or scaling laws [35]. It does not require any such kind of knowledge about microscopic interactions. Scaling theory classifies the polymer solutions based on various concentration regimes, each of which is characterized by specific scaling behaviors:

1. **Scaling in the dilute regime**: In this case, the polymer coils do not overlap, and particularly, in a high-grade solvent, the polymer chain generally swells due to favorable polymer-solvent interactions. It was observed that the size of the polymer chain, i.e., the radius of gyration (R), scales with the number of monomers (N), and the relation is expressed as follows:

$$R{\sim}a\ N^{\nu} \tag{3.26}$$

   where a denotes the size of the monomer and ν has a value close to 0.59.

2. **Scaling in the semi-dilute regime:** Here, the polymer coils can overlap without having any chain entanglement in the polymer solution. The intermolecular interactions make the polymer chain convert into a network-like structure. When the concentration of the polymer reaches beyond the overlap concentration c^*, the chains start to interpenetrate, forming a solution containing entangled fragments. The correlation length scale (ξ) signifies the effective distance between the polymer segment and other chains. This is represented as follows:

$$\xi{\sim}c^{\frac{3\nu}{3\nu-1}} \tag{3.27}$$

   The different macroscopic parameters like osmotic pressure (Π), viscosity (η), and diffusion coefficient all obey scaling laws:

$$\Pi{\sim}K_B T c^{\frac{3\nu}{3\nu-1}} \tag{3.28}$$

$$\eta{\sim}c^{k} \tag{3.29}$$

   where in eq. (3.29), the term k depends specifically on the regime and solvent quality.

3. **Concentrated regime:** In this case, entanglements and interactions increase to a greater extent. The overlap concentration (c^*) is the intersection between the semi-dilute state and the concentrated regime, which is expressed as follows:

$$c^*{\sim}\frac{N}{R^3}{\sim}N^{(1-3\nu)} \tag{3.30}$$

3.4.3.2 Applications of Scaling Theory

Scaling theory has made a notable contribution to the field of polymer solutions and has been successfully utilized to explain:

(i) Light scattering data of polymer solutions.
(ii) Determination of osmotic pressure in polymer solutions.

(iii) Measurement of viscosity and diffusion coefficients.
(iv) Application in polymer brushes and polyelectrolyte solutions.

3.4.4 Reptation Theory

Polymer chains can easily move in dilute solutions, and their motion and related dynamics have been well explained by the Rouse model or the Zimm model. Semi-dilute and concentrated solutions of polymer are extremely viscous; they can behave like elastic rubber at a particular range of time and frequency. Concentrated polymer solutions and the respective melts are a multiple-particle complex system. The movement of the polymer chains is constrained in concentrated polymer solutions as each of the polymer chains is entangled with various other chains. One of the most widely used theoretical models for polymer dynamics in such constrained environments is the reptation model, which was initially introduced by Pierre-Gilles de Gennes [34] in the 1970s and later developed by Doi and Edwards [36] in the 1980s. The issue was resolved by Doi-Edward's reptation model, assuming that a polymer chain in an entangled system moves (reptates) in a hypothetical tube formed by the constraints of other polymer chains (Figure 3.8) [37]. The polymer chain moves along the axis of the tube back and forth but does not cross the boundaries of the tube due to entanglements. This constrained motion is called reptation (derived from the word "reptile"), which is similar to the slithering of snakes passing through one another.

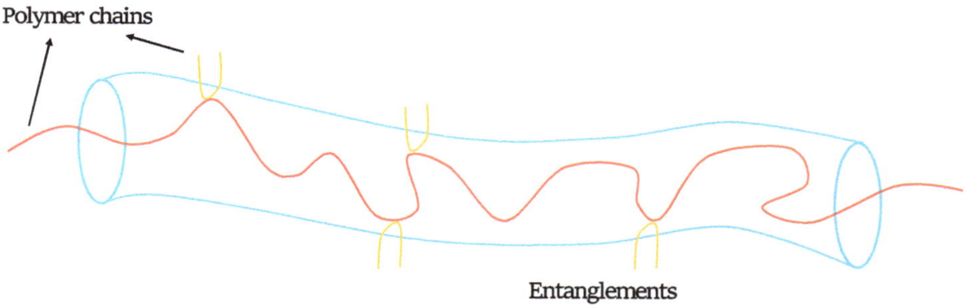

Figure 3.8: Tube model – the effect of polymer chain entanglements.

3.4.4.1 Reptation Time

The motion of the polymer chain along the tube is one-dimensional, and the most important quantity is the reptation time [38] or relaxation time (τ), the time taken by the chain to travel out of the tube. According to this theory, if there is an entangled system, then reptation time τ is directly proportional to the cube of molecular mass as follows:

$$\tau \propto M^3 \tag{3.31}$$

Let the length of the tube be L through which the polymer chain moves. Considering the overall motion of the polymer chain, the tube mobility (μ_{tube}) is given by the following equation:

$$\mu_{tube} = \frac{v}{f} \tag{3.32}$$

where v is the velocity of the polymer chain pulled by the force f. The movement of polymer chains inside the tube appears to be a Brownian type of motion, which is identified as the diffusion coefficient of the tube D_{tube}, utilizing Einstein's relation. Let the mobility of the polymer segment be μ_{seg}, then the mobility of the tube (μ_{tube}) having N (degree of polymerization) number of segments will be given by the following equation:

$$\mu_{tube} = \frac{\mu_{seg}}{N} \tag{3.33}$$

where μ_{tube} is inversely proportional to the degree of polymerization (N). According to the well-known Einstein's relation, the coefficient of diffusion D_{tube} for the confined tube is expressed as follows:

$$D_{tube} = k_B T \mu_{tube} \tag{3.34}$$

where the constant term k_B represents the Boltzmann constant and T represents the usual absolute temperature.

The mean squared displacement $s(t)^2$ due to one-dimensional Brownian motion is given as follows:

$$s(t)^2 = 2D_{tube}t \tag{3.35}$$

Using eq. (3.34) in the above equation, we get the following:

$$s(t)^2 = 2(k_B T)t\mu_{tube} \tag{3.36}$$

Therefore, the required time that a polymer chain takes to travel through the confined tube of length L is given by the following equation:

$$t = \frac{L^2}{2k_B T\ \mu_{tube}} \tag{3.37}$$

It was observed that the above time (t) is comparable to the reptation time (τ); therefore, we can write the following equation:

$$\tau \propto \frac{L^2}{\mu_{tube}} \tag{3.38}$$

Now, we know that the length of the tube (L) is directly proportional to the degree of polymerization (N), and μ_{tube} is inversely proportional to N.

So, the relation becomes:

$$\tau \propto N^3 \text{ and } \tau \propto M^3 \tag{3.39}$$

3.4.4.2 Important Assumptions Made in the Reptation Theory

(i) Entanglement of polymer chains with other neighboring chains, and for a short period, the polymer chains wiggle within the confined tube.
(ii) The polymer chains cross along the axis of the tube in one-dimensional motion.
(iii) Polymer chains do not cross each other, and the boundary of the imaginary tube is due to the constraints caused by the tube.

3.4.4.3 Limitations and Modern Approach

(i) The reptation model is less effective in the case of short polymer chains.
(ii) It considers the tube as a static object, but actually, it is formed due to the motion of the surrounding chains.
(iii) Molecular dynamics simulations and single-molecule experiments have proven to be more sophisticated in the field of polymer solution dynamics, helping researchers to contribute beyond the assumptions of reptation theory.

3.5 Industrial and Biological Applications of Polymer Solutions

Polymer solutions are important in many industrial and biological applications because of their unique physicochemical properties, versatility, and tunability. Processing and functionalization can be done easily in their solution form, which is not possible in the solid state. Various factors, including temperature and pH, as well as their interactions with other chemicals, can significantly alter the nature of the polymer [39]. The diverse applications of polymer solutions are possible only because of their unique nature and adaptability, and that is the reason it is a crucial material in many industries. The properties of polymer solutions can be altered by using various solution processing techniques, including solvent casting and phase inversion [40]. Figure 3.9 depicts the various factors that influence polymer solution applications.

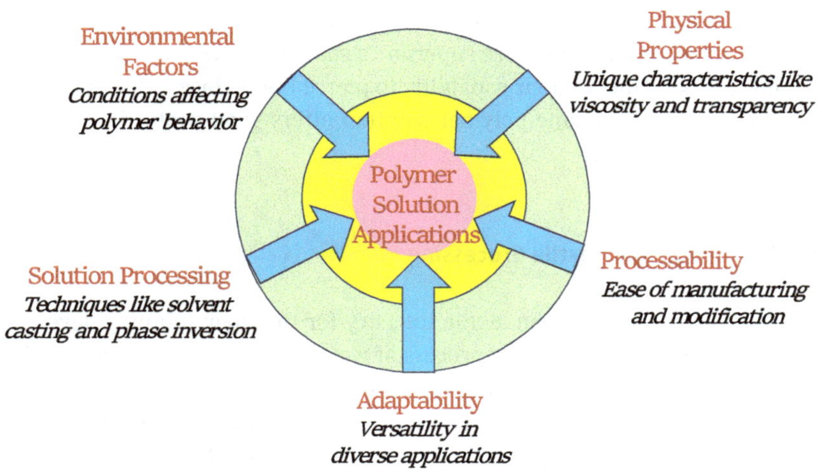

Figure 3.9: Factors influencing polymer solution applications.

3.5.1 Industrial Applications

3.5.1.1 Coatings and Surface Treatments

Polymer solutions are extensively used in various industries such as paints, varnishes, and specialized coatings. The solvent evaporates when polymer solutions are applied, leading to polymer films that make available functional properties. Acrylic solutions are very durable, can retain color, and offer many other environmental benefits in comparison to other substitutes [41]. The usage of polyurethane and epoxy solutions in the manufacturing of coating materials is prominent in the machinery and infrastructure. The viscosity of these solutions can be controlled by using polymer concentration and molecular weight distribution, which is a primary requirement in the spray coating or dip coating. Fluoropolymer solutions are very important for electronic protection because they offer a hydrophobic coating [42]. The manufacturing of transparent electrodes for touchscreen and photovoltaic devices can be achieved by using a conductive polymer solution [43].

3.5.1.2 Adhesives and Sealants

Polymer solutions are widely used in the adhesive industry, which manufactures materials that can be used to bond things efficiently. The functionality and molecular weight of the polymer solutions regulate their resistance properties. Typically, structural adhesives start off as reactive polymer solutions that are without difficulty cured after application, including cyanoacrylates, polyurethanes, or epoxies [44]. Adhesives made from polymer solutions are superior to solid adhesives as they establish strong mechanical locking due to their capability to pierce the surface of the substrate before polymerization [45]. Silicone polymer solution offers good protection against moisture along with robustness.

3.5.1.3 Fiber Production and Textile Processing

The usage of polymer solutions in the textile industry for the fiber spinning developments is very common these days. In the process of wet spinning, cellulose can be dissolved in an appropriate solvent, for example, *N*-methylmorpholine *N*-oxide [46]. Polyamides, polyesters, or polyacrylonitriles (PANs) are widely used polymer solutions in electrospinning [47]. Many textile finish treatments are achieved by employing polymer solutions to impart the desired properties to the textiles (e.g., water repellence, flame resistance, and antimicrobial activity) [48]. These textile finishes encompass the surface fabric of the textile, modifying desired properties lacking interrupting or limiting the comfort of wearing the fabric while preserving its breathability. The relationship of industry, types of polymers, and the applications of polymer solutions is given in Table 3.2.

Table 3.2: Polymers in various industries, their polymer types, and applications.

Industry	Polymer types	Applications	References
Biomedical	Polylactic acid (PLA), polyglycolic acid (PGA), and polycaprolactone	Tissue engineering, drug delivery, implants, and medical devices	[49]
Packaging	Polyethylene (PE), polyethylene terephthalate (PET), polypropylene (PP), and polyhydroxyalkanoates (PHA)	Food packaging, bottles, containers, and sustainable packaging	[50]
Automotive	Acrylonitrile butadiene styrene (ABS), polycarbonate (PC), polyamides (PA), and polyurethane (PU)	Dashboards, bumpers, interior parts, and lightweight components	[51]
Water treatment	Polysulfone (PSF), polyethersulfone (PES), polyvinylidene fluoride (PVDF), and polyamide (PA)	Membranes for filtration, desalination, and wastewater treatment	[52]
Electronics	Polyimides (PI), polydimethylsiloxane (PDMS), liquid crystal polymers (LCP), and fluoropolymers	Flexible displays, circuit boards, sensors, and wearable electronics	[53]

Table 3.2 (continued)

Industry	Polymer types	Applications	References
Construction	Polyvinyl chloride (PVC), expanded polystyrene (EPS), polyurethane foams, and fiber-reinforced polymers (FRP)	Insulation, pipes, window frames, and structural reinforcement	[54]
Renewable energy	Polyethylene naphthalate (PEN) and thermoplastic polyurethane (TPU)	Solar panels and wind turbine blades	[55]
Agriculture	Polyethylene (PE), polyvinyl chloride (PVC), biodegradable polyesters, and starch-based polymers	Greenhouse films, mulch films, irrigation systems, and controlled-release fertilizers	[56]
3D printing	Polylactic acid (PLA), acrylonitrile butadiene styrene (ABS), and thermoplastic polyurethane (TPU)	Rapid prototyping, custom parts, medical models, and manufacturing	[57]

3.5.2 Uses of Polymer Solutions in Plastics, Textiles, and Coatings

Polymer solutions are used as key intermediates in the manufacture and processing of plastics, textiles, and coatings, particularly across many sectors. Their excellent properties, processability, and ability to make films and structures are all critical for modern manufacturing. This chapter reviews the range of polymer solutions in the three major areas mentioned above. Table 3.3 shows the relationship between the characteristics of polymer solutions and various fields, such as biomedical, plastics, textiles, coatings, and environmental.

Table 3.3: Polymer solution applications in diverse fields.

Characteristic	Plastics, textiles, and coatings	Biomedical	Conducting and smart	Environmental
Processing	Electrically cross-linking into uniform films with controlled thickness	Drug-loaded particles or implants	Conductive coatings, flexible displays, solar cells, wearable sensors	Networks for water absorption
Application	Decorative finishes, corrosion protection, functional surfaces	Controlled drug delivery systems	Flexible displays, solar cells, and wearable sensors	Water management and removal of particles
Property control	Rheological properties for application	Polymer-drug-solvent interactions for release	Chain conformation and charge density for conductivity	Chain conformation and charge density for flocculation

3.5.3 Plastics Manufacturing and Processing

3.5.3.1 Solution Processing of Thermoplastics

Solution casting methods are also capable of constructing particular plastics from polymer solutions. For instance, the melt processing of high-performance engineering plastics, such as polyimides (which have poor processability owing to their thermal and mechanical properties), can be made possible by dissolving in a solvent (e.g., *N*-methylpyrrolidone or dimethylacetamide) [58]. After the polymer solution is cast into a film or coating, the solvent is removed by drying, demonstrating superior thermal stability and mechanical properties. Those derivatives, for example, cellulose acetate and cellulose nitrate, are commonly processed in solution form only. These are also used as supports for photographic films, cigarette filters, and some special membranes. The solvent casting enables a more precise control of film thickness and optical clarity according to the requirement, which is not possible using melt processing. Polymer solutions are also used for the manufacture of microporous plastic membranes, using phase inversion techniques. Polysulfone, polyethersulfone, or polyvinylidene fluoride dissolved in solvents like dimethylformamide can be cast into films and immersed in non-solvent baths, creating precisely controlled porous structures used in filtration, separation, and battery separator applications [59].

3.5.3.2 Polymer Blending and Compatibilization

Solution blending is a better method compared to melt blending to fabricate polymer alloys with better properties. Polymers can be molecularly mixed when they are dissolved in a single solvent, a mixing that is difficult to achieve during melt processing. This is especially useful for heat-sensitive or widely different processing temperature polymers [60]. Compatibilizers, which are usually additions of either block or graft copolymers, can be used in polymer solutions to increase the miscibility of incompatible polymer blends [61]. Compatibilizers contain segments that interact positively with both components, which reduce the interfacial tension and mechanical properties of the blend.

3.5.3.3 Recycling and Upcycling

Solvent-based recycling methods use a polymer solution to purify and extract plastics from mixed waste streams. Through selective dissolution, other polymers can be isolated, contaminants can be removed, and high-value materials such as engineering plastics can be obtained from complex durable products. This allows for better recycling of plastics that have properties closer to virgin materials than typical mechanically based recycling methods [62].

3.5.4 Textile Applications

3.5.4.1 Fiber Formation

The manufacture of synthetic fibers has a great reliance on the polymer solutions as intermediates. In the wet spinning method used to manufacture polymers such as PAN (in the case of acrylic fibers) and cellulose (for rayon and lyocell), these polymers are dissolved in a solvent [63]. The solution is extruded through a spinneret, allowing the polymer to combine with a coagulation bath for solvent exchange to occur and precipitate the polymer into continuous filaments. The process of dry spinning uses polymer solutions of cellulose acetate or spandex (polyurethane) in volatile solvents. During dry spinning, polymer solutions are extruded through spinnerets into warm air, where they evaporate, leaving behind a solid fiber structure with controlled morphology and properties [64]. Electrospinning is a sophisticated fiber formation method that only relies on polymer solutions. Once high voltage is applied to a polymer solution, it is capable of producing ultrafine fibers with nanometer diameters. The resulting nanofibers provide textiles with extraordinary surface area for filtration, protective cloth, and medical textiles [65].

3.5.4.2 Textile Finishing and Modification

Functional finishes for textiles also often use polymer solutions to apply specific properties. Water repellent treatments use fluoropolymer solutions to create durable hydrophobic surfaces on textiles [66]. Antimicrobial finishes use solutions of quaternary ammonium polymers that can be durable to protect fabrics from microorganisms. Flame-retardant treatments typically contain phosphorus in polymer solutions, and upon combustion, create a barrier. It is vital for textiles for use in public areas, transport, and protective clothing. Cross-linkable resin solutions have changed the way wrinkle-resistant finishes are applied to easy-care cotton fabrics [67], particularly dimethyloldihydroxyethyleneurea, which has played a key role in the development of easy-care cotton fabrics. The resin solution travels by capillary action within the fiber structure. Curing the resin results in the formation of cross-links between the resin and the acetate or cotton fabric, resulting in a finished version of finished fabric containing no more than 1% of the total weight of the resin. In essence, cross-linking stabilizes the dimensions of the swollen fiber, so that the finished cotton fabric is much more resistant to dimensional change that occurs during washing and drying.

3.5.4.3 Technical Textiles

There have been considerable advances in medical textiles in terms of the solution capabilities of polymers. In particular, wound dressings usually consist of hydrogel-forming polymer solutions, which are used to create a moist environment for healing [68]. Controlled drug release has been generated by incorporating pharmaceutical agents with biodegradable polymer solutions, which are then processed into fibers or coatings. Protective clothing uses solution technology to create barriers against chemical and biological agents. Solution-cast fluoropolymers or polyurethanes are used to create selectively permeable membranes in protective clothing to eliminate agents from reaching the wearer while allowing water vapor to escape to keep the wearer comfortable [69].

3.5.5 Coatings Industry

1. **Industrial protective coatings:** Corrosion-resistant coatings are mainly based on polymer solutions of epoxies, polyurethanes, or acrylics. These formulations provide effective protection for infrastructure, marine vessels, and industrial equipment from harsh environmental conditions. Spray coating and dip coating are the two applications that can be achieved by altering the viscosity of the polymer solutions [70]. Vinyl esters or phenolics offer a highly crosslinked framework, due to which they are used in the manufacturing of chemical-resistant linings for tanks and pipes. These materials tend to form a water-resistant barrier against chemicals while keeping their structural form intact. The polymers based on silicone or fluoropolymer solutions are capable of retaining their properties at 650 °C and higher temperatures [71].
2. **Architectural and decorative coatings**: One of the chief clients of polymer solutions in consumer goods is aquatic latex paints [72]. As water evaporates, aqueous dispersions form continuous sheets that offer protective and aesthetically pleasing wall, ceiling, and facade coatings [73]. High-performance wood coatings are polymer solutions of polyurethane, alkyl, or acrylic resins, which are formulated for durability while revealing the beauty of the substrate. The solvent can affect the degree of penetration, drying time, and appearance. Textured and specialty finishes often employ thixotropic polymer (a fluid that gets lesser viscous when mechanical stress applied on it and return to its original viscosity when at rest) solutions, some with associative thickners. Bis thixotropic (opposite to thixotropic) solutions behave like a gel while being stored, then they flow during application. Decorative applications may include stippling, brush marks, or an appearance of artificial stone [74].
3. **Functional coatings**: Publications on antimicrobial coatings in healthcare environments typically cover polymer solutions containing quaternary ammonium

compounds or silver nanoparticles [75]. Antimicrobial coatings are an example of systems that release active agents in a controlled manner for more permanent protection against pathogenic microorganisms [76]. Easy-clean coatings are based on either fluoropolymer or silicone solutions, which create low-surface-energy interfaces that have stain-resistant surface attributes, which also allow for cleaning. Easy-clean coatings occur in numerous applications from architectural facades to consumer electronics [77]. Conductive coatings are solutions of inherently electrically conductive polymers such as poly(3,4-ethylenedioxythiophene) (PEDOT):poly (styrene sulfonate) (PSS), or polymer composites containing conductive fillers. Conductive coatings are utilized to make transparent electrodes for touch screens, electromagnetic interference shielding, and antistatic protection [78]. The growth in polymer solutions, in conjunction with the polymer sector resilience for innovative products, has significantly grown the stage across the plastic, textile, and coating industries. Any polymer solution can be used to produce legitimate, sophisticated materials through its development and processing technologies. The wide-ranging use of polymer solutions in plastic, textile, and coating industries is shown in Figure 3.10.

3.5.6 Biomedical Applications of Polymer Solutions

Polymers solutions are very critical for the biomedical fields due to their adaptability, biocompatibility, and functionality. Polymeric solutions play a key role in drug delivery systems, tissue engineering, diagnostics, and treatment [79]. Biomedical applications of polymer solutions have been summarized in Table 3.4.

3.5.7 Drug Delivery Systems

Polymer solutions are capable of providing complex mechanisms that show controlled drug release and controlled drug delivery. Water-soluble polymers such as PEG or poly(*N*-vinylpyrrolidone) can act as carriers for pharmaceutical agents by improving solubility and pharmacokinetic characteristics. Additionally, PEGylation can enhance drugs with increased circulation time in the bloodstream, reduced immunogenicity, and increased drug stability. Biodegradable polymer solutions, such as PLGA (Polylactoic–co–glycolic acd) and polycaprolactone, allow for the development of nanoparticles and microparticles that can encapsulate therapeutic agents [80]. Delivery systems can act as a barrier to sensitive drugs from degradation while offering sustained or triggered release profiles. Altering the polymer's molecular weight, crystallinity, and composition will provide precision in degradation and provide scientists with the opportunity to deliver drugs from days to months [81]. In situ, forming injectable systems are polymer based formulations that exhibit sol–gel phase transition following

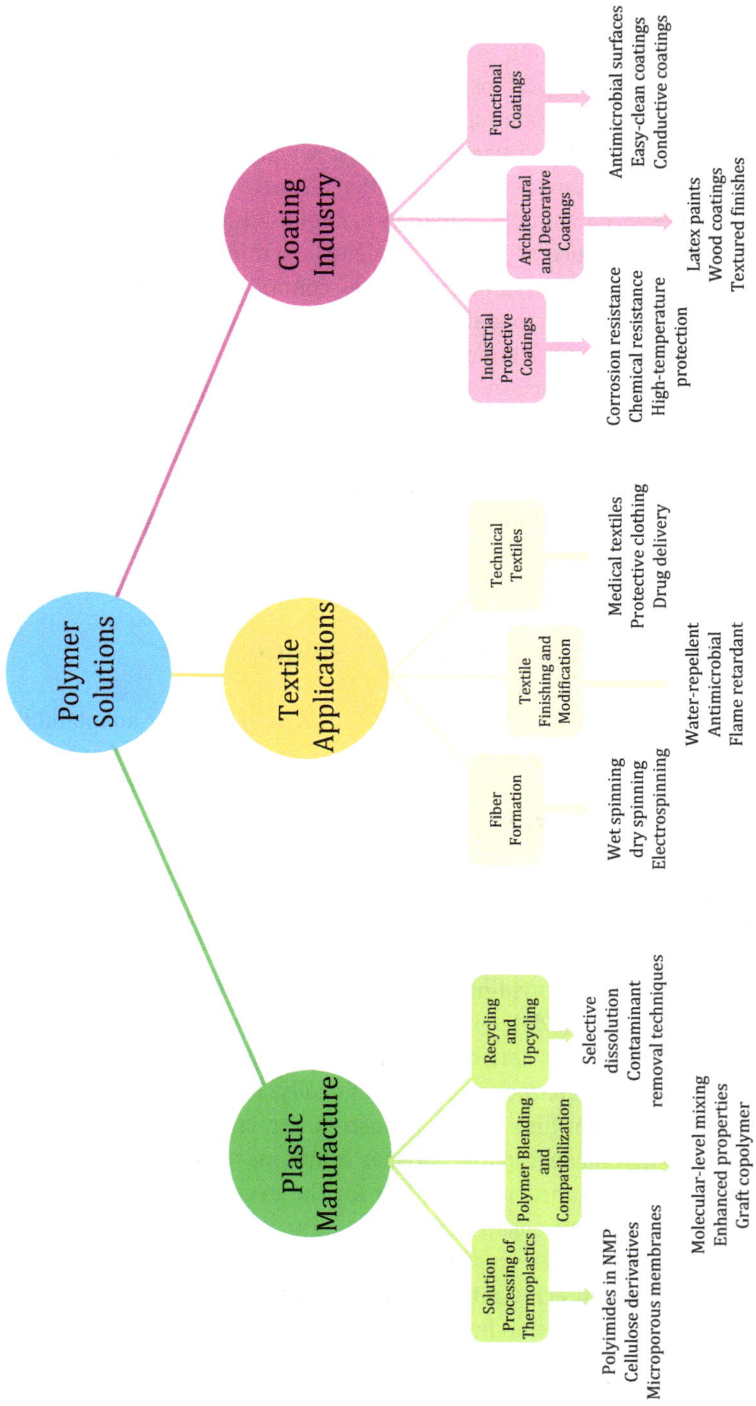

Figure 3.10: Use of polymer solutions in plastic, textile, and coating industries.

Table 3.4: Biomedical applications of polymer solutions.

Characteristic	Drug delivery systems	Tissue engineering	Medical imaging and diagnostics	Surgical materials and wound management
Drug carriers	Water-soluble polymers improve solubility	Natural polymers mimic the extracellular matrix	Gadolinium chelates enhance MRI contrast	Chitosan solutions control bleeding
Delivery systems	Biodegradable polymers encapsulate agents	Injectable hydrogels repair tissues	Nanoparticles combine imaging techniques	PEG sealants prevent leakage
Material properties	Thermosensitive polymers form hydrogels	Bioinks enable 3D bioprinting	Conductive polymers detect biomarkers	Alginate solutions form moisture-retentive gels
Therapeutic effect	N/A	N/A	N/A	Growth factors promote healing

administration [82]. Thermo-responsive polymers like PNIPAM and several poloxamer solutions (which are aqueous mix of block copolymers) are liquid at room temperature but transition to hydrogels at physiological temperature [83]. This effect allows for delivery using minimally invasive methods with subsequent extended local drug release at targeted anatomical sites [84].

3.5.7.1 Tissue Engineering and Regenerative Medicine

Polymers in solution are also converted to scaffolds that support cell growth for tissue regeneration. Natural polymers, including collagens, fibrins, and hyaluronic acid, are solutions that are capable of being manufactured into structures that mimic the characteristics of native extracellular matrix. Many of these solutions can be freeze-dried into porous scaffolds or electrospun into fibrous networks that guide cellular organization in tissues [85]. Injectable hydrogel precursor solutions provide minimally invasive approaches to repair tissues. These formulations can be delivered through small needles and cross-linked in situ afterwards with light, ionic, or enzymatic methods [86]. Hydrogels provide 3D environments to encapsulated cells and allow nutrients to diffuse while also integrating with surrounding tissues [87]. Three-dimensional bioprinting strongly relies on bioink formulations to make tissue structures. Bioink formulations consist of polymer solutions containing cells and bioactive molecules for controlled deposition to modify architecture. The formulations must have optimized rheological properties that ensure printability and cell viability [88]. The ability to have soluble polymers keep their rheological properties consistent with the filament stabilization during processing will depend upon polymer concentration and molecular weight, and ideally, using thixotropic agents after dilution to give suitable printing characteristics [89].

3.5.7.2 Medical Imaging and Diagnostics

Polymer solutions have been extensively used in the medical imaging field. Using the linear polymers such as PEG-bound gadolinium chelates, to boost contrast in magnetic resonance imaging (MRI) images with lesser toxicity. These macromolecular contrast agents are able to offer longer retention and targeting capabilities. Several polymer solutions serve as the basis for diverse coated nanoparticle solutions that combine numerous characteristics of dissimilar imaging modalities, such as computed tomography, fluorescence imaging, and MRI, to deliver multimodal imaging vehicles. Self-assembly nanotechnology in block copolymer solutions could result in micelles that combine imaging agents and stealth to encompass circulation times. Many simple devices that are built with sensing elements are often functional polymer solutions [90]. Polypyrrole and polyaniline can be converted into thin films by dissolving them in a suitable solvent, and the resultant polymer can be used as a biological sensor [91]. These biological sensors allow the sensitive and rapid detection of disease biomarkers.

3.5.7.3 Surgical Materials and Wound Management

Adhesive polymer solutions cover the need for tissue modification through adhesion in a surgical setting [92]. Chitosan solutions provide hemostasis and adhere to tissues in moist environments, facilitating bleeding control [93]. PEG-based surgical sealants cross-link upon application to create water-tight barriers to stop air and/or fluid leakage at surgical sites. Advanced wound dressings contain polymer solutions that create the ideal environment for wound healing. Alginate solutions form gels that maintain moist wound healing by retaining wound exudate. Antimicrobial and/or antiseptic polymers such as polyhexamethylene biguanide can be incorporated into the polymer solution to provide anti-infection/treat a wound infection [94]. Thermoresponsive polymer dressings undergo temperature-driven phase changes to go from adherent to non-adherent, allowing for atraumatic dressing changes [95]. Polymer solutions characterized by growth factors and antimicrobial peptides allow for bioactive wound dressings that can stimulate healing [96]. They can be electrospun or cast into films and allow a sustained release of therapeutic molecules applied directly to the wound site [97]. The ongoing growth of polymer solution technologies for biomedical applications represents increasingly complex strategies for the treatment of diseases, repair of tissues, and diagnoses [98]. As we learn more about polymer-biological interactions, these potentially wide-ranging forms and applications of polymers will be at the forefront of biomedical advances [99].

3.6 Conducting and Smart Polymer Solutions

The electrical properties, optical properties, and responsive properties of inorganic materials come together in conducting and smart polymers. Due to this unique blend, they are highly versatile for advanced technological applications across a wide range of industries [100].

3.6.1 Conducting Polymer Solutions

Conducting polymers or intrinsically conductive polymers have extended π-conjugated systems providing electron conduction along polymer chains [101]. These included polyacetylene, polypyrrole, polythiophene, polyaniline, and PEDOT [poly(3,4–ethylenedioxythiophene). Usually, in their neutral forms, the conductors have poor solubility and processability; however, through chemical or electrochemical oxidation (p-doping) or chemical or electrochemical reduction (n-doping), charge carriers and counterions are introduced, thus increasing the conductivity and solubility in various suitable solvent systems [102]. The solution properties of conducting polymers depend critically upon the polymer backbone structure, type of dopant, oxidation state, and the solvent in question [103].

3.6.1.1 Processing and Fabrication Methods

Solution processing opens up many fabrication routes for conducting polymers that would not be available through any traditional means of melt processing [104]. Spin coating can be used to fabricate thin, uniform films in a controllable thickness down to the nanometer scale. The film morphology, however, can be affected directly by the viscosity and concentration of the polymer solution; as the molecular weight of a polymer increases, a more robust film can typically be achieved, but with trade-offs in processability [105]. To get optimal print quality, some common surfactants aim to alter the surface tension and evaporation. Electrospinning is an exceptional method that yields conductive polymers into nanofibers. The viscosity, concentration, and conductivity of the polymer solution dictate many properties such as fiber diameter, morphology, and fiber alignment [106].

3.6.1.2 Applications of Conducting Polymer Solutions

To manufacture transparent conductive films, PEDOT:PSS solutions are used to get touch screens, organic light-emitting diodes, and photovoltaic devices. Moreover, the enhancement of its conductivity through post-deposition with polar solvents and/or acids is accomplished by altering the conformation of the polymer chain and improving the transfer of electrical charge from chain to chain [107]. Flexible electronics often utilize solution-processed conducting polymers that can maintain conductive properties through mechanical strain. Producing wearable sensors, rollable displays, and conformable circuit components that are intended to take the shape of non-planar surfaces is possible with these conducting polymers. Adding a solvent or a plasticizer to the solution, even a high-boiling solvent, can further enhance the mechanical properties of these materials without having a detrimental effect on the electrical properties. Conducting polymer solutions are also used as active materials for electrochemical devices such as supercapacitors, batteries, and electrochromic displays [108].

3.6.2 Smart Polymer Solutions

A smart polymer solution is a polymer dissolved in a solvent (often water) that can reversibly change its properties in response to external stimuli such as temperature, pH, light, electric or magnetic fields, ionic strength, or specific chemicals.

3.6.2.1 Stimuli-Responsive Mechanisms

Smart polymers exhibit significant changes in properties in response to external stimuli such as temperature, pH, light, electric fields, or other specific chemical triggers [109]. In solution, these changes are often observed as conformational changes, self-assembly, or phase changes that alter solubility, viscosity, or optical properties [110]. Temperature-responsive polymers, such as PNIPAM, show LCST behavior, where the polymer remains soluble below a critical temperature but precipitates above the critical temperature. The underlying mechanism of thermal phase separation is due to the balance between hydrogen bonding with water and hydrophobic interactions among polymer chains [111]. pH-responsive polymers contain acidic (e.g., carboxylic acid) or basic (e.g., amine) groups that ionize over different ranges of pH, causing dramatic changes in solubility [112]. For example, when polyacrylic acid solutions are at a high pH, they expand

Table 3.5: Types of smart polymers: composition, response mechanism, and applications.

Polymer type	Composition	Response mechanism	Applications	References
Thermo-responsive polymers	Poly(*N*-isopropylacrylamide) (PNIPAM), poly(*N*-vinylcaprolactam) (PVCL), and poly(ethylene glycol) (PEG) derivatives	Exhibit lower critical solution temperature (LCST) or upper critical solution temperature (UCST) behavior	Drug delivery systems, tissue engineering scaffolds, cell culture surfaces, and smart textiles	[114]
pH-responsive polymers	Poly(acrylic acid) (PAA), poly(methacrylic acid) (PMAA), chitosan, and poly(2-(dimethylamino) ethyl methacrylate) (PDMAEMA)	Functional groups that accept or release protons in response to pH changes	Oral drug delivery, cancer therapy, biosensors, and membrane technology	[115]
Photo-responsive polymers	Azobenzene derivatives, spiropyran/spirooxazine compounds, coumarin, and diarylethene	Light-induced isomerization, cleavage of photolabile groups, or cross-linking	On-demand drug release, optical data storage, smart surfaces, and photopatterning	[116]
Electro-responsive polymers	Polyaniline (PANI), polypyrrole (PPy), poly (3,4-ethylenedioxythiophene) (PEDOT), and poly(vinyl alcohol) (PVA) with conductive fillers	Conformational changes, swelling/deswelling, or actuation in response to electric fields	Controlled release systems, soft robotics, artificial muscles, and smart windows	[117]
Magneto-responsive polymers	Polymer matrices (e.g., PVA, PDMS, and PAAm) with embedded magnetic nanoparticles (Fe_3O_4 and $CoFe_2O_4$)	Alignment, movement, or heating in response to magnetic fields	Remotely controlled drug delivery, hyperthermia cancer treatment, soft actuators, and MRI contrast agents	[118]
Self-healing polymers	Dynamic covalent bonds (Diels-Alder adducts, disulfide bonds), non-covalent interactions (hydrogen bonding, π–π stacking, and metal-ligand coordination)	Reversible bond formation at damage sites	Electronic skins, protective coatings, wearable devices, and soft robotics	[119]

Table 3.5 (continued)

Polymer type	Composition	Response mechanism	Applications	References
Shape memory polymers	Thermoplastic polyurethanes, epoxy resins, poly(lactic acid) (PLA), and semicrystalline polymers with crystalline and amorphous phases	Temporary shape fixation and recovery triggered by stimuli	Minimally invasive medical devices, aerospace deployable structures, smart textiles, and 4D printing	[120]
Glucose-responsive polymers	Phenylboronic acid derivatives, glucose oxidase (GOx)-containing polymers, concanavalin A-modified polymers	Specific binding to glucose or enzymatic glucose detection	Insulin delivery, glucose sensing, closed-loop diabetes management	[121]
Enzyme-responsive polymers	Peptide-functionalized polymers, polysaccharides (hyaluronic acid, chitosan), and polyesters with enzymatically cleavable linkages	Degradation or conformational change triggered by specific enzymes	Tumor-targeted drug delivery, tissue engineering, biosensing, and wound dressings	[122]
Multi-responsive polymers	Block copolymers with different responsive segments and hybrid materials combining organic polymers with inorganic components	Multiple stimuli trigger distinct or synergistic responses	Theranostics, programmable materials, advanced drug delivery, and smart coatings	[123]

due to repulsion between the negatively charged carboxylate groups, and they collapse at low pH due to protonation of the carboxylate groups [113]. Various smart polymers, along with their applications and response mechanisms, are given in Table 3.5.

Photosensitive polymers incorporate either chromophores or photocleavable functional groups that change position and/or structure following exposure to light [124]. For example, utilizing numerous light wavelengths, azobenzene-containing polymers in solution can establish reversible trans-cis isomerization in those directions, altering their solubility, viscosity, and self-assembly characteristics [125].

3.6.2.2 Solution Behavior and Processing

Various smart polymers show very complex behavior and need to be processed because they can be responsive to the concentration of the stimulus. Block copolymer solutions have the capability to self-assemble into nanostructures with very diverse morphologies (micelles, vesicles, and worms), reliant on solvent quality, block ratio, and environmental factors. Each of these self-assemblies can revert into their transient state upon stimulus, which helps enable encapsulation and continued controlled release of their molecules [126]. When processing smart polymer solutions, it is also important to consider all the mentioned responsive attributes and construct strategies that allow modification to occur at the time of deposition. Surface patterning strategies often rely on precipitation techniques of stimulus molecules to generate defined structures in response to the stimulus. Alternatively, cross-linking reactions may also be activated in situ to solidify any transient solution structures into permanent networks after deposition [127].

3.6.2.3 Applications of Smart Polymer Solutions

Many drug delivery systems use smart polymer solutions to make targeted or controlled drug delivery. Thermosensitive polymers, which transition from a sol to a gel at human body temperature (37 °C), can be used to create injectable, sol-gel formulations that form in situ, enabling continuous local drug delivery. A dual-responsive (e.g., pH and temperature) system may target specific biological environments, such as tumor tissue, which are generally more acidic (lower pH) and warmer (higher temperature) [128]. Soft actuators and artificial muscles utilize smart polymer solutions as precursors to hydrogels that contract or expand due to specific stimuli. Sensing platforms often utilize smart polymer solutions as recognition elements. Molecularly imprinted polymers made from solution precursors contain binding sites that are complementary to a specific analyte. When exposed to the target molecules, these materials exhibit detectable changes in optical properties, mass, or electrical characteristics [129].

3.7 Future Directions

Finally, the convergence between the fields of conducting and smart polymer technologies offers particularly exciting potential opportunities. While the fundamental building blocks of conducting and smart polymers are examined here, the electrical conductivity combined with stimulus responsiveness creates materials that may be electronically addressed to elicit a physical or chemical change locally. There can be more complex polymer architectures by upgrading our knowledge regarding structure-property relationships. Dendritic and bottlebrush polymers in solution form a branched framework at the macromolecular level, resulting in their excellent responsive properties. The conducting and smart polymers are refined with innovative methods that make them very pivotal for electronics, health care, and energy storage devices. There are several implications in the production of smart materials using a polymeric solution. The enthalpy and entropic contributions to Gibbs free energy are the exclusive property of the polymer solutions that make them better than their competitors. The Flory-Huggins theory, scaling concepts, and reptation models are extensively discussed in this chapter to achieve good control over polymeric solutions. The industrial method of preparation of various consumer products by using solution-based polymers is now very common. The advancements in the field of the biomedical industry, i.e., tissue engineering, drug delivery, and diagnostic applications, are possible only because of the high-end properties of polymeric solutions. The pioneering advancements in smart polymers are beneficial for the growth of electronic devices and various sensing materials. In the context of future revolution, it seems prospective that polymer solutions will be unified with nanotechnology and artificial intelligence-guided materials design to achieve some particularly promising outcomes. With ongoing advances in polymer solution science, we can anticipate new applications in the fields of medicine, energy, and environmental sustainability. It is unquestionably the elementary principles outlined in this chapter that will lay the foundation for all this progress to come in the future.

Practice Questions

1. What is the principle of the DLS method?

2. What is the hydrodynamic radius? Write the expression for the nonspherical and spherical molecules.

3. What is the principle of KB theory? What are KBIs?

4. Write the full form of the following terms: (i) PLGA, (ii) PNIPAM, (iii) PFG-NMR, (iv) UCST, and (v) LCST.

5. What are "radius of gyration" and "refractive index increment?"

6. Explain the effect of polymerization in "good solvent" and "bad solvent."

7. Why does the diffusion coefficient (D) of a particle decrease when either the viscosity of the medium increases or the size of the particle increases?

8. In the Flory-Huggins mean-field theory, what is the physical significance of the interaction parameter (χ)? How does its value influence the miscibility of a polymer in a given solvent?

9. Why does the size of a polymer coil (radius of gyration) in a good solvent generally increase with molecular weight according to a power-law relationship, and how does this reflect solvent quality?

10. Distinguish between the three different regimes of polymer solution based on their concentration.

11. Describe the Rouse model and the Zimm model in the context of the viscoelastic properties of both dilute and semi-dilute polymer solutions.

12. What is reptation theory, and why is it named "reptation?" Explain the theory with the help of the tube model.

13. What is scaling theory? Explain the scaling involved in the dilute and semi-dilute regimes of polymer solution.

14. The viscosity of water is 0.894 cP at 25 °C and 0.353 cP at 80 °C. How much faster is the diffusion at 80 °C compared with 25 °C for a particle suspended in water?

15. A certain spherical particle is suspended in water at an ambient temperature of 298 K (25 °C). The dynamic viscosity of water at this temperature is given as 0.89×10^{-3} Pa s. The radius of the particle is given as 100 nm. Using the Stokes-Einstein equation (3.9), calculate the diffusion coefficient (D) of the particle in $m^2\ s^{-1}$. (Given: Boltzmann constant $k_b = 1.38 \times 10^{-23}$ J K^{-1}.)

16. A nanoparticle has a measured diffusion coefficient of 2.5×10^{-11} $m^2\ s^{-1}$ in a given solvent at 310 K. The dynamic viscosity of the solvent is known to be 0.001 Pa · s. Using the Stokes-Einstein equation (eq. (3.9), determine the radius of the particle (R_s) in nanometers. (Given: Boltzmann constant $k_B = 1.38 \times 10^{-23}$ J K^{-1}.)

17. What is the primary mechanism behind the formation of polymer films in coating applications?

18. Define smart polymers and provide two examples of external stimuli they respond to.

19. Analyze the relationship between polymer concentration, molecular weight, and viscosity in coating applications. How does this relationship affect the choice of application method?

20. Compare and contrast the mechanisms of action for pH-responsive and temperature-responsive polymers. Why might a dual-responsive system be advantageous in drug delivery?

21. Evaluate the role of compatibilizers in polymer solution blending. How do they improve the properties of incompatible polymer blends?

22. Compare the advantages and limitations of solution-cast membranes versus melt-processed membranes for filtration applications.

23. Explain the mechanism of PEGylation in drug delivery systems. Discuss how this process improves circulation time, reduces immunogenicity, and enhances drug stability.

24. Explain how conducting polymers achieve electrical conductivity through doping processes. Discuss both p-doping and n-doping mechanisms and their effects on solubility.

25. Provide a comprehensive explanation of how cross-linkable resin solutions create wrinkle-resistant finishes in cotton fabrics. Include the curing process and the effects on dimensional stability.

26. Explain the self-assembly behavior of block copolymer solutions and how different morphologies (micelles, vesicles, and worms) are formed based on environmental conditions.

27. In electrospinning, the solution conductivity affects fiber formation. If a PAN solution in DMF has a conductivity of 2.5 μS cm^{-1} at 12%w/w concentration, and the conductivity needs to be increased to 8.0 μS cm^{-1} by adding salt:(i) Calculate the conductivity increase factor.(ii) If the salt contributes 0.5 μS cm^{-1} per 0.1%w/w addition, how much salt should be added?

Hints for Practice Questions

Question No. 14:

According to the Stokes-Einstein equation, the diffusion coefficient of a sphere (D) is given by the following:

$$D = \frac{k_b \times T}{6 \times \pi \times n_s \times R_s} \tag{3.9}$$

The term ($k_b/6\pi R_s$) is a constant term for the same water particle.
Thus, eq. (3.9), at the two given temperatures, becomes:

$$\frac{D_{80}}{D_{25}} = \frac{T_{80} \times \eta_{25}}{T_{25} \times \eta_{80}}$$

Given:

$T_{25} = 25 + 273 = 298$ K

$T_{80} = 80 + 273 = 353$ K

$\eta_{25} = 0.894$ cP $= 0.894 \times 10^{-3}$ Pa s

$\eta_{80} = 0.353$ cP $= 0.353 \times 10^{-3}$ Pa s

Substitute the values:

$$\frac{D_{80}}{D_{25}} = \frac{353 \times 0.894 \times 10^{-3}}{298 \times 0.353 \times 10^{-3}}$$

$$\frac{D_{80}}{D_{25}} = \frac{315.58}{2\,105.19} = 3$$

Therefore, diffusion at 80 °C is about three times faster than at 25 °C.

Question No. 15:

Given:

Temperature: $T = 298$ K

Viscosity of water: $\eta_s = 0.89 \times 10^{-3}$ Pas

Stoke's radius: $R_s = 100$ nm $= 100 \times 10^{-9}$ m $= 1.0 \times 10^{-7}$ m

Boltzmann constant: $k_b = 1.38 \times 10^{-23}$ JK^{-1}

Substituting the above given values in eq. (3.9), we get the following equation:

$$D = \frac{1.38 \times 10^{-23} \times 298}{6 \times 3.14 \times 0.89 \times 10^{-3} \times 1.0 \times 10^{-7}}$$

$$= 2.45 \times 10^{-12}\ \text{m}^2\ \text{s}^{-1}$$

Question No. 16:

From eq. (3.9), $R_s = \dfrac{k_b T}{6\pi n_s D}$

Given:

Temperature: $T = 310$ K

Viscosity of solvent: $\eta_s = 0.001$ Pas

Diffusion coefficient of sphere: $D = 2.5 \times 10^{-11}$ m^2 s^{-1}

Boltzmann constant: $k_b = 1.38 \times 10^{-23}$ J K^{-1}

Substitute the values in the above equation, we get the following equation:

$$R_s = \frac{1.38 \times 10^{-23} \times 310}{6 \times 3.14 \times 0.001 \times 2.5 \times 10^{-11}}$$

$$= 9.08 \times 10^{-9}\ \text{m}$$
$$= 9.08\ \text{nm}$$

Question No. 27:

Given:

- Initial conductivity, κ_i = 2.5 μS cm^{-1}
- Target conductivity, κ_t = 8.0 μS cm^{-1}
- Salt effect: 0.5 μS cm^{-1} increase per 0.1%w/w salt added. (Assume linear response and that %w/w refers to the final solution; small mass-change effects ignored.)

(a) **Conductivity increase factor**

$$\text{Factor} = \frac{\kappa_t}{\kappa_i} = \frac{8}{2.5} = 3.2$$

Answer: (a) 3.2 times

(b) **Required salt addition**

1. Required increase:
   $\Delta\kappa = \kappa_t - \kappa_i$ = 8.0 − 2.5 = 5.5 μS cm^{-1}
2. Each 0.1% w/w salt → 0.5 μS cm^{-1}
   Number of 0.1% steps:
   5.5/0.5 = 11 steps
3. Total %w/w salt: 11 × 0.1% = 1.1% w/w
   Answer: (b) 1.1% w/w salt

References

[1] Ligia G, Deodato R. Polymer solution behavior: polymer in pure solvent and mixed solvent. In: Physicochemical Behavior and Supramolecular Organization of Polymers. Springer; 2009. pp. 1–42. https://doi.org/10.1007/978-1-4020-9372-2_1

[2] Gandhi KJ, Deshmane SV, Biyani KR. Polymers in pharmaceutical drug delivery systems: a review. Int J Pharm Sci Rev Res. 2012;14(2):57–66. Available from:: https://www.researchgate.net/publication/285986312_Polymers_in_pharmaceutical_drug_delivery_system_A_review

[3] Mousavi SM, Raveshiyan S, Amini Y, Zadhoush A. A critical review with emphasis on the rheological behavior and properties of polymer solutions and their role in membrane formation, morphology, and performance. Adv Colloid Interface Sci. 2023;319:102986. https://doi.org/10.1016/j.cis.2023.102986

[4] Salgado-Chavarría D, Palacios-Alquisira J. One hundred years of macromolecular chemistry. Educ Quím. 2021;32(1):20–30. https://doi.org/10.22201/fq.18708404e.2021.1.76662

[5] Rasmussen SC.. From polymer to macromolecule: origins and historic evolution of polymer terminology. 2020. https://doi.org/10.70359/bhc2020v045p091.

[6] Flory PJ. Principles of polymer chemistry. Ithaca (NY): Cornell University Press; 1953. Available from:: https://archive.org/details/principlesofpoly0000paul

[7] Huggins ML. Theory of solutions of high polymers. J Am Chem Soc. 1942;64(7):1712–1719. https://doi.org/10.1021/ja01259a068

[8] Huggins ML. Thermodynamic properties of solutions of long-chain compounds. Ann N Y Acad Sci. 1942;43(1):1–32. https://doi.org/10.1111/j.1749-6632.1942.tb47940.x

[9] Tanaka F. Polymer physics: applications to molecular association and thermoreversible gelation. Cambridge: Cambridge University Press; 2011. https://doi.org/10.1017/CBO9780511975691

[10] Su WF. Principles of polymer design and synthesis. Berlin: Springer; 2013. pp. 6–10. https://doi.org/10.1007/978-3-642-38730-2

[11] Holehouse AS., Pappu RV Collapse transitions of proteins and the interplay among backbone, sidechain, and solvent interactions. Annu Rev Biophys. 2018;47(1):19–39. https://doi.org/10.1146/annurev-biophys-070317-032838

[12] Van Krevelen DW., Te Nijenhuis K Properties of polymers: their correlation with chemical structure; their numerical estimation and prediction from additive group contributions. 4th ed. Amsterdam: Elsevier; 2009. Available from:: https://www.sciencedirect.com/book/9780080548197/properties-of-polymers

[13] Patel SS., Takahashi KM Polymer dynamics in dilute and semidilute solutions. Macromolecules. 1992;25(17):4382–4391. https://doi.org/10.1021/ma00043a022

[14] Polychronopoulos ND, Vlachopoulos J. Polymer processing and rheology. In: Functional polymers. Cham: Springer; 2019:Vol. 86. pp. 1–47. https://doi.org/10.1007/978-3-319-92067-2_4-1

[15] Johnson MF, Evans WW, Jordan I, Ferry JD. Viscosities of concentrated polymer solutions. II. Polyisobutylene. J Colloid Sci. 1952;7(5):498–510. https://doi.org/10.1016/0095-8522(52)90033-0

[16] Onogi S, Kimura S, Kato T, Masuda T, Miyanaga N. Effects of molecular weight and concentration on flow properties of concentrated polymer solutions. J Polym Sci Part C Polym Symp. 1967;15(1):381–406. https://doi.org/10.1002/polc.5070150134

[17] Andrade EDC. The viscosity of liquids. Nature. 1930;125(3148):309–310. https://doi.org/10.1038/125309b0

[18] McCrackin FL. Relationship of intrinsic viscosity of polymer solutions to molecular weight. Polymer. 1987;28(11):1847–1850. https://doi.org/10.1016/0032-3861(87)90289-8

[19] Teraoka I. An introduction to physical properties of polymer solutions. Hoboken (NJ): Wiley; 2002. https://doi.org/10.1002/0471224510

[20] PE R Jr. A theory of the linear viscoelastic properties of dilute solutions of coiling polymers. J Chem Phys. 1953;21(7):1272–1280. https://doi.org/10.1063/1.1699180

[21] Zimm BH. Dynamics of polymer molecules in dilute solution: viscoelasticity, flow birefringence, and dielectric loss. J Chem Phys. 1956;24(2):269–278. https://doi.org/10.1063/1.1742462

[22] Miller CC. The Stokes-Einstein law for diffusion in solution. Proc R Soc Lond A. 1924;106(740):724–749. https://doi.org/10.1098/rspa.1924.0100

[23] Dandeno JB. Osmotic Theories, with Special Reference to Van't Hoff's Law. Bulletin of the Torrey Botanical Club. 1909 Jun 1;36(6):283–98. https://doi.org/10.2307/2479368

[24] Flory PJ., Bueche AM Theory of light scattering by polymer solutions. J Polym Sci. 1958;27(115):219–229. https://doi.org/10.1002/pol.1958.1202711518

[25] Schärtl W. Light scattering from polymer solutions and nanoparticle dispersions. Berlin: Springer; 2007. https://doi.org/10.1007/978-3-540-71951-9
[26] Muthukumar M. Thermodynamics of polymer solutions. J Chem Phys. 1986;85(8):4722–4728. https://doi.org/10.1063/1.451748
[27] Elias HG. Thermodynamics of polymer solutions. In: Macromolecules. Vol. 3. Physical structures and properties. Weinheim: Wiley-VCH; 2008. pp. 297–364. https://doi.org/10.1002/9783527627233.ch10
[28] Nemirovsky AM, Bawendi MG, Freed KF. Lattice models of polymer solutions: monomers occupying several lattice sites. J Chem Phys. 1987;87(12):7272–7282. https://doi.org/10.1063/1.453320
[29] Taimoori M, Modarress H, Mansoori GA. Generalized Flory–Huggins model for heat-of-mixing and phase-behavior calculations of polymer–polymer mixtures. J Appl Polym Sci. 2000;78(7):1328–1340. https://doi.org/10.1002/1097-4628(20001114)78:7%3C1328::AID-APP30%3E3.0.CO2-2
[30] Fisher ME, Hiley BJ. Configuration and free energy of a polymer molecule with solvent interaction. J Chem Phys. 1961;34(4):1253–1267. https://doi.org/10.1063/1.1731729
[31] Clark EA, Lipson JEG. LCST and UCST behavior in polymer solutions and blends. Polymer. 2012;53 (2):536–545. https://doi.org/10.1016/j.polymer.2011.11.045
[32] Kirkwood JG, Buff FP. The statistical mechanical theory of solutions. I. J Chem Phys. 1951;19(6):774–777. https://doi.org/10.1063/1.1748352
[33] Newman KE. Kirkwood–Buff solution theory: derivation and applications. Chem Soc Rev. 1994;23 (1):31–40. https://doi.org/10.1039/CS9942300031
[34] de Gennes PG. Scaling concepts in polymer physics. Ithaca (NY): Cornell University Press; 1979. https://doi.org/10.1063/1.2914118
[35] Kosmas MK, Freed KF. On scaling theories of polymer solutions. J Chem Phys. 1978;69(8):3647–3659. https://doi.org/10.1063/1.437073
[36] Doi M, Edwards SF. The theory of polymer dynamics. Oxford: Clarendon Press; 1986. Available from: https://cds.cern.ch/record/346518/files/0198520336_TOC.pdf.
[37] Viovy JL, Rubinstein M, Colby RH. Constraint release in polymer melts: tube reorganization versus tube dilation. Macromolecules. 1991;24(12):3587–3596. https://doi.org/10.1021/ma00012a020
[38] Paul W, Binder K, Heermann DW, Kremer K. Dynamics of polymer solutions and melts. Reptation predictions and scaling of relaxation times. J Chem Phys. 1991;95(10):7726–7740. https://doi.org/10.1063/1.461346
[39] Kim H, Park J, Lee D. Environmental responsiveness of polymer solutions: a physicochemical perspective. J Polym Res. 2023;30(5):112. https://doi.org/10.1021/acs.iecr.4c03325
[40] Zhao L, Chen Y, Wang T. Advances in polymer solution processing techniques for scalable manufacturing. Mater Today Chem. 2025;30:101643. https://doi.org/10.48550/arXiv.2504.07834
[41] Haque A, Mondal S, Khan I, Bera S, Banerjee D. Acrylic-based water-borne coatings: advances in formulation and performance. Coatings. 2021;11(4):471. https://doi.org/10.3390/coatings11040471
[42] Martinez-Gomez A, Castillo-Ortega MM, Rodriguez-Felix F, Liu J, Vazquez-Torres H. Fluoropolymer-based hydrophobic coatings: design principles and emerging applications. ACS Appl Mater Interfaces. 2024;16(6):7528–7539. https://doi.org/10.1021/acsami.3c20075
[43] Kim J, Lee H, Park J, Cho S, Yoon M. PEDOT-based conductive polymers for flexible electronics: progress and challenges. Adv Funct Mater. 2022;32(11):2109095. https://doi.org/10.1002/adfm.202109095
[44] Chen Y, Wu X, Zhang L, Yang K, Lin M. Reactive polymer solutions for high-performance structural adhesives. Polym Chem. 2021;12(7):994–1010. https://doi.org/10.1039/D0PY01406K
[45] Patel P, Desai S, Panchal K, Trivedi R, Mehta C. Mechanism of adhesion in polymer-based adhesives: surface interactions and mechanical interlocking. J Mater Sci. 2022;57(19):12478–12501. https://doi.org/10.1007/s10853-022-07114-x

[46] Nguyen TH, Wen X, Gong Y, Wang C, Lee P. Cellulose dissolution and regeneration in sustainable fiber production: Technological advances and challenges. *Cellulose*. 2020;27(12):6837–6877. https://doi.org/10.1007/s10570-020-03257-9

[47] Jadhav A, Marathe R, Karthik C, Singh N, Mohanty S. Electrospinning of polymer solutions: Physics, processing, and applications. *J Mater Chem A*. 2022;10(3):1217–1260. https://doi.org/10.1039/D1TA08967C

[48] Rodríguez-Tobías H, Santiago-Valtierra G, Medellín-Rodríguez FJ, Avilés-Arellano LM, Neira-Velázquez G. Functional polymer coatings for advanced textile finishing: Antimicrobial and protective properties. *J Coat Technol Res*. 2024;21(1):41–68. https://doi.org/10.1007/s11998-023-00832-2

[49] Li J, Wu C, Chu PK, Gelinsky M. 3D printing of hydrogels: Rational design strategies and emerging biomedical applications. *Mater Sci Eng R Rep*. 2020;140:100543. https://doi.org/10.1016/j.mser.2020.100543

[50] Yousefi H, Su HM, Imani SM, Alkhaldi K, Filipe CDM, Didar TF. Intelligent food packaging: A review of smart sensing technologies for monitoring food quality. *ACS Sens*. 2019;4(4):808–821. https://doi.org/10.1021/acssensors.9b00240

[51] Zabihi O, Ahmadi M, Khayyam H, Naebe M. Fish DNA-modified clays: Towards highly flame-retardant polymer nanocomposites with improved interfacial and mechanical performance. *Sci Rep*. 2016;6:38194. https://doi.org/10.1038/srep38194

[52] Yu S, Pang H, Huang S, Tang H, Wang S, Qiu M et al. Recent advances in metal-organic framework membranes for water treatment: A review. *Sci Total Environ*. 2021;800:149662. https://doi.org/10.1016/j.scitotenv.2021.149662

[53] Li L, Han L, Hu H, Zhang R. A review on polymers and their composites for flexible electronics. *Mater Adv*. 2023;4(3):726–746. https://doi.org/10.1039/D2MA00940D

[54] Riahinezhad M, Hallman M, Masson JF. Critical review of polymeric building envelope materials: Degradation, durability and service life prediction. *Buildings*. 2021;11(7):299. https://doi.org/10.3390/buildings11070299

[55] Subadra SP, Griskevicius P. Sustainability of polymer composites and its critical role in revolutionising wind power for green future. *Sustain Technol Green Econ*. 2021;1(1):1–7. https://doi.org/10.21595/stge.2021.21974

[56] Yang Y, Yang J, Wu WM, Zhao J, Song Y, Gao L et al. Biodegradation and mineralization of polystyrene by plastic-eating mealworms: Part 2. Role of gut microorganisms. *Environ Sci Technol*. 2015;49(20):12087–12093. https://doi.org/10.1021/acs.est.5b02663

[57] Arefin AM, Khatri NR, Kulkarni N, Egan PF. Polymer 3D printing review: Materials, process, and design strategies for medical applications. *Polymers*. 2021;13(9):1499. https://doi.org/10.3390/polym13091499

[58] Yang G, Wu M, Lin W, Jiang B, Jiang Z. Processing of polyimide solutions for electronic applications. *J Appl Polym Sci*. 2019;136(48):48237. https://doi.org/10.1002/app.48237

[59] Forouharshad M, Putti M, Basso A, Prato M, Monticelli O. Biobased system composed of electrospun sc-PLA/POSS/cyclodextrin fibers to remove water pollutants. *ACS Sustain Chem Eng*. 2020;8(31):11611–11618. https://doi.org/10.1021/acssuschemeng.0c03581

[60] Kumar S, Samal SK, Dash P, Mohanty S. Solution-based polymer blending techniques: A comprehensive review. *Polym-Plast Technol Mater*. 2022;61(1):85–109. https://doi.org/10.1080/25740881.2021.1941243

[61] Vollmer I, Jenks MJ, Roelands MC, White RJ, Van harmelen T, De wild P et al. Beyond mechanical recycling: Giving new life to plastic waste. *Angew Chem Int Ed Engl*. 2020;59(36):15402–15423. https://doi.org/10.1002/anie.201915651

[62] Zhao X, Korey M, Li K, Copenhaver K, Tekinalp H, Celik S et al. Converting plastic waste into high-performance composites by solid-state shear pulverization. *J Clean Prod*. 2022;330:129883. https://doi.org/10.1016/j.jclepro.2021.129883

[63] Ozipek B, Karakas H. Wet spinning of synthetic polymer fibers. In: *Advances in filament yarn spinning of textiles and polymers*. Woodhead Publishing; 2014. pp. 174–186. https://doi.org/10.1533/9780857099174.2.174

[64] Xue J, Wu T, Dai Y, Xia Y. Electrospinning and electrospun nanofibers: Methods, materials, and applications. *Chem Rev*. 2019;119(8):5298–5415. https://doi.org/10.1021/acs.chemrev.8b00593

[65] Kenry K, Lim CT. Nanofiber technology: Current status and emerging developments. *Prog Polym Sci*. 2023;136:101623. https://doi.org/10.1016/j.progpolymsci.2022.101623

[66] Yildiz Z, Onen A, Gungor A. Preparation of flame-retardant polymer-based textiles and investigation of their properties. *J Ind Text*. 2021;50(9):1343–1389. https://doi.org/10.1177/1528083719858760

[67] Montazer M, Harifi T. *Nanofinishing of textile materials*. Woodhead Publishing; 2018. https://doi.org/10.1016/C2016-0-01870-X

[68] Gupta B, Jain R, Singh H. Fabrication of drug-loaded biodegradable nanofibers via the electrospinning technique for controlled drug release applications. *Int J Polym Mater Polym Biomater*. 2023;72(3):188–211. https://doi.org/10.1080/00914037.2021.1958419

[69] Lyon SB, Bingham R, Mills DJ. Recent progress in the understanding of corrosion control by coatings. *Prog Org Coat*. 2022;164:106765. https://doi.org/10.1016/j.porgcoat.2022.106765

[70] Ding Y, Liu M, Li S, Zhang S, Zhou WF, Wang B. Contribution of the side groups to the performance of a series of phenolic bio-based epoxy resins and their composites. *J Appl Polym Sci*. 2021;138(5):49817. https://doi.org/10.1002/app.49817

[71] Wu S, Yang S, Lu Y, Ge S. Waterborne silicone-modified polyurethane coatings: A critical review. *Prog Org Coat*. 2022;163:106644. https://doi.org/10.1016/j.porgcoat.2021.106644

[72] Overbeek A. Polymer heterogeneity in waterborne coatings. *J Coat Technol Res*. 2023;20(1):1–13. https://doi.org/10.1007/s11998-022-00629-7

[73] Aggarwal LK, Thapliyal PC, Karade SR. Acrylic-based waterborne coatings for exterior masonry surfaces – A review. Constr Build Mater. 2021;269:121338. 10.1016/j.conbuildmat.2020.121338

[74] Kostansek E. Associative polymer solutions and networks: Structure and rheology. J Coat Technol Res. 2020;17(3):523–549. 10.1007/s11998-019-00297-0

[75] Das I, Sonawane SH, Trivedi P. Development and characterization of antimicrobial polymer nanocomposite-based coatings on stainless steel and glass. Mater Today Proc. 2022;51:1214–1219. 10.1016/j.matpr.2021.06.042

[76] González-Henríquez CM, Sarabia-Vallejos MA, Rodriguez-Hernandez J. Advances in the fabrication of antimicrobial hydrogels for biomedical applications. Materials. 2019;12(4):641. 10.3390/ma12040641

[77] Milionis A, Sharma CS, Hopf R, Uggowitzer M, Bayer IS, Poulikakos D. Engineering fully organic and biodegradable superhydrophobic materials. Adv Mater Interfaces. 2020;7(24):2001268. 10.1002/admi.202001268

[78] Worsley R, Pimpolari L, McManus D, Ge N, Ionescu R, Wittkopf JA et al. All-2D material inkjet-printed capacitors: Toward fully printed integrated circuits. ACS Nano. 2019;13(1):54–60. 10.1021/acsnano.8b06464

[79] Tibbitt MW, Rodell CB, Burdick JA, Anseth KS. Progress in material design for biomedical applications. Proc Natl Acad Sci U S A. 2023;120(6):e2210817120. 10.1073/pnas.2210817120

[80] Makadia HK, Siegel SJ. Poly Lactic-co-Glycolic Acid (PLGA) as a biodegradable controlled drug delivery carrier. Polymers. 2022;14(2):290. 10.3390/polym14020290

[81] Washington KE, Kularatne RN, Karmegam V, Biewer MC, Stefan MC. Recent advances in aliphatic polyesters for drug delivery applications. ACS Biomater Sci Eng. 2021;7(7):2475–2505. 10.1021/acsbiomaterials.0c01610

[82] Dimatteo R, Darling NJ, Segura T. In situ forming hydrogels for cell delivery and tissue regeneration. Adv Drug Deliv Rev. 2022;181:114067. 10.1016/j.addr.2021.114067

[83] Alexandridis P, Lindman B. Amphiphilic block copolymers: Self-assembly and applications. Elsevier Sci. 2023; 10.1016/C2021-0-01938-X

[84] Hoare TR, Kohane DS. Hydrogels in drug delivery: Progress and challenges. Polymer. 2024;249:125558. 10.1016/j.polymer.2023.125558

[85] Narayanan LK, Thompson TL, Behr JM. Scaffolds for regenerative engineering. Regen Eng Transl Med. 2023;9(1):19–42. 10.1007/s40883-022-00259-w

[86] Parmar PA, Skaalure SC, Chow LW, St-Pierre JP, Stoichevska V, Peng YY et al. Temporally degradable collagen-mimetic hydrogels tuned to chondrogenesis of human mesenchymal stem cells. Biomaterials. 2022;121:203–216. 10.1016/j.biomaterials.2021.121068

[87] Kong HJ, Mooney DJ. Polysaccharide-based hydrogels in tissue engineering. Polymers. 2020;12 (10):2389. 10.3390/polym12102389

[88] Lee A, Dai G. Recent advances in 3D bioprinting and organ-on-a-chip technologies for improved drug development and precision medicine. Adv Sci. 2024;11(4):2305626. 10.1002/advs.202305626

[89] Hospodiuk M, Dey M, Sosnoski D, Ozbolat IT. The bioink: A comprehensive review on bioprintable materials. Biotechnol Adv. 2023;56:107789. 10.1016/j.biotechadv.2022.107789

[90] Dincer C, Bruch R, Costa-Rama E, Fernández-Abedul MT, Merkoçi A, Manz A et al. Disposable sensors in diagnostics, food, and environmental monitoring. Adv Mater. 2022;34(6):2104192. 10.1002/adma.202104192

[91] Gizzie NA, Meier AR, Zakeri B. Advances in biosensing platforms for point-of-care diagnostic applications. Curr Opin Biomed Eng. 2024;25:100462. 10.1016/j.cobme.2023.100462

[92] Li J, Celiz AD, Yang J, Yang Q, Wamala I, Whyte W. et al. Tough adhesives for diverse wet surfaces. Science. 2022;375(6578):292–297. 10.1126/science.abj9778

[93] Chenite A, Buschmann MD, Wang D, Chaput C, Kandani N. Rheological characterisation of thermogelling chitosan/glycerol-phosphate solutions. Carbohydr Polym. 2022;187:233–240. 10.1016/j.carbpol.2021.118616

[94] Kamoun EA, Kenawy ERS, Chen X. A review on polymeric hydrogel membranes for wound dressing applications: PVA-based hydrogel dressings. J Adv Res. 2023;28:31–44. 10.1016/j.jare.2022.08.016

[95] Zheng Y, Wang X, Ji S, Chen S, Jiang H. Thermoresponsive hydrogels in biomedical applications. Mater Chem Front. 2021;5(22):6053–6069. 10.1039/D1QM00785H

[96] Zhao X, Wu H, Guo B, Dong R, Qiu Y, Ma PX. Antibacterial anti-oxidant electroactive injectable hydrogel as a self-healing wound dressing with hemostasis and conductive properties. Biomaterials. 2023;273:121148. 10.1016/j.biomaterials.2022.121148

[97] Huang S, Fu X. Naturally derived materials-based cell and drug delivery systems in skin regeneration. J Control Release. 2022;306:187–202. 10.1016/j.jconrel.2021.07.015

[98] Prestwich GD, Healy KE. Why regenerative medicine needs an extracellular matrix. Matrix Biol. 2022;111:38–59. 10.1016/j.matbio.2022.03.003

[99] Khademhosseini A, Langer R. A decade of progress in tissue engineering. Nat Protoc. 2023;18 (1):139–157. 10.1038/s41596-022-00731-5

[100] Kumar S, Patel R, Nguyen NT. Solution-processable functional polymers: From synthesis to applications. Chem Rev. 2021;121(12):7145–7183. 10.1021/acs.chemrev.0c01291

[101] Li J, Zhang Q, Wang L. Mechanisms of charge transport in conducting polymers: Insights from solution-state characterization. Prog Polym Sci. 2020;104:101127. 10.1016/j.progpolymsci.2020.101232

[102] Nguyen TP, Yoon SH, Lee JW. Controlling the doping process in conductive polymers for improved stability and performance. Adv Funct Mater. 2019;29(22):1900501. 10.1002/adfm.201900501

[103] Chen X, Yang Y, Liu M. Molecular engineering of solution-processable conductive polymers: Strategies and emerging applications. Chem Soc Rev. 2022;51(3):1192–1232. 10.1039/D1CS00525A

[104] Zhang F, Chen L, Yang X. Solution processing techniques for conducting polymer thin films: Recent developments and future perspectives. Adv Mater. 2020;32(15):1904743. 10.1002/adma.201904743
[105] Jiang Y, Liu P, Yang Z. The role of molecular weight in the solution processing of conducting polymers. Macromolecules. 2021;54(8):3652–3664. 10.1021/acs.macromol.0c02725
[106] Wu X, Zhang Y, Chen P. Structure-property relationships in electrospun conducting polymer nanofibers. J Mater Chem A. 2022;10(9):4570–4589. 10.1039/D1TA09788J
[107] Patel DK, Sakhaei AH, Mirzaeifar R. Enhancement strategies for PEDOT: PSS-based thermoelectric materials: A comprehensive review. Adv Electron Mater. 2021;7(5):2000845. 10.1002/aelm.202000845
[108] Yang J, Liu Y, Chen X. Conducting polymer composites for electrochemical energy storage: Recent advances and perspectives. Adv Mater. 2019;31(45):1904539. 10.1002/adma.201904539
[109] Stuart MAC, Huck WTS, Genzer J. Emerging applications of stimuli-responsive polymer materials. Nat Mater. 2019;18(12):1172–1183. 10.1038/s41563-019-0446-9
[110] Dong L, Feng Y, Wang L. Functional supramolecular polymers: Structure, dynamics, and applications. Chem Soc Rev. 2020;49(23):8480–8540. 10.1039/D0CS00574F
[111] García-Fernández L, Cui J, Del Campo A. Understanding LCST transition mechanisms in polymer solutions: Beyond simplified models. Adv Funct Mater. 2022;32(4):2107783. 10.1002/adfm.202107783
[112] Kumar A, Montemagno C, Choi HJ. Smart pH-responsive polymers with tunable properties for biomedical applications. Biomacromolecules. 2020;21(9):3589–3609. 10.1021/acs.biomac.0c00669
[113] Chen W, Ma Y, Pan J. pH-responsive polymer hydrogels: Design strategies and applications. Adv Funct Mater. 2021;31(31):2102332. 10.1002/adfm.202102332
[114] Abuwatfa WH, Awad NS, Pitt WG, Husseini GA. Thermosensitive polymers and thermo-responsive liposomal drug delivery systems. Polymers. 2022;14(5):925. 10.3390/polym14050925
[115] Palanikumar L, Al-Hosani S, Kalmouni M, Nguyen VP, Ali L, Pasricha R et al. pH-responsive high stability polymeric nanoparticles for targeted delivery of anticancer therapeutics. Commun Biol. 2020;3:95. 10.1038/s42003-020-0817-4
[116] Giménez VMM, Arya G, Zucchi IA, Galante MJ, Manucha W. Photo-responsive polymeric nanocarriers for target-specific and controlled drug delivery. Soft Matter. 2021;17(38):8577–8584. 10.1039/D1SM00999K
[117] Carayon I, Gaubert A, Mousli Y, Philippe B. Electro-responsive hydrogels: Macromolecular and supramolecular approaches in the biomedical field. Biomater Sci. 2020;8(20):5589–5600. 10.1039/D0BM01268H
[118] Sharifianjazi F, Irani M, Esmaeilkhanian A, Bazli L, Asl MS, Jang HW et al. Polymer-incorporated magnetic nanoparticles: Applications for magnetoresponsive targeted drug delivery. Mater Sci Eng B. 2021;272:115358. 10.1016/j.mseb.2021.115358
[119] Zhu S, Liu Z, Li W, Zhang H, Dai G, Zhou X. Research progress of self-healing polymer materials for flexible electronic devices. J Polym Sci. 2023;61(15):1554–1571. 10.1002/pol.20230020
[120] Delaey J, Dubruel P, Van Vlierberghe S. Shape-memory polymers for biomedical applications. Adv Funct Mater. 2020;30(44):1909047. 10.1002/adfm.201909047
[121] Aktas Eken G, Huang Y, Prucker O, Rühe J, Ober C. Advancing glucose sensing through auto-fluorescent polymer brushes: From surface design to nano-arrays. Small. 2024;20(22):2309040. 10.1002/smll.202470170
[122] Guo F, Jiao Y, Du Y, Luo S, Hong W, Fu Q et al. Enzyme-responsive nano-drug delivery system for combined antitumor therapy. Int J Biol Macromol. 2022;220:1133–1145. 10.1016/j.ijbiomac.2022.09.176
[123] Kim J, G DA, Debnath P, Saha P. Smart multi-responsive biomaterials and their applications for 4D bioprinting. Biomimetics. 2024;9(8):484. 10.3390/biomimetics9080484
[124] Zhao H, Wang J, Zhou F. Light-responsive polymers: From mechanisms to applications. Prog Polym Sci. 2022;126:101509. 10.1016/j.progpolymsci.2022.101509

[125] Wang D, Xie Y, Wu S. Advances in azobenzene-based polymers as photoswitchable materials. Chem Soc Rev. 2023;52(5):1660–1687. 10.1039/D2CS00759F

[126] Patel K, Angelos S, Zink JI. Stimuli-responsive block copolymer assemblies for controlled cargo delivery. J Am Chem Soc. 2022;144(7):3126–3141. 10.1021/jacs.1c12549

[127] Yao X, Chen L, Ju J. Stimulus-triggered crosslinking strategies for smart polymeric materials. Chem Eng J. 2021;419:129506. 10.1016/j.cej.2021.129506

[128] Tan Z, Parisi C, Di Michele L. Recent progress in stimuli-responsive drug delivery systems for tumor targeting. Adv Healthc Mater. 2021;10(8):2001993. 10.1002/adhm.202001993

[129] Singh V, Zhu P, Sheng J. Functional mechanisms of molecularly imprinted polymers: From molecular recognition to sensing applications. Chem Soc Rev. 2021;50(21):11954–004. 10.1039/D1CS00484K

Subash Chandra Mohapatra and Vishnu Kumawat*

4 Colloidal Chemistry

Abstract: A colloid is a state of matter in which particles ranging in size from 1 to 100 nm form a heterogeneous mixture, and either do not diffuse or have an extremely low tendency to pass through a semipermeable/animal membrane. Colloids, a prominent feature of surface chemistry, provide a large surface area due to the dispersion of finely divided particles within a medium. This chapter serves as a contextual guide for readers to understand the methods of preparing colloids and categorizing colloid particles based on dispersion, particle structure, and the interaction of small particles with solvent molecules. The hydrophilic-lipophilic balance theory provides insight into the stability of emulsions and their applications, based on the relationship between the hydrophilic and hydrophobic components of surfactants. When colloid molecules get associated at a specific concentration (critical micelle concentration), they are called associated colloids, i.e. "micelles." Micelles are defined as the dynamic aggregation of amphiphilic molecules, such as soap, that have excellent mechanical properties. Micelles serve as an excellent platform for facilitating chemical reactions, increasing their speed by organizing reactant molecules at a molecular level. Incorporating amphiphilic molecules with well-defined structures into water can lead to remarkable selectivity and activity. These molecules are commonly utilized to mimic membranes for detecting membrane proteins and peptides, as well as for drug delivery.

4.1 Introduction

Colloid science is an exciting interdisciplinary field that brings together ideas from physics, biology, materials science, and other related disciplines. What truly captures our interest is the size of the particles involved, rather than their chemical formulation, whether organic or inorganic, their origins, such as biological or mineral sources, or their physical states, like single or multiple phases.

In 1861, Thomas Graham, during his research on the diffusion of various substances through vegetable and animal membranes, based on his findings, classified the different substances according to their diffusion properties [1]:

1. **Colloidal substances:** Noncrystalline substances that rarely diffuse or have a slight tendency to diffuse in solution. They do not pass through animal or vegetable membranes and exist in an amorphous form, such as gelatin, proteins, gum, and starch. The term "colloid" originates from the Greek word "kolla," meaning glue-like.

https://doi.org/10.1515/9783112208236-004

2. **Crystalloid substances:** Crystalloid substances, such as salts, sugar, acids, and bases, are crystalline and can pass quickly through animal and vegetable membranes.

Graham thought that the difference in the behavior of "crystalloids" and "colloids" was due to the particle size. After Graham classifies the substances, it was soon realized and demonstrated by many researchers that the classification cannot be rigidly based on their diffusion rate through animal membranes. It can no longer be accepted, as many crystalline substances can transform into colloids under appropriate conditions and means, regardless of their nature. It was understood that any substance, no matter what it is, can be transformed into a colloid, simply by breaking it down into tiny particles of colloidal size. Colloids do not form a distinct chemical class; instead, they represent a specific state of matter, similar to solids or liquids. This state is unique because particle size dimensions greatly influence its macroscopic properties.

For more than 100 years, chemists have been investigating the colloidal state of matter, which Ostwald referred to as a "world of neglected dimensions." In 1845, the dispersion of silver chloride and Prussian blue in an aqueous medium was explained by Selmi. Shortly after, Faraday investigated gold sols and concluded that this colloidal state is thermodynamically unstable, with stabilization occurring as a kinetic phenomenon. Some of Faraday's dispersions are still exhibited at the British Museum.

In the early twentieth century, science related to colloids brought together chemistry (preparative) and physics (theoretical). Einstein identified the relationship between the zigzag motion of colloidal particles and their diffusion coefficient, while Perrin utilized this bridge to determine Avogadro's number (N_A). Since then, it has connected multiple scientific fields, including biology and materials science. It has also been demonstrated that various so-called colloidal substances possess a crystalline structure, and conversely, many crystalline substances exhibit colloidal characteristics, as confirmed by X-ray analysis. For example:

(i) Soap (sodium salt of long-chain fatty acids) behaves as a colloid in water while remaining crystalline in benzene.
(ii) NaCl behaves like a crystalline substance in water, while displaying colloidal properties in benzene.

Given these research findings, the term "colloidal substance" is no longer valid and is thus discarded in favor of "colloidal state."

Now, the colloidal state of matter is defined as: "A heterogeneous system characterized by the microscopic dispersion of insoluble particles (1–100 nm) within particles of another substance." The colloidal system is a combination of two distinct phases: continuous and discontinuous. It is unlikely that the particles of the dispersed phase will form a single phase with those of the dispersion medium, as the dispersed phase mainly consists of large molecules or groups of smaller molecules. In the continuous phase, that is, the dispersion medium, tiny particles within the colloid size range (1–100 nm) of the other phase, called the discontinuous phase or dispersed phase, remain suspended.

A colloidal system has two main components:

1. **Dispersed phase**: The substance, often composed of solid particles and droplets, is dispersed in the medium and constitutes the dispersed phase.
2. **Continuous phase**: This is the medium in which the particles of the dispersed phase are spread. It could be anything, such as a liquid, gas, or solid.

Despite being a heterogeneous mixture, colloids can also appear uniform to the naked eye, which is why they can be easily mistaken for solutions. However, under a microscope, colloidal particles are visible. As illustrated in Figure 4.1, light traverses a true solution without scattering, whereas in a colloidal solution, it undergoes scattering.

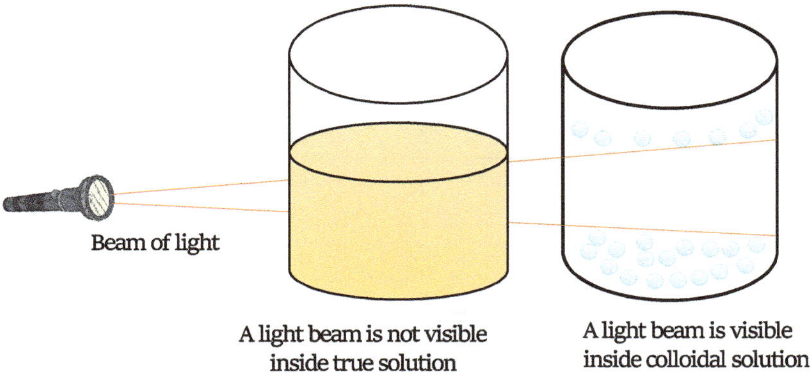
Beam of light
A light beam is not visible inside true solution
A light beam is visible inside colloidal solution

Figure 4.1: Observation: when a light beam is passed through a true solution and a colloidal solution.

The dispersed phase particles have a diameter of approximately 1 nm to 1 µm, bridging the size range between those in true solutions and suspensions. Colloidal particles do not settle even under the influence of gravitational force, primarily due to their small size, and cannot be separated by ordinary filtration. Instead, they require ultrafiltration techniques.

Colloids can exist in various forms, depending upon both the physical state of the dispersed phase and the dispersion medium, such as aerosols, emulsions, foams, sols, and gels. Their unique characteristics, such as the Tyndall effect (the scattering of light), Brownian motion (the random movement of particles), and electrophoresis, make them an essential study subject in scientific and industrial applications.

Colloids are of profound significance across various fields. For example, in the pharmaceutical industry, colloidal systems are utilized to improve the solubility and bioavailability of medicine. In human physiology, many biological fluids, like blood, lymph, and cytoplasm, exhibit colloidal properties, facilitating the transport of nutrients, hormones, and waste products. Industrially, colloids play a vital role in the manufacture of paints, inks, cosmetics, and food products.

4.1.1 Characteristics of Colloids

Dispersions can be broadly categorized into true solutions, colloidal solutions, and suspensions, based on particle size and other physical properties. True solutions consist of particles smaller than 1 nm, are homogeneous, pass through filter paper and semipermeable membranes, and do not scatter light. Colloids contain particles between 1 and 100 nm. They are heterogeneous, do not pass through semipermeable membranes, and scatter light (Tyndall effect). Suspensions have particles larger than 100 nm, are heterogeneous, can be filtered out, and their particles tend to settle over time. Different types of dispersion with their properties are presented in Table 4.1.

Table 4.1: A summary of the specifications of true solutions, colloidal solutions, and suspensions.

Properties →	**Visibility**	**Filtration**	**Diffusion rate**	**Appearance**	**Settling**
True solutions	Invisible	Easily filter through regular filter paper and animal membrane	Fast	Transparent and clear	Do not settle
Colloidal solutions	Invisible to the naked eye, the scattering of light is recorded under an ultramicroscope	Molecules such as these do not pass through an animal membrane, but they do pass through ordinary filter paper	Slow	Translucent	Do not settle
Suspensions	Visible with the naked eye	Do not pass	Do not diffuse	Opaque	Settle while standing

Colloidal systems occupy an intermediate state between true solutions and suspensions. The differences can be highlighted based on various parameters. Colloids can be differentiated from other solutions by several characteristics:

1. **Particle size:** Colloids consist of particles that are found in sizes between those of particles in a solution and a suspension, typically ranging from 1 nanometer to 1 micrometer. The particle size ranges of true solutions, colloidal solutions, and suspensions are summarized in Table 4.2 and can be easily understood with the help of Figure 4.2.
2. **Appearance:** Colloids are mainly translucent. True solutions are clear and transparent. Colloids are translucent due to the scattering of light. Suspensions are usually opaque and turbid.

Table 4.2: Variation in particle size.

Type of state	Size of particles
True solution/molecular solutions	<10 Å
Colloidal solutions	10–1,000 Å
Suspensions	>1,000 Å

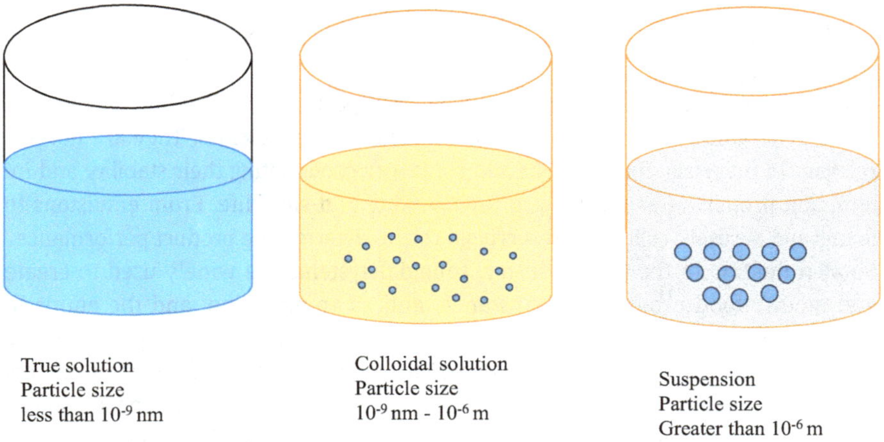

Figure 4.2: Different solutions and sizes of their particles.

3. **Dispersion**: Colloids are mixtures where one substance is evenly dispersed throughout another, made up of two phases: (i) the dispersed phase (particles), and (ii) the continuous phase (medium).
4. **Stability**: The particles in a colloid do not settle out over time, which distinguishes them from suspensions. The true solutions are stable indefinitely. Colloids are relatively stable as the particles do not settle readily. The suspensions are unstable, and the particles settle over time.
5. **Filtration:** In terms of filtration, these solutions behave differently. True solutions pass through all filters. However, colloids pass through filter paper but not through semipermeable membranes. Suspensions usually do not pass through regular filter paper due to their large particle size.
6. **Visibility:** Particles in true solutions and colloids are not visible under an optical microscope. Particles in suspensions are visible under a microscope or even to the naked eye.

These distinctions are crucial in choosing appropriate systems for applications such as drug delivery in pharmaceuticals, where colloids are preferred due to their stability and bioavailability. These properties underscore the practical utility of colloids.

For instance, in drug delivery systems, colloids enhance the stability and controlled release of active pharmaceutical ingredients. Similarly, colloidal gels are being explored for use in tissue engineering and regenerative medicine due to their biocompatibility and tunable physical properties.

4.1.2 Role of Colloidal Solutions in Various Fields

Colloidal systems are a cornerstone of both theoretical research and practical applications across many fields:

1. **Everyday applications:** Colloids are an integral part of everyday life. Colloids are found in products such as paints, food, cosmetics, and medicines. They are not just relevant in industrial formulations and consumer goods, citing their stability and interaction properties as essential to functionality and shelf life. From emulsions to foams and aerosols, colloids play a crucial role in determining product performance.
2. **Food industry**: In the food industry, colloidal systems are widely used to create and modify food products. For instance, milk is an emulsion, and the emulsification process is fundamental to the preparation of different milk products like butter, mayonnaise, and salad dressings. The colloidal nature of their ingredients determines their texture, appearance, and shelf life. Gels, such as those used in jellies, puddings, and confectionery, also rely on the colloidal behavior for their consistency and stability. Colloids are also used in creating foams, such as whipped cream and beer foam, which contribute to the texture of these products.
3. **Cosmetics and personal care**: In the cosmetics industry, colloids are employed in the formulation of many products like creams, lotions, shampoos, and other personal care products. The emulsions and gels in these products help to stabilize ingredients and improve the texture and application of the products. For example, colloidal silica is often used as a thickening agent in cosmetics.
4. **Paints and coatings**: The application of colloids in the paint industry enables the production of smooth, uniform coatings. Pigments are dispersed in a liquid medium (often water or oil) to form a colloidal system. The properties of the colloidal dispersion determine the application, coverage, and drying time of paints and coatings.
5. **Environmental protection**: Colloidal systems play a significant role in environmental science, especially in water treatment, soil studies, and pollutant remediation. According to Fanun (2014), colloids contribute to the removal of heavy metals, radioactive materials, and other contaminants through adsorption and surface interaction processes. For example, pollutants can adsorb onto colloidal particles in water, affecting their transport and bioavailability. Understanding these interactions is essential in pollution control and water purification technologies. Their ability to form stable dispersions allows for effective transport and treatment within environmental systems. The mobility of colloidal particles in groundwater can also influence the spread

of contaminants, making colloidal behavior a critical factor in assessing and managing environmental risks.

6. **Industry, materials science, and nanotechnology:** Colloidal systems are extensively used in various industrial processes. For example, the production of paints, inks, and coatings involves colloidal dispersions of pigments and other additives. In materials science, colloids are fundamental in fabricating nanostructures and hybrid materials. Antonietti and Göltner (1997) described how colloidal superstructures can be engineered via self-assembly processes, enabling the creation of highly symmetric nanomaterials. Colloids are used as templates or carriers for the synthesis of nanoparticles. These materials have unique properties, like enhanced conductivity, strength, and reactivity, and find applications in diverse fields, such as optics, catalysis, sensor technologies, electronics, energy storage, and drug delivery [2]. The properties of colloids such as stability, charge, and particle size determine the performance of these materials in terms of color, texture, durability, and distribution.
7. **Colloids in physics and model systems**: Colloids enable physicists to explore fundamental interactions at the mesoscopic level. Colloidal dispersions are employed to simulate atomic behaviors in controllable laboratory environments, particularly under external fields. This makes them invaluable in studying phase transitions, diffusion, and dynamic assemblies.

Colloidal systems are a cornerstone of both theoretical research and practical applications across many fields. Their versatility in mimicking natural systems, supporting industrial innovation, and providing solutions to environmental challenges illustrates their immense importance. As research advances, the scope and sophistication of colloid applications continue to expand.

4.2 Classification of Colloidal Systems

4.2.1 Classification Based on Two Physical States

Colloids consist of two components: the dispersed phase and the continuous phase. The various types of colloids, along with their respective examples, are summarized in Table 4.3.

1. **Sol (solid in liquid)**: In this type of colloid, solid particles are dispersed in a liquid medium. Examples include paints, inks, and blood plasma.
2. **Gel (liquid in solid)**: A gel is a system where a liquid is dispersed within a solid medium. Examples include jelly and toothpaste.
3. **Emulsion (liquid in liquid)**: This is a colloid in which liquid droplets are dispersed into another liquid. Examples include milk, mayonnaise, and lotions.

4. **Aerosol (solid or liquid in gas)**: Aerosols are colloids where solid or liquid particles are dispersed into a gas. Examples include fog, smoke, and mist.
5. **Foam (gas in liquid or solid)**: A foam is a substance where gas bubbles are dispersed within a liquid or solid. Examples include shaving cream, whipped cream, and beer foam.
6. **Solid Foam (gas in solid)**: This is a colloidal system in which gas is dispersed within a solid. Examples include sponge materials.

Table 4.3: Summary of colloid systems based on dispersion medium and dispersed phase.

S. no.	Dispersion medium	Dispersed phase	Name of colloid system	Examples
1.	Liquid	Solid	Sol	Pigmented ink, sediment, and paint
2.	Liquid	Liquid	Emulsion	Milk, hand cream, biological membranes, mayonnaise, latex, liquid biomolecular condensate, and oil on water
3.	Liquid	Gas	Foam	Soap suds, shaving cream, and Chantilly cream
4.	Gas	Solid	Solid aerosol	Smoke and dust storms
5.	Gas	Liquid	Liquid aerosol	Fog, clouds, mist, hairspray, and steam
6.	Gas	Gas	None	No colloid formation
7.	Solid	Solid	Solid sol	Cranberry glass and ruby glass
8.	Solid	Liquid	Gel	Jelly, gelatine, and agar
9.	Solid	Gas	Solid foam	Lava, pumice, and Styrofoam

4.2.2 Classification of Colloids

Generally, colloidal sols can be divided into two main categories, called "lyophilic and lyophobic." If water consists of a dispersion phase, it is classified as hydrophilic and hydrophobic. This difference arises from how the dispersion medium interacts with the dispersed phase. Lyophilic and lyophobic are Greek words: "lyo" (meaning to loosen/dissolve) and "phillos" (meaning showing affinity or loving), that is, liquid- or solvent-loving; and "phobic" (meaning hating or fear), i.e., liquid- or solvent-hating.

1. **Lyophilic colloids:** Molecules of these substances (like starch, gum, rubber, gelatine, etc.) which are large enough to meet the lowest limit of the colloidal sol formation, and can turn into the colloidal state when mixed with an appropriate dispersion medium, are referred to as lyophilic colloids or lyophilic sols. Since these substances are liquid-loving, they have a strong attraction force and affinity with the molecules of the dispersion medium. That is why they form a colloid sol di-

rectly by dissolving in a medium, so they are also known as "intrinsic colloids." These colloids are reversible, as removing the dispersion medium (such as water by evaporation) allows the sol to be prepared by simply adding more solvent to the residue. In other words, they can easily be reformed after coagulation. These colloids cannot be easily precipitated, and hence, they are relatively stable.

2. **Lyophobic colloids:** Molecules of these substances such as metal-sulfides and hydroxides, gold, and other metals are smaller than the lower limit of colloid formation. It is quite difficult, and not possible, to prepare a colloid by simply mixing these substances with the dispersion medium, because they do not meet the lower limit of the colloidal sol by aggregation of many individual molecules. Therefore, their colloidal sols require special methods for preparation. Since their colloidal sols are prepared only using indirect methods, they are classified as "extrinsic colloids." Lyophobic colloids have a fear of interaction with the solvent/medium; hence, they do not form stable colloids and readily precipitate/coagulate. Furthermore, their residue left after evaporation (when water is the dispersion medium) cannot be easily reformed into colloidal sols by ordinary means, which is why they are irreversible. The differences between lyophilic and lyophobic colloids, based on various properties, are presented in Table 4.4. This comparison highlights their distinct stability, methods of preparation, and behavior in solution.

Table 4.4: A summary of the properties of lyophilic and lyophobic colloids.

S. no.	Feature	Lyophilic colloid	Lyophobic colloid
1.	**Preparation**	Can be prepared by simple means of mixing	Needs special methods/ techniques
2.	**Affinity for solvent**	High-solvent-loving and hydrated	Low-solvent repelling and less hydrated
3.	**Reversibility**	Reversible and can be reconstituted	Cannot be reverted to the original
4.	**Stability**	High and resistant to coagulation	Less stable and coagulate quickly
5.	**Visibility**	Not visible either by the naked eye or an ultramicroscope	Visible under ultramicroscope
6.	**Viscosity**	Increases after the formation of sol	Remains the same as medium
7.	**Surface tension**	Lower than the dispersion medium	Nearly the same as medium
8.	**Examples**	Protein, gum, starch, etc.	Metal hydroxides and sulfides

4.2.3 Classification Based on Particle Interactions

In the formation of colloids, the size of the dispersed phase particles plays a decisive role. When a solid material is dispersed in a medium in which it is otherwise insoluble, and the resulting particles fall within the colloidal size range, the system is referred to as a colloidal dispersion. Common examples include gold sol and As_2S_3 dispersion in water. Colloids or colloidal dispersions can thus be classified according to whether the particle dimensions of the substances lie within the colloidal range:

1. **Multimolecular colloids:** These form when many atoms or tiny molecules (usually less than 1 nm in diameter) cluster together in a dispersion medium, creating particles of colloidal size. The dispersing behavior depends on the aggregating properties of the particles. For example, a gold sol may comprise particles of varying sizes, each composed of several atoms. Sulfur sol consists of particles containing about a thousand S_8 sulfur molecules, held together through van der Waals forces.
2. **Macromolecular colloids:** These colloids consist of molecules that are already large enough to fall within the colloidal size range. When certain substances contain large molecules, known as macromolecules, which have high molecular masses, they dissolve in an appropriate liquid to form a solution where the dispersed particles are within the colloidal size. Such substances are referred to as macromolecular colloids. Typically, these are polymers with very high molecular weights. Examples of natural macromolecules include starch, proteins, cellulose, enzymes, and gelatin. Examples of man-made macromolecules include polyethylene, nylon, polystyrene, and synthetic rubber. Macromolecular colloids are typically more stable and exhibit many characteristics similar to those of true solutions. Macromolecular colloids are defined as large molecules with sizes comparable to those of colloidal particles, and such dispersions are known as macromolecular colloids. Despite their colloidal nature, these solutions share many aspects with true solutions.
3. **Associated colloids:** At low concentrations, these substances behave like ordinary electrolytes. However, when their concentration exceeds a certain limit, the molecules aggregate to form larger units called micelles, which exhibit colloidal properties. Thus, associated colloids act as normal electrolytes in dilute solutions but display colloidal behavior at higher concentrations due to micelle formation [3]. This phenomenon occurs only above a characteristic temperature, known as the Krafft temperature (T_k), and above a minimum concentration, referred to as the CMC (Critical micelle concentration). Typical examples of associated colloids include soaps and synthetic detergents. The distinctions among multimolecular colloids, macromolecular colloids, and associated colloids are summarized in Table 4.5.

Table 4.5: Difference between different types of colloids.

Multimolecular colloids	Micromolecular colloids	Associate colloids
Many molecules or atoms aggregate to form multimolecular colloids with a diameter of less than 1 nm.	These are molecules of enormous size.	Many ions aggregate in a concentrated solution to form these colloids.
Their molecular masses are not very high.	These have high molecular masses.	They also have high molecular masses.
Weak van der Waals forces hold these colloids together.	Since they have longer chains, the van der Waals forces are comparatively stronger.	The higher the concentration, the greater the van der Waals forces, so the forces are stronger here.

4.3 Properties of Colloids

Colloidal systems show a set of characteristic properties that clearly set them apart from true solutions and coarse suspensions. These properties arise from the small size of dispersed particles and their interaction with the medium. Among the most important are optical, mechanical, and electrical properties, each of which plays a vital role in determining the behavior and practical applications of colloids. These distinctive features can be better understood by examining them individually, beginning with their optical properties:

1. **Optical properties**: One of the defining characteristics of colloids is their ability to scatter light. When a ray of light passes through a colloid solution, the particles make the light scatter, making the path of the beam of light visible. This does not occur in true solutions, where the solute particles are too small to scatter light. When a powerful converging beam of light is passed through a colloidal solution kept in the darkness, the beam's path becomes illuminated with a bluish glow. This phenomenon is known as the Tyndall effect (1869), and the illuminated path is referred to as the Tyndall cone, represented by Figure 4.3. This effect occurs because particles of colloidal size scatter light, which is fascinating to understand. Essentially, colloidal particles scatter ultraviolet and visible light because their size matches the wavelength of the light, causing illumination. It was observed during the experiment that the area of scattered light is greater than the area occupied by the particles themselves. Colloidal particles appear as bright spots because they are positioned at right angles to the light beam. Thus, the Tyndall effect is described as the scattering of light caused by particles within the colloidal range.

   The Tyndall effect is not displayed by the particles (ions or molecules) of true solutions, as they are too small to scatter light. Therefore, the Tyndall effect helps to differentiate between these two solutions. This concept has also led to the development of an ultramicroscope, an instrument used to detect particles of colloidal size. Additionally, the Tyndall effect confirms that colloidal systems are heterogeneous.

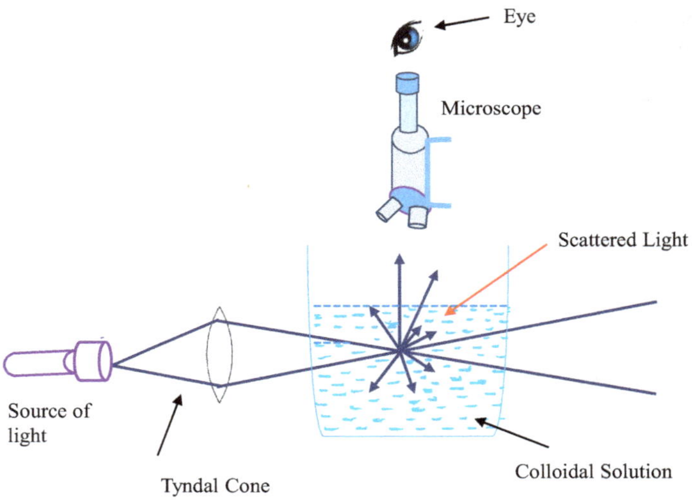

Figure 4.3: A representation of the Tyndall effect.

2. **Mechanical properties**: Colloidal particles are in continuous, random motion because they collide with molecules in their surrounding medium. This motion, known as Brownian motion, helps to keep the particles dispersed and prevents them from settling out. The molecules in the continuous medium are kinetically energetic, which causes this phenomenon. An important property, namely Brownian movement, is exhibited by the colloidal particles. When analyzed under an ultramicroscope, these particles are seen to move in a constant zigzag pattern. The credit for discovering this important property was given to botanist Robert Brown in 1827. During his research, he found that suspended pollen grains in water move randomly in a zigzag manner. The movement was named Brownian movement after him. It can be described as the continuous, irregular, or zigzag movement of colloidal particles within a dispersion medium, as shown in Figure 4.4.

   Cause of Brownian movement: Brownian movement occurs from the uneven collisions between colloidal particles and the moving particles of the dispersion medium. These molecules constantly collide with the particles from all directions, transferring momentum. Since these collisions are not equally distributed in all directions, they produce a net force along a specific direction, which leads to further collisions with the medium's molecules, causing the particle to change direction in a zigzag fashion. The process continues, leading to a random zigzag motion of the colloidal particle. The Brownian movement decreases as the size of

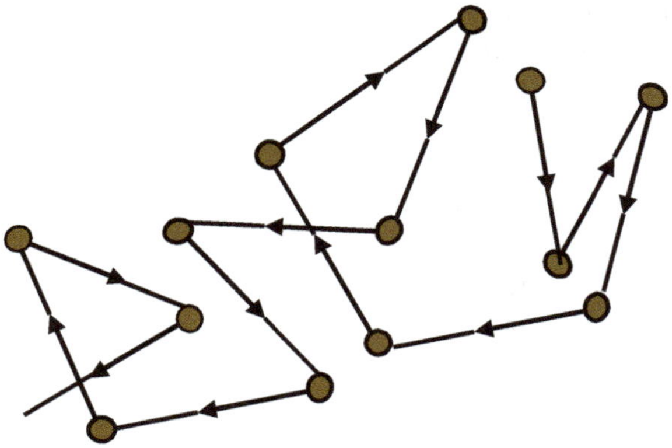

Figure 4.4: Brownian (zigzag motion).

the colloidal particle increases. That is why suspensions do not show this kind of movement. Brownian motion is essential for stabilizing a sol because it counteracts gravitational forces on colloidal particles, stopping them from settling.

3. **Electrical properties:** A defining property of colloidal solutions is that dispersed particles carry a definite electrical charge, which plays a central role in their stability. In a colloidal system, all particles acquire the same type of charge, while the dispersion medium carries an equal but opposite charge. Thus, although the overall sol is electrically neutral, the uniform charge on colloidal particles prevents aggregation by generating electrostatic repulsion.

For instance, in a silver iodide sol (AgI), the charge depends on the relative concentrations of ions present during preparation:

(i) When $AgNO_3$ is in excess, AgI particles adsorb Ag^+ ions and become positively charged; the counter ions in the medium are NO_3^-.

(ii) When KI is in excess, AgI particles adsorb I^- ions and become negatively charged; the counter ions in the medium are K^+.

This selective adsorption of ions determines the charge on colloidal particles and prevents them from settling, thus stabilizing the sol. The arrangement of ions in AgI sol, under conditions of excess $AgNO_3$ and excess KI, illustrating the development of positive and negative sols, respectively, is presented in Figure 4.5. Similarly, in a ferric hydroxide sol, the $Fe(OH)_3$ particles are positively charged, and the surrounding medium holds the counter negative ions, ensuring stability.

Thus, colloidal sols can be classified based on their charge as either positively or negatively charged. Examples of these include:

(i) **Positively charged sols**: Hydroxide sols of metals such as $Fe(OH)_3$, $Al(OH)_3$, and $Cr(OH)_3$, TiO_2 sols, sols of hemoglobin, and methylene blue (basic dye).

(ii) **Negatively charged sols**: Colloidal sols of metals, such as Cu, Ag, Au, and Pt, sulfide sols of metals, such as As_2S_3 and CdS, starch sols, and Congo red (acidic dye).

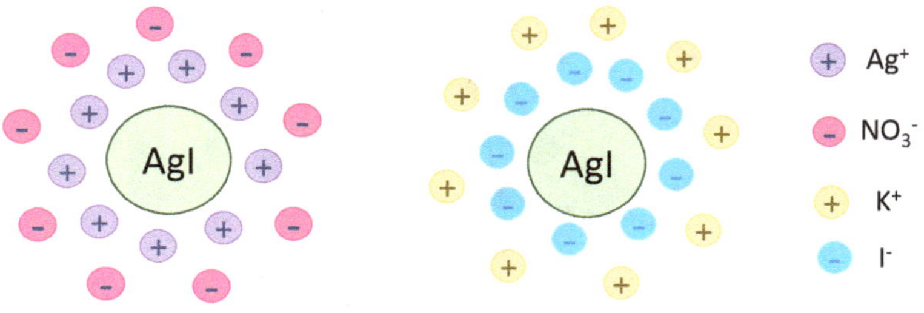

Figure 4.5: A representation of charged colloidal sols.

4.4 Preparative Methods for Colloidal Solutions

The particle size of the dispersed phase is a crucial factor in the preparation of colloidal solutions. It serves as a connecting link between a suspension and a true solution. Colloidal solutions are characterized by particles that are larger than those in a true solution but smaller than those in a suspension. The typical size range of colloidal particles lies between 10 and 2,000 Å. This intermediate particle size is the key feature that imparts unique properties to colloids and guides the methods used for their preparation.

4.4.1 Preparation of Lyophilic Sols

Lyophilic sols are colloidal systems in which the dispersed phase has a strong affinity for the dispersion medium. Because of this natural attraction, the particles of the dispersed phase readily intermix with the medium, resulting in the spontaneous formation of a stable colloidal sol without the need for any external stabilizer. The preparation process is generally simple and slightly exothermic due to the interaction between the solute and solvent molecules. Common examples of lyophilic sols include gum sols, starch sols, egg albumin sols, gelatin sols, and rubber in benzene. These systems are typically reversible in nature, meaning that if the solvent is removed (e.g., by evaporation), the sol can be reconstituted by simply adding the dispersion medium back.

4.4.2 Preparation of Lyophobic Sols

Unlike lyophilic sols, lyophobic sols cannot be prepared by simple mixing of the dispersed phase with the dispersion medium, as there is little or no natural affinity between them. As a result, direct dissolution does not yield a colloidal solution. Furthermore, the particle size of individual entities does not fall within the colloidal range, and hence, special techniques and stabilizing agents are required for their preparation. The methods employed for preparing lyophobic sols can be broadly classified into two categories:
1. Dispersion methods
2. Condensation methods

Both approaches ensure that particles in the range of 10–2,000 Å are obtained, and stabilizers are often added to prevent coagulation, thereby imparting stability to the sol.
1. **Dispersion methods:** In dispersion techniques, larger solid particles are broken down mechanically or physically into extremely fine particles that fall within the colloidal size range. These methods rely on external forces to reduce particle size and usually require stabilizers to prevent coagulation. These methods employ mechanical, electrical, or ultrasonic forces, and stabilizing agents are often added to prevent the newly formed particles from aggregating.
   (i) **Mechanical dispersion:** In this technique, the coarse particles of the dispersed phase are broken down mechanically to achieve colloidal dimensions. A colloid mill is commonly employed for this purpose. The substance is first mixed with the dispersion medium and an appropriate stabilizing agent, and then passed through the mill. Inside the mill, rapidly rotating discs (operating at nearly 7,000 revolutions per minute) exert intense shearing forces on the mixture, as illustrated in Figure 4.6. These forces grind the particles to sizes that fall within the colloidal range, leading to the formation of a sol. One drawback of this method is that it does not always yield particles of uniform sizes. Nevertheless, it is widely applied in industries for the preparation of paints, inks, varnishes, and dyes.
   (ii) **Ultrasonic dispersion:** In this method, high-frequency sound waves (around 20,000 Hz) generated by an ultrasonic device are used to break down coarse particles into colloidal dimensions. The technique is especially effective for substances of low mechanical strength, such as graphite, gypsum, and resins. The process works on the principle of cavitation. Piezoelectric oscillations convert electrical energy into mechanical vibrations, which produce alternating compressions and rarefactions in the liquid medium. This results in the formation of microscopic bubbles or cavities that rapidly collapse under external pressure. The collapse of these cavities releases energy, which breaks down larger particles into smaller ones. Since the process is exothermic, it assists in further reducing particle size and promotes uniform dispersion. Stabilizing agents are gen-

erally added to prevent the freshly produced colloidal particles from coagulating, thereby ensuring the stability of the sol. Ultrasonic dispersion has significant advantages: it allows precise control over particle size, avoids contamination from mechanical grinding, and produces sols with high uniformity. It is widely used in materials science, pharmaceuticals, and nanotechnology. A typical example is the preparation of mercury sols using this technique.

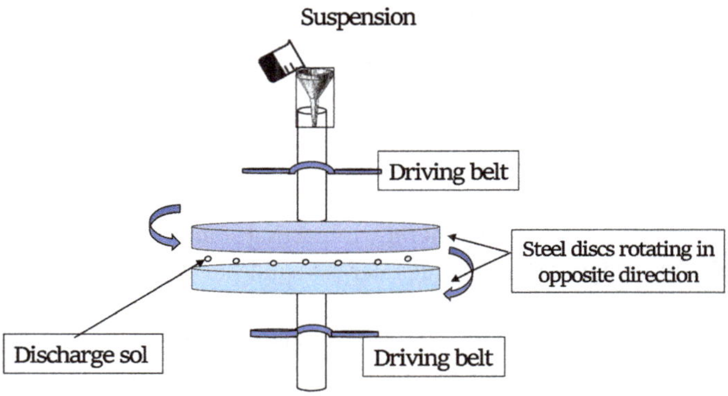

Figure 4.6: A pictorial representation of the working model of the colloidal/disc mill used in the dispersion method.

(iii) **Electrical disintegration method (Bredig's arc method)**: Bredig's arc method (1898) is a classical technique used for preparing colloidal solutions of metals such as gold, silver, and platinum. In this method, two electrodes made of the desired metal are immersed in water containing a small amount of alkali, which acts as a stabilizer. The liquid is maintained at a low temperature (around 0 °C) to minimize evaporation losses.

When an electric arc is struck between the electrodes, the intense heat produced vaporizes the metal. The vapors then cool rapidly in the dispersion medium, condensing into particles within the colloidal size range, as shown in Figure 4.7. The addition of stabilizing agents, such as potassium hydroxide, prevents coagulation of the freshly formed particles, thereby yielding a stable sol. This process is considered a combination of dispersion and condensation. Dispersion occurs because the metal breaks down into vapor, and condensation occurs because the vapor reaggregates into colloidal particles. Bredig's method is particularly effective for producing stable metal sols.

(iv) **Peptization**: In this process, newly formed precipitates are subjected to disintegration into colloidal-sized particles using a peptizing agent, which is typically an electrolyte sharing a common ion with the precipitate. This common ion causes the electrolyte to adsorb onto the precipitate surface, thereby imparting an electric charge to the particles. The resulting electrostatic repulsion between similarly charged particles prevents their coagulation, allowing them to remain dispersed in the medium in the colloidal state. This technique is particularly useful for preparing colloidal solutions of metal hydroxides (e.g., ferric hydroxide and aluminum hydroxide) and metal oxides (e.g., stannic oxide). For instance, when hydrochloric acid is added to a freshly prepared precipitate of aluminum hydroxide, the Al^{3+} ions generated are adsorbed on the surface of the precipitate, converting it into a stable sol of aluminum hydroxide, as depicted in Figure 4.8:

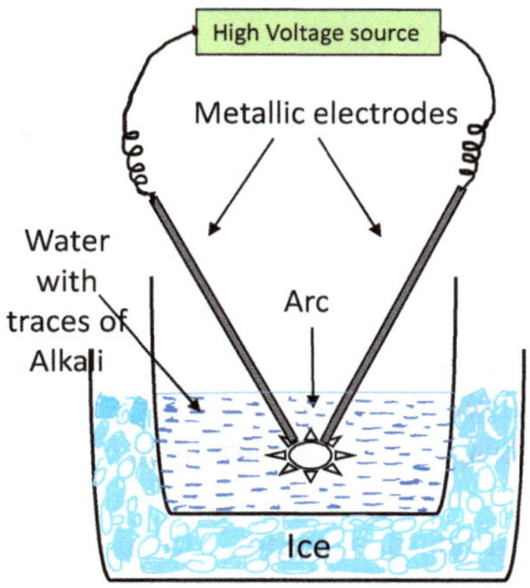

Figure 4.7: Bredig's arc electrical disintegration method.

$$Al(OH)_3 + HCl \longrightarrow AlCl_3 + H_2O$$

$$AlCl_3 \longrightarrow Al^{3+} + 3Cl^-$$

$$Al(OH)_3 + Al^{3+} \longrightarrow \{Al(OH)_3 . Al^{3+}\}$$

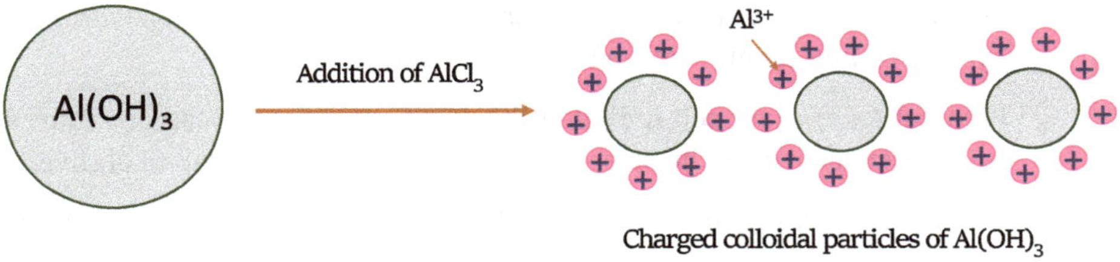

Figure 4.8: Preparation of colloidal solution of $Al(OH)_3$.

Similarly, the addition of $FeCl_3$ electrolyte to fresh precipitate of Ferric hydroxide produces a ferric hydroxide sol, as shown in Figure 4.9:

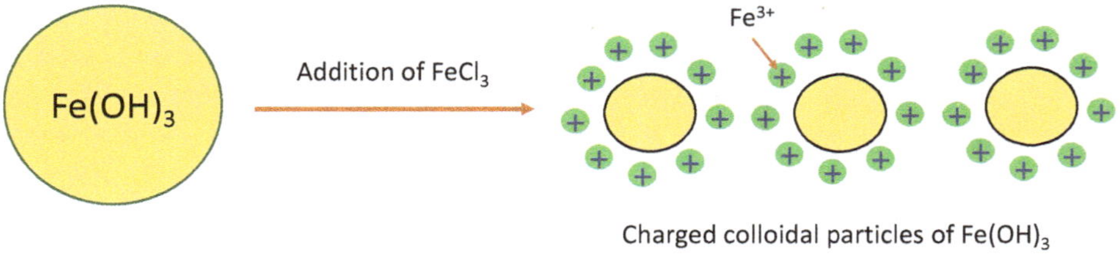

Figure 4.9: Preparation of a colloidal solution of $Fe(OH)_3$.

$$Fe(OH)_3 + Fe^{3+} \rightarrow \{Fe(OH)_3 . Fe^{3+}\}$$

2. **Aggregation or condensation methods:** In condensation techniques, very small atoms, ions, or molecules combine through chemical or physical processes to form larger aggregates of colloidal dimensions (10^{-9} to 10^{-6} m). These reactions are carefully carried out under controlled conditions so that the size of the resulting particles remains within the colloidal range. Through these processes, stable sols of various metal oxides, hydroxides, and sulfides can be obtained. The main chemical reactions employed in condensation methods include:
   (i) **Double decomposition reaction:** Colloidal sols can also be prepared by double decomposition, in which two soluble reactants interact to produce an insoluble substance of colloidal dimensions. A typical example is the preparation of arsenious sulfide (As_2S_3) sol. When a dilute aqueous solution of arsenious oxide (As_2O_3) is treated with hydrogen sulfide (H_2S) gas, a colloidal sol of As_2S_3 is formed:

$$As_2O_3 + H_2S \rightarrow As_2S_3(sol) + 3H_2O$$

   (ii) **Reduction:** Colloidal sols of several metals can be prepared by reducing their dilute salt solutions with suitable reducing agents. Metals such as gold, silver, platinum, palladium, iridium, copper, and lead are commonly obtained by this technique. The choice of reducing agent depends on the metal salt used. Typical reducing agents include formaldehyde, hydrazine, phenylhydrazine, hydroxylamine, tannic acid, and stannous chloride.

   Preparation of silver sol: A dilute solution of silver nitrate can be reduced using formaldehyde in the presence of water to yield a colloidal sol of silver:

$$2AgNO_3 + HCHO + H_2O \rightarrow 2Ag(sol) + HCOOH + 2HNO_3$$

   Preparation of gold sol: Gold sol may be prepared by reducing a dilute solution of gold chloride ($AuCl_3$) with reducing agents such as formaldehyde or hydrazine:

$$2AuCl_3 + 3HCHO + 3H_2O \longrightarrow 2Au(sol) + 3HCOOH + 6HCl$$

$$2AuCl_3 + 2NH_2NH_2 \longrightarrow 4Au(sol) + 3N_2 + 12HCl$$

Preparation of gold sol: Alternatively, gold sol can also be obtained by reducing gold chloride with stannous chloride ($SnCl_2$):

$$2AuCl_3 + 3SnCl_2 \longrightarrow 2Au(sol) + 3SnCl_4$$

(iii) **Oxidation reaction**: In this technique, colloidal sols are obtained by oxidizing a suitable substance under controlled conditions. A common example is the preparation of sulfur sol by oxidizing hydrogen sulfide gas. When H_2S gas is passed through an aqueous solution of nitric acid (HNO_3) or bromine water, sulfur is liberated in colloidal form:

$$2H_2S + HNO_3 \longrightarrow 2S(sol) + 2H_2O + NO$$

$$2H_2S + Br_2 \longrightarrow 2S(sol) + 2HBr$$

(iv) **Hydrolysis process:** Colloidal sols of certain metal hydroxides can be prepared by hydrolyzing the salts of heavy metals under controlled conditions. A classic example is the preparation of ferric hydroxide sol. When a small quantity of ferric chloride ($FeCl_3$) solution is added to boiling water, extensive hydrolysis occurs, producing a red-brown colloidal sol of ferric hydroxide:

$$FeCl_3 + 3H_2O \xrightarrow{\Delta} Fe(OH)_3(sol) + 3HCl$$

(v) **Exchange of solvent:** In this method, a solution of a substance in a solvent, in which it is soluble, is added to a second solvent, in which the substance is insoluble. This leads to the precipitation of the substance as colloidal particles, forming a stable sol. Example: sulfur and phosphorus are soluble in alcohol. When their alcoholic solutions are poured into water, where they are insoluble, colloidal sols of sulfur or phosphorus are formed.

(vi) **Excessive cooling:** Colloidal solutions of ice in organic solvents, such as ether, can be prepared by rapidly cooling a mixture of water and the organic solvent. The sudden drop in temperature causes the formation of finely divided ice particles dispersed in the organic medium, resulting in a stable colloidal system.

4.5 Hydrophilic-Lipophilic Balance (HLB)

In 1949, Griffin introduced the concept of the hydrophilic-lipophilic balance (HLB) to describe the balance between the hydrophilic (water-loving) and lipophilic (oil-loving) portions of a surfactant molecule. This concept plays a crucial role in surface chemistry, particularly in understanding how surfactants interact with water (hydrophilic) or oil (lipophilic). The HLB value is expressed on a numerical scale ranging from 0 to 20 to predict the behavior of nonionic surfactants for a desired emulsion type. The HLB principle is complemented by Bancroft's rule, which states that "The phase in which the surfactant is more soluble tends to become the continuous phase of the emulsion." These concepts assist in choosing surfactants to achieve the desired emulsion [4]. According to Bancroft's rule, the nature of the continuous phase in an emulsion depends on the solubility of the surfactant. If the surfactant is more soluble in water, it promotes the formation of an oil-in-water (O/W) emulsion. Conversely, if the surfactant is more soluble in oil, it favors the formation of a water-in-oil (W/O) emulsion. Thus, the type of emulsion is largely governed by the surfactant's solubility characteristics rather than the relative amounts of oil and water present. Thus, the HLB system provides a practical guide to select the appropriate surfactant for a desired emulsion type.

Over the last 10 years, scientists have expanded their use in areas such as mixing liquids that usually do not mix (emulsification), delivering drugs in the body, food production, cosmetics, and small-scale technologies known as nanotechnology.

HLB value for nonionic surfactants can be calculated using the following quation:

$$\mathrm{HLB} = 20 \times \left(\frac{M_\mathrm{H}}{M_\mathrm{H} + M_\mathrm{L}}\right) = 20 \times \left(\frac{M_\mathrm{H}}{M_\mathrm{T}}\right) \tag{4.1}$$

where M_H is the mass of the hydrophilic group, M_L is the mass of the lyophilic group, and M_T is the total molecular mass of the surfactant molecule ($M_\mathrm{H} + M_\mathrm{L}$).

Let us understand with the help of an example. If a hydrophilic group of a surfactant has a molecular weight of 50 and the molecular weight of the surfactant molecule is 250, its HLB would be calculated as:

$$\mathrm{HLB} = \left(20 \times \left(\frac{50}{250}\right)\right) = 4$$

Typically, surfactant analysis data provide the best basis for evaluating the HLB value. When the hydrophilic segment of an emulsifier is composed solely of ethylene oxide units, the HLB value can be calculated using the following equation:

$$\mathrm{HLB} = \frac{\%\ \text{of ethylene oxide}}{5} \tag{4.2}$$

This simplified equation is particularly useful for polyoxyethylene-based nonionic surfactants, where the hydrophilic contribution arises exclusively from the ethylene oxide groups.

The formula for emulsifiers, including esters, fatty acids, and polyhydroxy groups, is given as follows:

$$\text{HLB} = \left(20 \times \left(1 - \frac{\text{Saponification value of fatty acid (S)}}{\text{Acidity value of given fatty acid (A)}}\right)\right) \tag{4.3}$$

For example, for polysorbate:

$$\text{Saponification value of fatty acid (S)} = 45$$

$$\text{Acidity value of given fatty acid (A)} = 275$$

$$\text{HLB} = 20 \times \left(1 - \frac{45}{275}\right) = 16.73$$

Mixing two or more emulsifiers makes it possible to calculate the overall HLB value of the blend by taking a weighted average of the individual HLB values. The general formula is given as follows:

$$\text{HLB}_{\text{mixture}} = f_1(\text{HLB})_1 + f_2(\text{HLB})_2 + f_3(\text{HLB})_3 + \cdots \tag{4.4}$$

where f_1, f_2, etc. are weight fractions of the emulsifiers in the mixture, and $(\text{HLB})_1$, $(\text{HLB})_2$, etc. are the HLB values of the individual emulsifiers.

For example, when a Polysorbate-80 (HLB = 15.0) is mixed with Sorbitan oleate (HLB = 4.3), containing 75% and 25%, respectively, the HLB value is calculated using the following method:

$$\text{Polysorbate-80} = f_1 \times n(\text{HLB})_1 = 75\% \times 15.0 = 11.25$$

$$\text{Sorbitan oleate} = f_2 \times n(\text{HLB})_2 = 25\% \times 4.3 = 1.075$$

$$\text{HLB} = 12.325$$

By adjusting the ratio of emulsifiers, formulators can fine-tune the HLB to match the required value for a stable emulsion, ensuring both efficiency and long-term stability.

4.5.1 Davies' HLB (Group Contribution Method)

This method determines the HLB of a surfactant molecule by summing the contributions from its hydrophilic and lipophilic components. The formula used to find out the HLB is given as follows:

$$\text{HLB} = 7 + \sum(\text{Number of hydrophilic groups}) - \sum(\text{Number of lyophilic groups}) \tag{4.5}$$

4.5.2 Significance of the HLB Principle

The HLB principle is significant and has become essential because it provides guidelines for choosing appropriate surfactants to create stable emulsions. It helps in classifying the right surfactant and determining the correct quantity of emulsifier needed to achieve a stable O/W emulsion. An emulsion is a colloidal system in which both the dispersed phase and the dispersion medium are liquids that do not naturally mix. The size of suspended particles in the emulsions is usually larger than in sols, and these particles are sometimes visible under a microscope.

Emulsions are mainly classified into two categories:
1. Oil-in-water (O/W), where oil particles are dispersed in an aqueous medium
2. Water-in-oil (W/O), where water molecules are dispersed in a liquid (e.g., oil)

The type of emulsion obtained is decided by the relative amounts of the two liquids, with the excess liquid acting as the dispersion medium. For example, if oil is in excess, the emulsion is W/O; if water is in excess, it is O/W. Common examples include: (1) milk, which is an O/W emulsion with liquid fat molecules dispersed into water; and (2) cod-liver oil emulsion, where water droplets are dispersed into oil. Since emulsions are mixtures of two immiscible liquids, they are usually unstable and tend to separate upon standing. To create stable emulsions, small amounts of other substances, known as emulsifying agents or emulsifiers, are added.

4.5.3 The HLB Scale and Its Applications

As stated before, it is a numerical scale ranging from 0 to 20, as shown in Figure 4.10, used to predict the behavior of nonionic surfactants in emulsions, where 0 indicates fully lipophilic and 20 indicates fully hydrophilic. To determine the HLB value of a nonionic surfactant with a known composition, it is first necessary to understand the concept of the hydrophilic group. The HLB scale provides a context guide to understand how surfactants work to form different emulsions (W/O and O/W). The relationship between HLB values and the type of emulsion formed is summarized in Table 4.6.

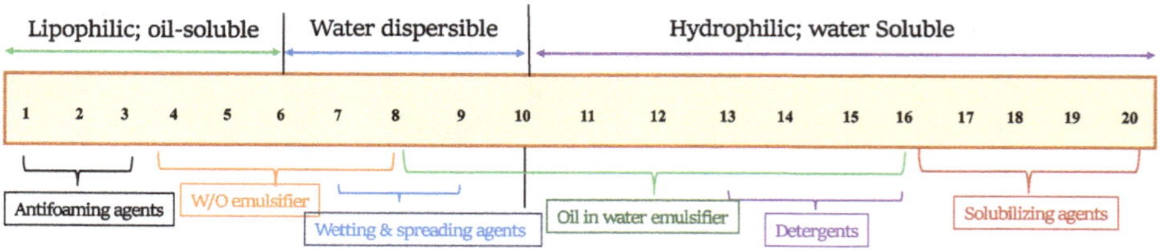

Figure 4.10: HLB scale.

Table 4.6: HLB values and the types of emulsion.

S. no.	HLB value range	Solubility description	Application as an emulsion
1.	<3	Do not have dispersibility in water	None: antifoaming agents
2.	3–6	Low dispersibility	Poor emulsion
3.	7–9	Milky dispersion is obtained after shaking	W/O emulsion: wetting agents
4.	8–10	Stable milky dispersion	O/W-type emulsion
5.	10–13	Translucent to clear solution	O/W-type emulsion: detergent
6.	>13	Clear solution	O/W-type emulsion: solubilizers

If HLB is less than 10, it is lipid-soluble (insoluble in water); and if HLB is more than 10, it is water-soluble (insoluble in lipid).

In simple terms, a lipophilic emulsifier has an HLB value below 10, whereas a hydrophilic emulsifier has a higher HLB number. That is, the HLB number increases with increasing polarity and hydrophilicity.

4.5.4 Application of the HLB Principle Across the Industries

1. **Selection of surfactants in pharmaceuticals and cosmetics:** The HLB scale is widely used in the pharmaceutical and cosmetic sectors for choosing appropriate surfactants. It helps in improving the dissolution of poorly water-soluble drugs and in preparing emulsions for creams, ointments, and lotions. Surfactants selected based on their HLB values impart the desired stability, consistency, and texture to the formulations.
2. **Food industry:** Similar to pharmaceuticals, the HLB principle is applied to select suitable bioemulsifiers or biosurfactants that stabilize emulsions in food products [5]. This ensures proper texture, mouthfeel, and extended shelf life.
3. **Chemical and cleaning industry:** In chemical processes, surfactant selection guided by the HLB scale plays a role in cleaning, wetting, and emulsification. For detergents and household cleaners, the correct HLB value enhances cleaning performance by stabilizing emulsions and enabling efficient suspension and removal of dirt or grease particles.
4. **Role of surfactants in emulsification:** Emulsions are heterogeneous systems composed of two immiscible liquids, such as oil and water that are thermodynamically unstable. To prevent phase separation, special agents, known as emulsifiers or surfactants, are used. Being amphiphilic molecules, surfactants have both hydrophilic and lipophilic segments, which allow them to reduce interfacial tension and promote mixing of immiscible liquids. This results in the formation of micelles, which help stabilize emulsions.

Surfactants form a protective layer around the dispersed phase droplets, preventing them from coagulating or merging, thereby maintaining emulsion stability. Depending on the application, different classes of surfactants are chosen, for instance, nonionic surfactants are often used in food formulations, while anionic surfactants are common in household cleaning agents and personal care products. Their amphiphilic nature makes them indispensable in a wide range of emulsions, including detergents, shampoos, lotions, lubricants, and paints.

4.6 Micelle Formation

Surface science investigates the unique behavior of amphiphilic molecules, especially their micellar and thermodynamic properties. Surfactants, being amphiphilic, accumulate at the interface between two immiscible phases, thereby reducing interfacial tension.

When the concentration of these amphiphilic molecules reaches a particular threshold, known as the CMC, they self-assemble into organized aggregates called micelles. Such colloids are referred to as associated colloids. A micelle can be described as a dynamic cluster of surfactant molecules, with hydrophobic tails oriented inward and hydrophilic heads facing the surrounding aqueous medium [6, 7].

Micelle formation occurs only above a certain concentration (CMC) and a minimum temperature, termed as the Kraft temperature or critical micelle temperature. This self-assembly imparts special properties to surfactant systems, such as enhanced solubilization, detergency, and stability, commonly demonstrated in soap and detergent solutions.

4.6.1 Amphiphilic Molecules/Surfactants

Paul Winsor, an architect, introduced the term amphiphilic, about 50 years ago, to describe the dual nature of surfactant molecules. The word is derived from the Greek roots “amphi” (meaning both) and “philos” (meaning love or affinity). Amphiphilic molecules contain two distinct parts, as outlined in Figure 4.11:

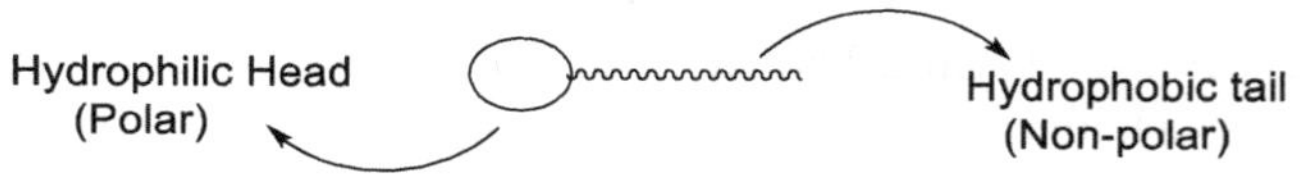

Figure 4.11: A pictorial representation of an amphiphilic molecule.

(i) A hydrophobic tail is a nonpolar or water-repelling group, consisting of a long hydrocarbon chain that avoids water.
(ii) A hydrophilic head is a polar or water-loving group, consisting of an ionic or nonionic group, which interacts well with water or other polar solvents.

This structural duality enables surfactants to accumulate at interfaces, where they reduce both surface tension and surface energy. Owing to this property, they are commonly referred to as surface-active agents. Surfactants can be categorized mainly into three types as follows, each with separate properties and applications:

1. Ionic surfactants: (i) anionic and (ii) cationic
2. Nonionic surfactants
3. Amphoteric

1. **Ionic surfactants:** An ionic surfactant is a surface-active agent that dissociates in water to form charged head groups, giving either cationic or anionic character.
   (i) **Anionic surfactants**: These amphiphilic molecules have a negatively charged hydrophilic group and dissociate in water to form an amphiphilic anion and a cation, generally belonging to alkali metals, as shown in Figure 4.12. They have a wide range of applications in the cleaning industry and are used in shampoos, detergents, and other products designed to remove greasy or sticky dirt from clothes. Examples include carboxylates, sulfates, betains, and sulfonates.

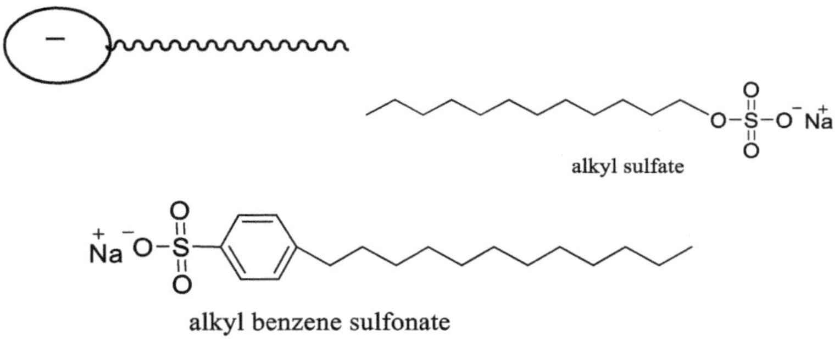

Figure 4.12: A pictorial representation of an anionic surfactant molecule.

   (ii) **Cationic surfactants**: The hydrophilic head is positively charged in this type of surfactant. Cationic surfactants have bactericidal, fungicidal, or other sanitizing properties that create a cationic disinfectant film on the surface [8]. Since hard surfaces are negatively charged, cationic surfactants are not suggested to be used to clean hard surfaces. Because of this, cationic surfactants are incompatible with interacting with anionic surfactant molecules, so they have limited scope as dirt cleaning agents, like anionic surfactants.

   Cationic surfactants are also used as antistatic agents to make the material's surface slightly more conductive, absorbing moisture from the air. In this process, the material's surface interacts with the hydrophobic part, and the cationic part of the surfactants absorbs the moisture from the air. Examples include quaternary ammonium salts such as cetyltrimethylammonium bromide (CTAB), depicted in Figure 4.13.

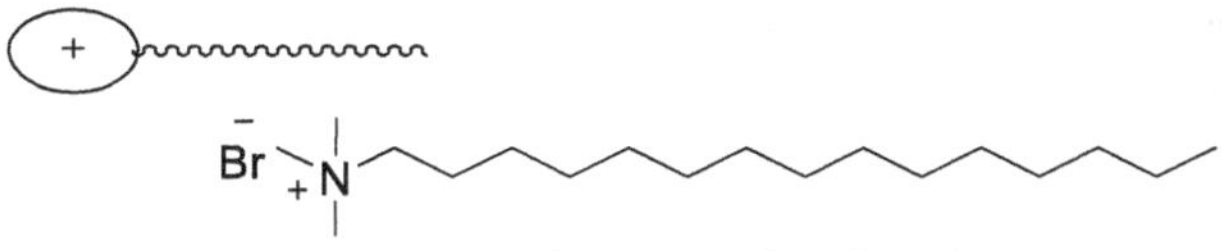

Figure 4.13: A pictorial representation of a cationic surfactant molecule.

2. **Nonionic surfactants:** There is no charge on the hydrophilic head of the surfactant's molecule, as shown in Figure 4.14. Despite the absence of any charge, they still maintain their polarity by keeping the oxygen-rich segment at one end (hydrophilic) while the large organic segment at the other end. The oxygen-rich component enables the molecule to interact with water through hydrogen bonding. These surfactants are highly versatile and utilized in pharmaceuticals, cosmetics, and household cleaning products. Examples include alcohol ethoxylates, alkoxylates, and ethylene oxide/propylene oxide block copolymers.

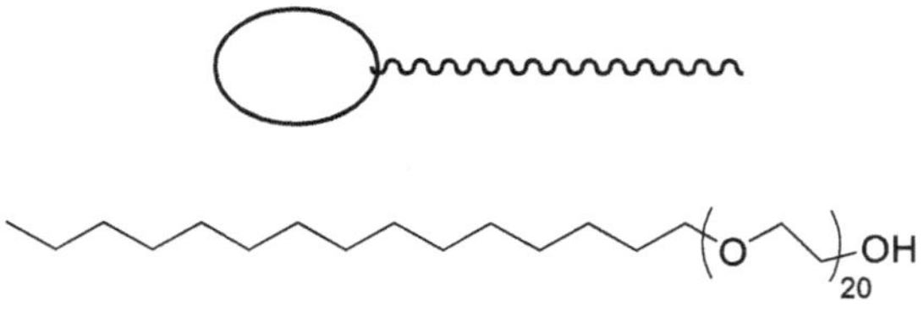

Figure 4.14: A pictorial representation of a nonionic surfactant molecule.

3. **Amphoteric surfactants:** These surfactants have both kinds of charges, positive and negative, within a molecule in their hydrophilic head, i.e., net charge is zero, as shown in Figure 4.15. Based on the pH of the solution, these surfactants are capable of functioning as both anionic and cationic surfactants, demonstrating dual functionality, i.e., they can change their charge based on the pH of the medium. This feasibility to change their nature makes them a valuable and versatile surface-active agent, having a wide range of applications. In an acidic medium ($pH < 7$), the anionic part combines with H^+ and gets neutralized, hence, only a positive charge remains on the molecule, acting as a cationic surfactant. On the other hand, in a basic medium ($pH > 7$), the cationic group is attracted to the OH^- ion, which neutralizes the positive charge, turning them into anionic surfactants. Examples include amino acid derivatives and cocamidopropyl betaine (CAB).

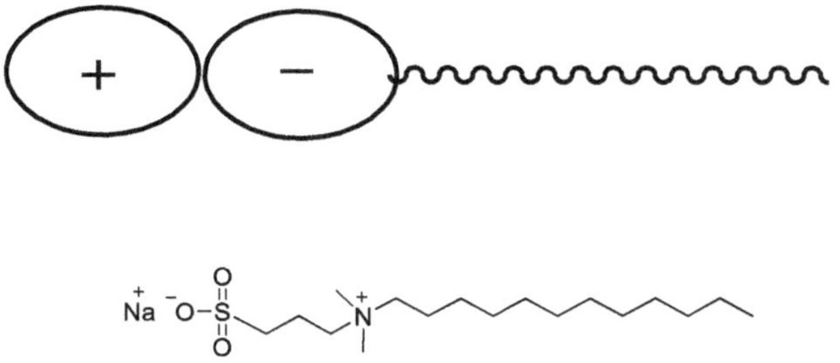

trialkylammoniopropanesulfonate

Figure 4.15: A pictorial representation of an amphoteric surfactant molecule.

4.6.2 Types of Micelles

Micelles are broadly classified into two categories based on the orientation of their hydrophilic and hydrophobic segments:

1. **Regular micelles:** These micelles are typically formed in aqueous solutions. In this arrangement, the hydrophobic tails of the amphiphilic molecules aggregate inward, away from water, forming the micellar core, whose pictorial representation is shown in Figure 4.16 for better understanding. Meanwhile, the hydrophilic heads remain exposed to the surrounding aqueous medium, forming the outer shell. Thus, the structure consists of a nonpolar core and a polar exterior, which stabilizes the micelle in water.

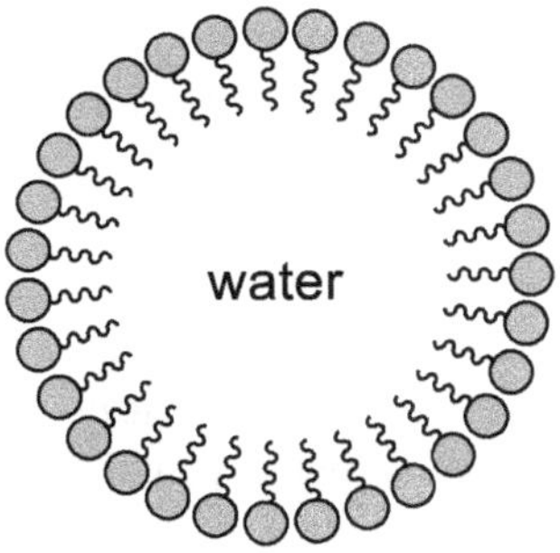

Figure 4.16: A pictorial representation of the structure of a micelle in an aqueous medium.

2. **Reverse micelle:** These micelles are formed in nonaqueous or nonpolar solvents. In this case, the hydrophilic heads of the amphiphilic molecules are directed inward, forming the micellar core, while the hydrophobic tails extend outward into the surrounding nonpolar medium. Thus, the core of the micelle is polar (hydrophilic), and

the shell is nonpolar (hydrophobic). This arrangement allows surfactants to stabilize polar solutes in organic solvents. Such structures are illustrated in Figure 4.17.

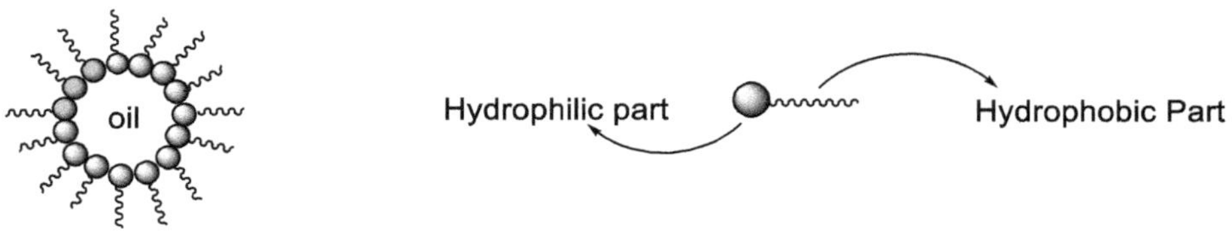

Figure 4.17: A pictorial representation of the structure of a reverse micelle.

4.6.3 Mechanism of Micelle Formation

Earlier, it was mentioned that micelles are associated colloids formed by the self-assembly of amphiphilic molecules, and their formation is a fundamental phenomenon in colloid and surface chemistry. This process is primarily driven by the dual affinity of surfactants, which possess both hydrophilic (polar) and hydrophobic (nonpolar) components.

Stepwise mechanism of micelle formation:

1. **At low concentrations:** When surfactant molecules are present in very small amounts, they remain dispersed as individual units in the aqueous medium. At this stage, the hydrophilic head interacts strongly with water, while the hydrophobic tail remains exposed to the solvent, which increases the free energy of the system.
2. **Approaching the CMC**: As the concentration of surfactant molecules gradually increases, they reach a threshold concentration known as the CMC. Beyond this point, the surfactant molecules begin to associate, orienting themselves to reduce the unfavorable contact between their hydrophobic tails and water. This aggregation leads to the spontaneous formation of micelles, thereby lowering the overall free energy of the system.
3. **Aggregation and micelle shape**: Initially, micelles tend to form small, nearly spherical structures near the CMC. With further increase in concentration, the shape may transform into more complex structures, such as lamellar sheets or cylindrical micelles, depending on the type of surfactant and environmental conditions.

4.6.3.1 Forces Governing Micelle Formation

Micelle formation is controlled by two opposing forces:

(i) Attractive forces: Hydrophobic tails cluster together inside the micelle due to van der Waals interactions, shielding them from water.

(ii) Repulsive forces: The polar heads, being hydrophilic, remain oriented toward water, creating electrostatic or hydrogen-bonding interactions that stabilize the micelle and prevent uncontrolled growth.

This balance results in an energetically favorable, ordered structure where hydrophobic segments are sequestered in the interior, while hydrophilic groups form the external shell.

4.6.3.2 Dynamic Nature of Micelles

Micelles are not static structures, they exist in a dynamic equilibrium. Surfactant molecules can continuously associate with or dissociate from the micellar aggregate, depending on factors such as concentration, temperature, and pressure. This reversible behavior highlights the flexible yet stable nature of micelle formation.

4.6.3.3 Framework of Micelle

(i) Core: The hydrophobic part arranges itself inside the micelle, generating a nonpolar center.
(ii) Shell: The hydrophilic part keeps its head facing outward, facilitating interaction with water or polar solvents. This arrangement stabilizes the micelle by minimizing the energy of the structure.
(iii) Shape: Spherical micelles are widely accepted in aqueous solutions. However, depending on the type of surfactant and environmental conditions, they may also be found in other shapes, such as lamellar or cylindrical.

4.6.4 Thermodynamics of Micelle Formation

Since micellization is a spontaneous process, which is governed by free energy change, i.e., ΔG.

As per Gibbs-Helmholtz equation:

$$\Delta G = \Delta H - T\Delta S \tag{4.6}$$

A spontaneous process results in a decrease in Gibbs free energy. To decrease Gibbs free energy and make ΔG negative, a balance is required between the change in enthalpy and the entropy change involving hydrophobic moieties. Since micelle formation occurs above the CMC, the change in entropy plays an important role because the change in enthalpy for the same is slightly positive ($\Delta H \sim 2\text{–}3$ kJ mol^{-1}).

Generally, the Gibbs free energy involved in micellization can be estimated as:

$$\Delta G_{\text{micellization}} = -RT \times \ln(\text{CMC}) \tag{4.7}$$

The hydrophobic part of amphiphilic molecules self-assembles in a manner that minimizes contact with water, which favors an increase in entropy ($\Delta S \sim 135–140$ kJ mol^{-1} K^{-1}), and it helps decrease Gibbs free energy, i.e., ΔG is negative.

4.6.5 Important Aspects of Micelle Formation

1. **Krafft temperature**: This temperature ensures the solubility of surfactant molecules in water, at which the concentration of surfactant molecules approaches CMC and leads to the aggregation of amphiphilic molecules.
2. **Reversible process**: Micelle formation is a reversible process. When the surfactant's concentration is lower than the CMC or the temperature drops below the Krafft temperature, the micelles revert to individual molecules.
3. **CMC**: It is defined as the minimum concentration of amphiphilic (surfactant) molecules needed for the formation of micelles. Below the CMC, no micelles are formed. CMC is a fundamental concept for understanding the chemistry of colloids, especially in the context of micelle formation and the study of surfactants. It is the minimum concentration of amphiphilic molecules required for a spontaneous transition from an individual molecule to self-assembly in aqueous solution [5]. It represents the threshold concentration of surfactant molecules above which the existence of an individual unit in aqueous solution does not occur. Surface tension is the key property that changes with the concentration of surfactant molecules, and it is interesting to study how it changes below and above the CMC.

4.6.6 Surface Tension of Surfactants

Surfactants are surface-active agents that activate the surface by lowering the surface tension. When surfactants are introduced to the system, they initially start to partition into the surface, lowering the surface energy by removing the hydrophobic part of the surfactant molecule from interacting with water, and thereby reducing the interfacial tension.

As the concentration of surface-active agents increases, a greater surface area is covered, further decreasing the surface tension or surface energy, leading to the aggregation of these agents into a self-assembled structure. The aggregation of surfactant molecules becomes the reason for a further decrease in surface energy by minimizing the unfavorable contact of their hydrophobic group with water. At the CMC, the subsequent addition of a surfactant molecule will increase the number of micelles. This variation of surface tension with concentration at various stages is shown in Figure 4.18.

Below the CMC, the surface tension-lowering effect is prominent and increases with an increase in concentration of surfactant molecules. Above CMC, the surface tension-lowering effect is minimal or constant, as the surface is almost covered, and micelle formation starts.

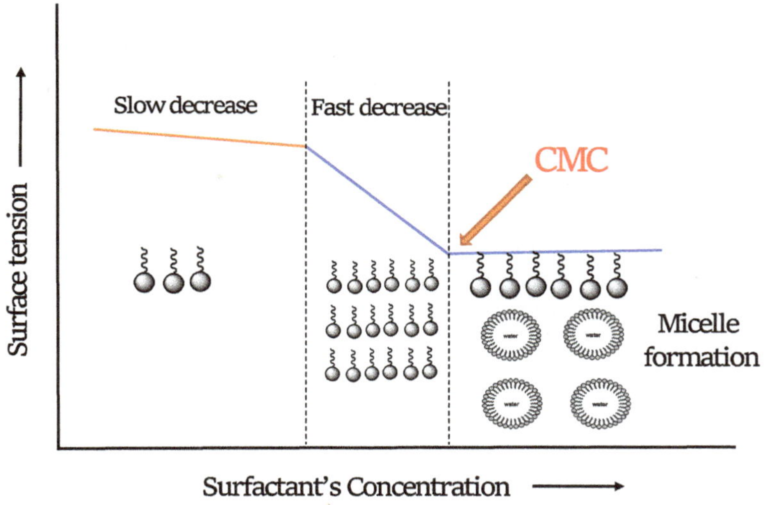

Figure 4.18: Variation in surface tension with surfactant concentration.

4.6.6.1 Significance of CMC

1. At the CMC, a significant surface tension reduction is observed, allowing surfactant molecules to have maximum surface activity.
2. Among its importance is that it is the threshold concentration for micelle formation, thus playing an important role in various applications such as solubilization of substances with a higher degree of hydrophobicity, cleaning, and emulsification.

4.6.7 Factors Affecting CMC

1. **Chemical structure of the amphiphilic molecule**: The length of the hydrophobic tail and the hydrophilic head affect the CMC, depending on their interaction with solvent molecules. A longer hydrophobic/water repelling tail favors higher hydrophobic interaction, which helps to reduce CMC. The longer hydrophilic/water-loving part, in the case of ionic surfactant, increases the required number of surfactant molecules to achieve CMC.

2. **Temperature:** The effect of temperature on CMC varies with the type of surfactants. Generally, in the case of a nonionic surfactant, an increase in CMC is observed with a rise in temperature because it enhances the hydrophobic interactions among organic moieties, and disrupts the hydrogen bonding interaction of the hydrophilic head with water/solvent. On the other hand, in ionic surfactants, a slight decrease in the CMC is observed with an increase in temperature. At lower temperatures, micelle formation is driven by changes in entropy, whereas at higher temperatures, it is governed by enthalpy.
3. **Presence of electrolytes:** The presence of electrolytes diminishes the value of the CMC of a particular surfactant in each solvent. This is because the electrolyte shields the ionic charges on hydrophilic heads, leading to a reduction in electrostatic repulsion between surfactant molecules. This facilitates the aggregation of amphiphilic molecules at a lower concentration [9].
4. **Solvent properties:** Solvent properties, such as polarity, fluidity, dielectric constant, hydrogen-bonded structure, hydrophobicity, and lipophilicity, significantly influence surfactant self-aggregation in aqueous solutions. Through the combination of suitable solvent components, these properties can be controlled and varied accordingly. In general, it can be said that the CMC increases with the increase in polarity of the solvent, and decreases with a decrease in polarity of the solvent. The surfactant molecules with a polar head are more soluble in polar solvents, which lead to a rise in the CMC value. In contrast, its opposite is true for surfactant molecules with nonpolar heads. Similarly, solvents with high dielectric constant (water/methanol/ethanol/ethylene glycol, etc.) increase the CMC since they are more likely to be present in bulk, while solvents with low dielectric constant (*n*-propanol/*n*-butanol) promote self-aggregation, which results in a reduction in CMC.

4.6.8 Methods of Determination of CMC

The determination of the CMC is crucial for understanding the behavior of surfactants, as it marks the onset of micelle formation. Various experimental techniques are employed to evaluate this parameter, based on the changes in physical and chemical properties of surfactants at concentrations around the CMC (Figure 4.19):

1. **Surface tension measurement:** Surface tension decreases steadily as the concentration of surfactant molecules increases, since surfactants accumulate at the air-water interface. Once the CMC is reached, the surface becomes saturated with surfactant molecules, and additional surfactants begin to form micelles in the bulk solution. Beyond this point, further increase in concentration produce little or no change in surface tension. The plateau observed in a surface tension versus concentration graph, as shown in Figure 4.19, indicates the CMC.

2. **Conductivity measurement:** For ionic surfactants, conductivity initially increases linearly with concentration due to the presence of free ions acting as charge carriers. At the CMC, a deviation from linearity is observed because monomers start aggregating into micelles, reducing the number of free ions contributing to conductivity. The break in the conductivity versus concentration plot (Figure 4.19) identifies the CMC.
3. **Spectroscopic methods:** Spectroscopic techniques, such as UV-visible or fluorescence spectroscopy, are often employed with probe molecules (e.g., dyes). These probes interact differently with surfactant molecules before and after micelle formation. At the CMC, a distinct shift in absorbance or emission wavelength is observed, as the dye becomes solubilized in the micellar core. This change is used to accurately determine the CMC.

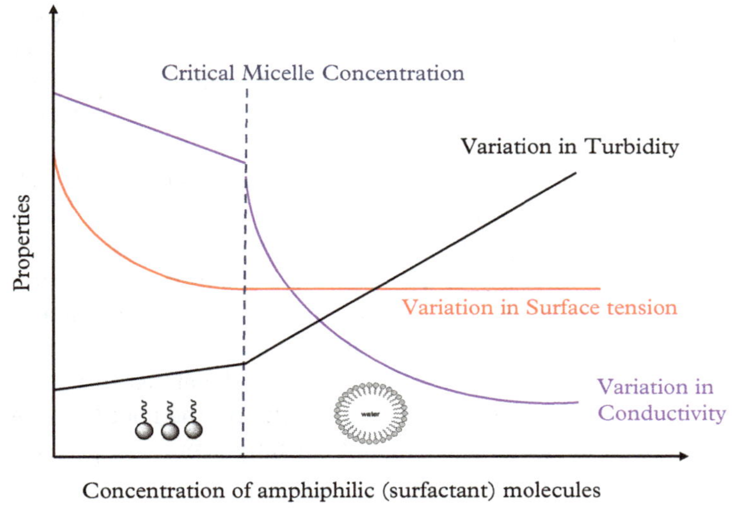

Figure 4.19: Change in various physical properties during micellization.

4. **Light scattering:** Dynamic light scattering (DLS) is used to measure the size and distribution of aggregates in solution. At concentrations below the CMC, only monomers are detected, whereas above the CMC, larger aggregates corresponding to micelles are observed (Figure 4.19). The appearance of these aggregates at a specific concentration provides an estimation of the CMC.

4.6.9 Application of Micelles

The CMC is a vital parameter in surface chemistry. It enables us to understand the nature of amphiphilic molecules and their ability to form micelles. It optimizes the use of surfactants in various industries, including cosmetics, pharmaceuticals, nanotechnology, and environmental science. Due to their unique properties, micelles are extensively used in several applications:

1. **Detergents and cleaning**: The hydrophobic tails of micelles trap oil and grease during washing by trapping them in the micelles, which enables their removal in water. To clean, soaps and detergents create emulsions in which dirt is suspended in water.
2. **Drug delivery systems**: Various challenges associated with drug administration can be overcome using polymeric micelles [10], i.e., aggregation colloids formed in solution by self-assembly of amphiphilic polymers. The low solubility of drugs in water and poor permeability across biological barriers are among these challenges. Hydrophobic (lipophilic) drugs can be used as drug carriers by being encapsulated and protected within the micelles, which improves their bioavailability and solubility. Micelles, especially polymeric micelles, due to their high stability and their core-shell structure (10–100 nm), can be transformed into targeting species to deliver drugs to the desired location in the body and minimize the side effects of drugs [2].

   For a micelle, the overall process as a drug delivery agent is based on the following features:
   (i) It should improve the stability of the encapsulated drug and enhance the solubility of the poorly water-soluble drugs.
   (ii) It should increase the targeting and make drug delivery more precise.
   (iii) It should have the right control over the release rate of a drug. Micelles provide an excellent platform for chemical reactions, enhancing the rate of chemical reactions by systematically assembling the reactant molecules on a molecular scale. Adding amphiphilic molecules with well-defined structures in water can result in incredible selectivity and activity. The micelles have been used extensively as membrane mimics to identify membrane proteins and peptides, as well as to deliver drugs.
3. **Cosmetics**: This also pertains to the cleaning action of micelles. Most skincare products utilize micelles to gently remove dirt and oil.
4. **Environmental science**: Micelles are employed to remove pollutants from water and soil. They also assist in oil spill remediation by breaking down and solubilizing hydrophobic contaminants.

This chapter has explored the multifaceted world of colloids, covering their nature, classification, and essential properties, along with the techniques employed in their preparation. Special emphasis was placed on the HLB and micelle formation, both of which are central to understanding the behavior of surfactants and their industrial relevance. The concepts discussed not only highlight the scientific significance of colloidal systems but also illustrate their wide-ranging applications in everyday life and technology. With this foundation, students are now equipped to connect the theoretical aspects of colloid chemistry with practical applications, paving the way for a deeper study of surface chemistry in the upcoming chapters.

Practice Questions

1. Give an example of an emulsifying agent other than soap that stabilizes O/W emulsions.

2. Why is milk considered to be a stable emulsion?

3. How do sol, gel, and emulsion differ from each other?

4. Provide some examples of emulsions you encounter in everyday life.

5. Explain why the term "colloidal state" is preferred over "colloidal substance" to describe the same substance acting as both a crystalloid and a colloid.

6. State the key condition required for a substance to exist in a colloidal state.

7. Explain the underlying reasons for the greater stability observed in lyophilic colloids compared to lyophobic colloids.

8. Define critical micelle concentration and Kraft temperature.

9. In the peptization method, under what conditions will $Al(OH)_3$ sol acquire a positive or negative charge?

10. Define surface tension and explain how it changes during micelle formation.

11. Describe in brief reversible and irreversible sols with examples.

12. Define Brownian motion. Explain the cause and discuss its significance.

13. What is the "Tyndall effect"? Explain.

14. What is a micelle? Describe the process of micellization.

15. What are the macromolecular, multimolecular, and associated colloids? Discuss their essential features.

16. What are the following methods for the preparation of colloidal sols:
    a. Bredig's arch method
    b. Mechanical dispersion by the colloidal disc method

17. Write a short note on the following:
    a. Significance and application of colloids.
    b. Methods for preparation of colloidal solutions
    c. Critical micelle concentration
    d. Thermodynamics of micellization

18. What is the colloidal state of matter? Discuss important features of different types of solutions, such as true, colloidal solutions, and suspensions.

19. Differentiate between lyophilic and lyophobic colloids and describe their significance and applications across various industries.

20. What is HLB? Explain.

References

[1] Thomas Graham; X. Liquid diffusion applied to analysis. *Phil. Trans. R. Soc.* 31 December 1861; (151): 183–224. https://doi.org/10.1098/rstl.1861.0011

[2] Ghezzi M, Pescina S, Padula C, Santi P, Del Favero E, Cantù L et al. Polymeric micelles in drug delivery: An insight of the techniques for their characterization and assessment in biorelevant conditions. *J Control Release*. 2021 Apr;332:312–336.

[3] Leermakers F, Eriksson JC Lyklema H. 4 Association colloids and their equilibrium modelling. In: *Fundamentals of Interface and Colloid Science*. 2005.

[4] Hydrophilic-Lipophilic Balance – An overview | ScienceDirect Topics [Internet]. [cited 2025 Jul 3]. Available from: https://www.sciencedirect.com/topics/agricultural-and-biological-sciences/hydrophilic-lipophilic-balance

[5] Baccile N, Seyrig C, Poirier A, Alonso-de Castro S, Roelants KW, SL ASS-A *Interfacial Properties, Interactions with Macromolecules and Molecular Modelling and Simulation of Microbial Bio-based Amphiphiles (Biosurfactants). A Tutorial Review.*

[6] Moroi Y. Micelle Formation. In: *Micelles*. Boston, MA: Springer US; 1992. pp. 41–96.

[7] Micelle – An overview | ScienceDirect Topics [Internet]. [cited 2025 Jul 7]. Available from: https://www.sciencedirect.com/topics/medicine-and-dentistry/micelle.

[8] Massarweh O, Abushaikha AS The use of surfactants in enhanced oil recovery: A review of recent advances. *Energy Rep.* 2020 Nov;6:3150–3178.

[9] Demissie H, Duraisamy R. Effects of electrolytes on the surface and micellar characteristics of Sodium dodecyl sulphate surfactant solution. *J Sci Innov Res.* 2016;5(6):208–214. Available from: www.jsirjournal.com

[10] Das PJ, Kumawat V, Singla M, Singh MR, Chandra Mohapatra S. Self-aggregation behavior of well-defined PEO-based amphiphilic tri-block copolymers and their application as metal detectors. *J Macromol Sci A*. 2024 Dec 3;61(12):1085–1094.

For Further Reading

[1] Butt HJ, Graf K, Michael K *Physics and chemistry of interfaces*. Wiley-VCH; 2006. pp. 269–277.

[2] Bose A, Roy Burman D, Sikdar B, Patra P Nanomicelles: Types, properties and applications in drug delivery. *IET Nanobiotechnol*. 2021 Feb 2;15(1):19–27.

[3] Lbr. P, Sharma LR, Pathania MS *Principles of physical chemistry*. Vishal Publishing Co.; 2004.

[4] Krister H, Shah DO, Schwuger MJ *Handbook of applied surface and colloid chemistry*. Wiley; 2002.

[5] Paul H, Raj R *Principles of Colloid and Surface Chemistry*. Third Edition, Revised and Expanded, 3rd Edition. CRC Press; 2016. p. 672.

[6] Solubility of Things. Colloidal Systems: Definition and Characteristics [Internet]. [cited 2025 Aug 26]. Available from: https://www.solubilityofthings.com/colloidal-systems-definition-and-characteristics.

[7] ACME Research Solutions. Role of surfactants in emulsion formation [Internet]. 2023 May 16 [cited 2025 Aug 26]. Available from: https://acmeresearchlabs.in/2023/05/16/role-of-surfactants-in-emulsion-formation/.

[8] Read Chemistry. Applications of Colloids [Internet]. 2024 May 18 [cited 2025 Aug 26]. Available from: https://readchemistry.com/2024/05/18/applications-of-colloids.

[9] Solubility of Things. Colloidal Systems: Definition and Characteristics [Internet]. Solubility of Things; [cited 2025 Aug 26]. Available from: https://www.solubilityofthings.com/colloidal-systems-definition-and-characteristics.

[10] Read Chemistry. Applications of Colloids [Internet]. 2024 May 18 [cited 2025 Aug 26]. Available from: https://readchemistry.com/2024/05/18/applications-of-colloids/.

[11] Kheterpal SC *Pradeep's physical chemistry*. 2nd ed. Vol. 1. New Millennium ed. Jalandhar: Pradeep Publications; 2004. ISBN: 9789352111305.

[12] Kk S, Lk. S *A textbook of physical chemistry*. New Delhi: Vikas Publishing House Pvt. Ltd; 2016. pp. 501–514 ISBN: 9789352590421, 9352590422.

[13] Bahl A, Bahl BS, Tuli GD *Essentials of Physical Chemistry*. Ltd, New Delhi: S. Chand and Company; 2014. pp. 886–872. ISB: 9355010605, 9789355010605.

[14] Raj G *Advanced physical chemistry*. 35th ed. Meerut: Goel Publication House; 2009. ISBN 9789389685343.

[15] Salager J-L *FIRP BOOKLET # E201-A FIRP BOOKLET # E201-A. SURFACTANTS in AQUEOUS SOLUTION*. Version # 1 (01/30/1993); Edited and published by: Laboratorio FIRP Escuela de Ingenieria Quimica, Universidad De Los Andes Mérida 5101 Venezuela.

Richa Tyagi and Rajni Grover*

5 Properties of Colloids

Abstract: Colloids are formed when the dispersed phase is heterogeneously distributed in the dispersion medium at the microscopic level. They are vital to many industrial, scientific, and environmental applications. The stability of a colloidal dispersion is altered by changing the interactions between charged colloid particles in the presence of electrolytes. When electrolytes are introduced, coagulation can take place, in certain instances, causing colloidal particles to cluster and precipitate out of the dispersion. Alternatively, in other instances, electrolytes can reinforce their stability by surrounding particles and creating layers that shield them from aggregation. The electric double layer, another building block concept, accounts for the stability of the colloidal system. Due to adsorption or dissociation of ions, a colloidal particle suspended in a liquid medium generally acquires a charge. A layer of counterions is attracted to this charged surface and forms the Stern layer, and then a diffuse layer of loosely held counterions is formed, i.e. the Gouy-Chapman layer. The entire charge distribution from the superposition of these layers stabilizes the colloid by inhibiting its aggregation. The colloidal dispersion stability depends considerably upon the zeta potential, which is an approximation for the effective charge of a moving colloidal particle. Colloids have a wide range of applications in various fields. This chapter will try to give an exhaustive overview of the basic properties and applications of colloids, giving an insight into their fundamental position in science and technology.

5.1 Introduction

Colloidal systems are created through the uneven distribution of the dispersed phase within the dispersion medium at the microscopic level. Colloidal systems are crucial in numerous industrial, scientific, and ecological applications. Because of the composition and interaction between the dispersed phase particles and the dispersion medium, these systems have unique physical and chemical properties. The behavior of colloidal particles with respect to light demonstrates important optical properties, for example, light scattering and the Tyndall effect, which characterize colloids in comparison to suspensions and true solutions [1]. Kinetic properties, for instance, diffusion and Brownian motion, are related to the freedom of movement of the colloidal particles and random collisions with the dispersion medium molecules. The electric double layer, yet another cornerstone, accounts for the stability of the colloidal system. Zeta potential and streaming potential also play a part in colloidal dispersion strength, which is a calculation of the effective charge on a moving colloidal particle. A second form of colloid, in which there is a solid particulate dispersed in a liquid

https://doi.org/10.1515/9783112208236-005

phase but potentially exhibiting settling over time, is referred to as a suspension. In contrast, the colloids, when two immiscible liquids are dispersed with some stabilizers, are referred to as emulsions. Colloids are used in a wide variety of applications across various disciplines. Throughout this chapter, a comprehensive examination of basic properties and uses of colloids will be performed, which will explain their vital importance in scientific research.

5.2 Optical Properties of Colloids

Because the colloidal particles tend to be between 1 and 1,000 nm in size, they are small enough to stay suspended in the medium but large enough to effectively hinder the passage of light [2]. Due to how the colloids interact with light in a unique manner, optical properties are the basis to describe colloidal systems. These interactions yield valuable information about the particle shape, size, and concentration, and even surface chemistry. Such optical properties provide a valuable tool for discriminating between colloids, suspensions, and true solutions. The colloidal particles are sufficient to absorb, scatter, and in some cases, reflect light, giving rise to optical effects that do not occur in true solutions. When a beam of light is transmitted through a colloidal solution, it does not move in a straight line as it would in a true solution; rather, it collides with the particles of the dispersed phase in colloidal systems and scatters. Particle dimensions and the refractive index difference between the dispersed phase and the dispersion medium are the key parameters that change the optical properties and the appearance of the colloidal system. The phenomenon provides a valuable tool to analyze and characterize colloids, as well as to make them more visually identifiable. One of the most popular demonstrations of light scattering is the Tyndall effect. Moreover, free ions in metallic colloids, on interaction with light, can produce vibrant and tunable hues through "surface plasmon resonance." This resonance effect is highly sensitive to the surrounding environment. Based on these distinctive optical characteristics, we have numerous analytical methods, particularly turbidimetry, UV-visible spectroscopy, and dynamic light scattering. Understanding these characteristics is therefore essential for basic research and a variety of applications in materials science, chemistry, and biology.

5.2.1 Tyndall Effect in Colloids

The Tyndall effect is one of the most visually captivating scientific phenomena observed in colloidal solutions that highlights the interaction between light and dispersed phase particles. Also, some fine suspensions and turbid solutions are optically active. When a strong beam of light is passed through a colloidal solution and is observed at a right angle, the path of the light becomes visible as a hazy cone or luminous beam, as can be visualized in Figure 5.1. This happens because the colloidal particles absorb part of the incident light and subsequently scatter it in various directions. When the dispersed phase particles in a colloid are sufficiently large to interact with light, then, due to this scattering, the system may even appear cloudy or turbid under ordinary illumination. In contrast, true solutions like salt solution or vinegar do not exhibit this effect. Because of the small size of solute particles, when a beam of light passes through a true solution, it remains invisible when observed at right angles. This distinction makes the Tyndall effect a useful technique for distinguishing between colloidal systems and true solutions. The Tyndall effect is a phenomenon we often observe in everyday situations, such as the hazy visibility of a film projector's light beam in a smoke-filled theatre or when the car headlights form glowing beams on dusty roads at night. These effects arise as a result of the scattering of light by colloidal dust or smoke particles suspended in the air medium.

Figure 5.1: Tyndall effect illustrating that the path of the light becomes visible when it passes through the colloidal solution. Figure generated and edited using Canva AI Image Generator [Internet]. Canva Pty Ltd; Available from: https://www.canva.com/ai-image-generator/.

5.2.2 Factors Influencing the Tyndall Effect

1. **Particle size and shape**: The size and geometry of the dispersed phase particles determine the effectiveness of light scattering in a colloidal system. The size of the dispersed phase particles must be within the colloidal range, which is usually 1–1,000 nm, in order to have the Tyndall effect visualized. Particles smaller than this range may not scatter light efficiently, while those that exceed the upper limit might settle out of the dispersion medium, leading to precipitation or turbidity, rather than stable colloidal behavior.
2. **Concentration of dispersed particles**: A higher concentration of colloidal particles increases the number of scattering centers in the dispersion medium and enhances the intensity of the Tyndall effect. However, extremely high concentra-

tions may lead to particle collisions, resulting in aggregation or coagulation, which may diminish the visibility of the effect or destabilize the colloid.

3. **Optical properties of the dispersion medium**: When a beam of light travels through different materials, the way it bends or refracts determines the magnitude of the Tyndall effect. Thus, the extent of light scattering is strongly determined by the difference in refractive indices of the dispersed phase and the dispersion medium [1]. If the refractive indices are too similar, the dispersed light might be insufficient or insignificant. On the other hand, effective scattering and an intensified Tyndall effect result from larger differences in refractive indices.
4. **Wavelength of incident light**: The wavelength of incident radiation influences the scattering of light by colloidal particles, which is governed by the Rayleigh scattering principle. Shorter wavelengths, such as violet or blue light, are scattered more strongly than longer wavelengths, such as red light. This is why blue light tends to spread out in different directions more easily. This explains the natural phenomena, like the blue hue of the sky during the day.

5.2.3 Applications of the Tyndall Effect

The Tyndall effect has several practical applications, both in the scientific domain and in everyday life. One may have witnessed the scattering of sunlight by morning mist or airborne particles that make the rays visible, particularly in humid and hilly regions, as can be visually experienced through Figure 5.2(a). Likewise, smoke machines are intentionally employed in concerts, theatre, and stage shows to make light beams visible, which enhance visual effects, as shown in Figure 5.2(b). The Tyndall effect can also be seen in something as simple as a laser pointer in a dusty environment, as the suspended dust particles scatter the light and make the beam visible.

In both science and industry, this effect plays a significant role in tools used to measure the size and movement of particles in colloidal suspensions. It also aids in environmental health assessment by monitoring the air quality index and by detecting airborne pollutants, including aerosols and fine dust particles. Since only colloids scatter light, the Tyndall effect is a quick method used in laboratories to distinguish colloids from true solutions. Some diagnostic tests, for example, the cerebrospinal turbidity test, are performed by observing the scattering of light by bacteria or viruses. Additionally, in pharmaceutical research, this visual effect is often used to assess the uniformity of micelles and polymeric nanoparticles in a drug delivery system.

5.2.4 Limitations of the Tyndall Effect

In spite of numerous visual applications of the Tyndall effect, it does have certain practical limitations. Although it is a useful technique to show the presence of colloidal particles in a solution, it is not a precise measurement tool as it does not give accurate information about the number and dimensions of the particles. Furthermore, it is not entirely reliable for the identification of colloids, since some fine suspensions, like flour diluted with water, can also scatter light and create confusion. The effect also depends on the concentration of particles; if the solution is too dilute, there may not be enough particles to scatter light visibly, and if it is too concentrated, the particles might cluster together or settle down, making uneven scattering. The difference in the refractive index of the dispersed phase and the dispersion medium must also be comparable to that of the wavelength of the incident light. Observation conditions required for this effect are a dark background and a focused, narrow light beam (such as from a laser). In a bright environment, or with a poor setup, the Tyndall effect might not be noticeable at all.

Figure 5.2: (a) Sunlight through dense forest creating Tyndall effect. Figure Sunbeams Shine Sun Forest. Source: Pixabay [Internet]. 2014 Nov 22. Available from: https://pixabay.com/photos/sunbeams-shine-sun-forest-540589/. (b) Lights on stage creating Tyndall effect. Figure Stage light show performance. Source: Pixabay [Internet]. 2017. Available from: https://pixabay.com/photos/stage-lightshow-show-performance-2223130/.

5.3 Kinetic Properties of Colloids

Colloidal systems display distinctive kinetic behavior, like diffusion, Brownian motion, and electrophoresis, all of which are a result of dispersed-phase particle movement and their interaction with the surrounding dispersion medium. All these properties depend on a number of variables, like medium viscosity, the size of the dispersed phase particles, and the system temperature. Obtaining a sharp grasp of these kinetic properties is critical to utilize them properly in both scientific investigation and an extensive variety of industrial uses.

5.3.1 Brownian Movement

One of the most characteristic kinetic features of colloidal systems is the Brownian movement. This is the random, continuous, and zigzag motion of particles of the dispersed phase that are suspended in the dispersion medium. This is due to the molecular kinetic nature of the particles. Small particles of the dispersion phase are in continuous random motion due to their thermal energy. These smaller and faster-moving molecules of the surrounding dispersion medium incessantly strike the dispersed phase particles and impart tiny impulses. Since collisions occur from all directions, as can be clearly seen in Figure 5.3, this bombardment generates a random, irregular path of dispersed phase particles. Thus, the Brownian effect is not the impact of collisions amongst the dispersed phase particles, but rather is due to unequal hits from the dispersion medium particles.

Botanist Robert Brown first observed the effect in 1827 when he noticed that the pollen grains were moving in water under a microscope. At that time, the reason for this motion wasn't clear, but later, Albert Einstein explained that it was caused by collisions with surrounding water molecules. Brownian motion is essential to comprehending the kinetic theory of gases.

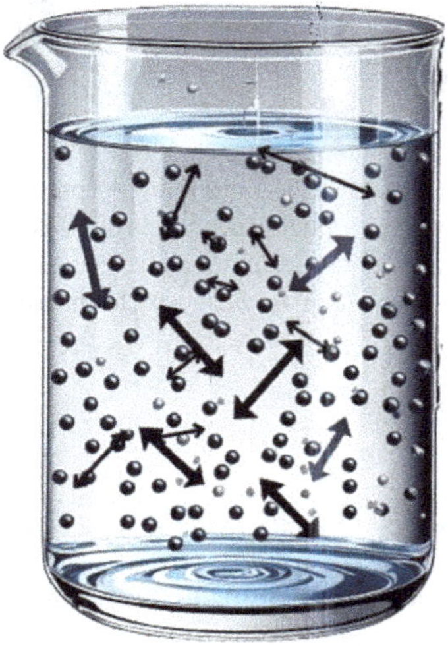

Figure 5.3: Brownian motion in colloidal systems, which shows continuous and random motion of dispersed phase particles. Figure generated and edited using Canva AI Image Generator [Internet]. Canva Pty Ltd; Available from: https://www.canva.com/ai-image-generator/.

5.3.2 Factors Influencing the Brownian Motion

1. **Influence of particle size**: The colloidal particles' random motion is inversely proportional to the particle size. Smaller size and lower mass of colloidal particles enable them to move around faster and at random due to a more sensitive effect by nearby dispersion phase molecules.
2. **Viscosity of the dispersion phase:** The forces between the particles of the dispersion medium also affect the movement of particles of the dispersed phase. When the dispersion medium of the colloidal system is more viscous, it prevents the free and continuous flow of the particles. This decreases the intensity of Brownian motion and influences the stability of the system.
3. **Dispersion medium density**: The density of the colloidal solution is also one of the most important things to take into consideration. It affects the balance between the random Brownian motion and gravitational settling of colloidal particles. When the dispersion medium is denser, it facilitates the Brownian motion of colloidal particles and allows them to suspend them more effectively. By doing this, the Brownian motion hinders sedimentation and helps maintain the stability of the colloidal system.
4. **Temperature**: The kinetic energy of the system's molecules rises with a rise in temperature, causing more intensive and active collisions with the colloidal particles, thus rendering the Brownian motion more intense and observable.

5.3.3 Significance of Brownian Motion

Brownian motion is the foundation for many scientific hypotheses and is evidence of molecular movement, supporting the kinetic theory of matter and the theory of statistical mechanics. It has the potential to influence the movement and distribution of active drug-delivery particles. It also accounts for the mechanism of diffusion of small aerosols and dust particles in the atmosphere. It also helps maintain the stability of colloidal systems. Since these random motions are contrary to the settling tendency of particles under gravity, they effectively prevent sedimentation, and therefore, maintain colloidal dispersions uniformly mixed in the long term.

From a mathematical standpoint, Einstein's diffusion theory can be used to characterize Brownian motion, which expresses the "average squared displacement" of a particle over time as:

$$\langle x^2 \rangle = 2 \times D \times t \tag{5.1}$$

where $\langle x^2 \rangle$ represents the one-dimensional mean square displacement, t is the elapsed time, and D represents the diffusion coefficient. The diffusion coefficient D can be further defined by the Stokes-Einstein equation:

$$D = \frac{k \times T}{6\pi\eta r} \tag{5.2}$$

where k represents the Boltzmann's constant (1.38×10^{-23} J K^{-1}), η is the viscosity of the medium, T is the absolute temperature in Kelvin, and r represents the radius of the colloidal particle.

According to this relation, larger particles in a fluid with higher viscosity and lower temperature tend to move slowly with a lower diffusion coefficient value. This, in turn, gives a lesser mean square displacement. Similarly, smaller particles at high temperature give a greater mean square displacement value. This random and continuous motion helps keep the colloidal particles uniformly dispersed, thereby minimizing the chances of sedimentation and aggregation.

Other considerable kinetic properties of colloids, apart from Brownian motion, are sedimentation, diffusion, osmotic pressure, ultracentrifugation, and viscosity. All these properties reflect the continuous movement of colloidal particles. These are not unique to colloids (they occur in molecular systems too), but their behavior in colloids is different due to the size and interactions of colloidal particles. By combining all these kinetic properties, one can get a more comprehensive understanding of colloidal systems. Diffusion and viscosity are important factors in colloidal dispersions because they define the stability of the system in a variety of applications by influencing how the particles move and interact within the medium. Regulation of viscosity in colloidal products, such as paints or food emulsions, guarantees appropriate texture and usefulness [3]. Colloidal diffusion plays a crucial role in biological transport, food stability, drug delivery systems, and the textile dyeing industry by ensuring the uniform distribution of particles. It has many environmental applications, including pollutant analysis and water treatment, where diffusion influences the movement and removal of colloidal impurities [4]. Sedimentation describes the influence of gravity on the dispersed phase particles. It can be minimized by random continuous collisions between the particles, i.e. Brownian motion, which makes the colloidal solution stable. Similar to true solutions, colloidal systems also exhibit osmotic pressure [5]. However, it is relatively low compared to that of a true solution because the number of particles per unit volume is much less. One of the most useful methods to determine the colloidal particle mass is the measurement of osmotic pressure. The osmotic pressure exerted by proteins, also called oncotic pressure, plays a crucial role in maintaining the balance of fluids between the tissues and the bloodstream. All these kinetic properties provide support for the dynamic nature of colloidal particles, which aids in explaining their stability and behavior under various circumstances.

5.4 Electrical Properties of Colloids

Colloidal dispersions are characterized by their unique electrical properties, which stem from the existence of charges on the surfaces of the particles in the dispersion medium. These charges significantly contribute to the stability of the system through the creation of repulsive forces, which withhold the particles from aggregating. Additionally, the behavior of charged colloidal particles in electric fields outside the system generates various significant electro-kinetic phenomena. A thorough knowledge of the electrical properties of colloids is required for theoretical, as well as practical purposes, in the areas of paints, pharmaceuticals, water purification, and materials science. The electric charge on the dispersed particles results from numerous factors, namely, adsorption and ionization of the surface groups. Such colloid charges account for various significant electric properties, namely, electrophoresis, electro-osmosis, and the electrical double layer.

5.4.1 Origin of Surface Charge on Colloidal Particles

The particles in a colloidal system acquire charge through various means:
1. By chemical reactions: Surface charge arises from chemical processes occurring during preparation, such as the adsorption of chloride ions on the surface of gold particles during reduction with chloroauric acid.
2. Ionization of surface groups: Some colloidal particles may have ionizable groups that can dissociate in the medium, contributing to their surface charge.
3. Selective ion adsorption: Particles may preferentially adsorb ions from the surrounding medium, for example, arsenous sulfide particles typically adsorb sulfide ions (S^{2-}), resulting in a negative charge, while ferric hydroxide particles often adsorb ferric ions (Fe^{3+}), leading to a positive charge.

The presence of the charges is crucial for maintaining the stability of the colloidal solution, as it creates repulsive interactions that prevent particles from aggregating.

5.4.2 Electrical Double Layer

Once the particles become charged, they attract counterions from the medium. Due to the electrostatic forces generated between the ions, they tend to associate with the surface of the colloids. This cluster of ions is not in the form of a single uniform layer, but rather forms an electrical double layer. The structure of the electric double layer consists of a compact, tightly bound layer of counterions, which is called the Stern layer, and a region where counterions are more loosely held and gradually merge with the bulk solution. The red particles are adjacent to the blue layer, as shown in

Figure 5.4. The region, which lies immediately next to the fixed layer, is called the diffuse layer. These two layers in combination do not create a physically rigid boundary, but represent just the arrangement of different charged particles in a dispersion medium. The electrical double layer ensures electrostatic balance and contributes significantly to colloidal stability by creating repulsive barriers between particles.

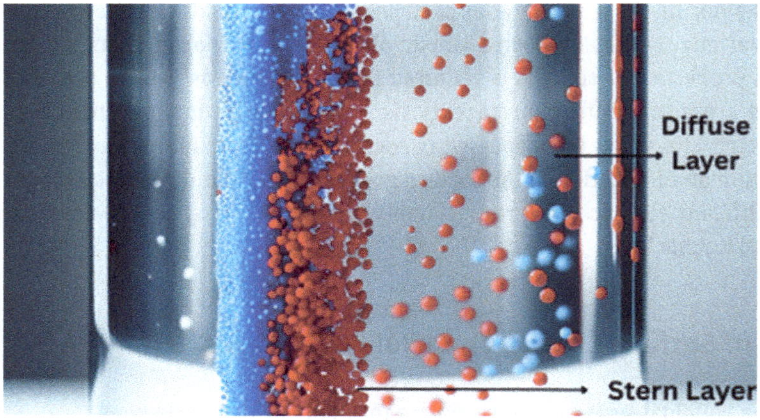

Figure 5.4: Electric double layer, where blue and red represent different charged particles. Figure generated and edited using Canva AI Image Generator [Internet]. Canva Pty Ltd; Available from: https://www.canva.com/ai-image-generator/.

5.4.3 Gouy-Chapman Model

The Gouy-Chapman model, developed independently by Gouy (1910) and Chapman (1913), extends Helmholtz's idea of the electrical double layer by introducing the concept of a diffuse layer. According to this theory, the distribution of ions near a charged surface is not rigidly fixed but governed by a dynamic balance between electrostatic attraction to the charged surface and thermal motion (Brownian movement) of the ions in solution.

In this model, counterions (oppositely charged to the surface) tend to accumulate close to the particle surface, while coions (similarly charged) are repelled. The result is a diffuse cloud of ions that gradually neutralizes the surface charge.

The electrostatic potential (ψ) in the diffuse layer decreases exponentially with increasing distance (x) from the particle surface, following the Poisson-Boltzmann equation:

$$\psi(x) = \psi_0 e^{-\kappa x} \tag{5.3}$$

where ψ_0 is the surface potential, and κ^{-1} is the Debye length, representing the thickness of the double layer.

The Debye length depends on the ionic strength (I) of the medium and is given by:

$$\kappa^{-1} = \left(\frac{\varepsilon kT}{2N_A e^2 I}\right)^{1/2} \tag{5.4}$$

where ε is the dielectric constant of the medium, k is the Boltzmann constant, T is the absolute temperature, N_A is the Avogadro's number, e is the electronic charge, and I is the ionic strength of the solution.

5.4.3.1 Key Features of the Model

- At low ionic strength, the double layer is extended (larger Debye length), leading to stronger electrostatic repulsion between particles and greater colloidal stability.
- At high ionic strength, the double layer is compressed, reducing electrostatic repulsion and promoting coagulation or flocculation.

Thus, the Gouy-Chapman model provides a more realistic description of colloidal stability compared to the simple Helmholtz model, as it accounts for the role of ion mobility and thermal agitation in the diffuse region [3].

5.4.4 Effect of pH and Electrolytes

The pH influences the ionization of the surface groups. It modifies the surface charge and the double layer. Each system has an isoelectric point, the pH at which the zeta potential is zero and stability is minimum. Also, on introducing electrolytes into a colloidal system, the stability of colloids is affected by the compression of the double layer and the reduction of the zeta potential [6]. An example is adding sodium chloride to a negatively charged colloid, which introduces Na^+ ions that neutralize surface charges, promoting coagulation.

5.4.5 Zeta Potential

Zeta potential is the electrical potential at the boundary, where the immobile fluid layer attached to the particle meets the surrounding dispersion medium. This potentially detaches the loosely bound diffuse layer from the tightly associated stern layer of counterions in the dispersion medium [1]. Thus, it is the zeta potential that reflects the actual surface charge experienced by the moving colloidal particles in the dispersion medium. The magnitude of this potential is the key parameter in assessing colloidal stability. Higher absolute zeta potential values (typically above ±30 mV) correspond to stronger repulsion and greater stability, while lower or nearly zero values of zeta potential indicate weak repulsions and dominant van der Waals attractive forces, leading to aggregation or coagulation of the colloidal system [7].

5.4.6 Factors Influencing Zeta Potential

There are several factors that affect the zeta potential of colloidal particles:

1. **pH:** The surface charge of the particle varies with pH, influencing zeta potential. Each colloid has an isoelectric point at which the zeta potential equals zero and stability is lowest, and there is a maximum tendency for particle agglomeration.
2. **Ionic strength**: The presence of salts condenses the double layer, decreasing zeta potential and facilitating aggregation [8].
3. **Ion valency**: Monovalent ions are not as effective in neutralizing the surface charge of colloidal particles as ions with greater valency, e.g. Ca^{2+} and Al^{3+}. This results in a reduction in zeta potential, thus decreasing the repulsive forces and promoting particle aggregation.
4. **Surfactants and additives**: Adsorption of surfactants and additives can modify the surface charge and zeta potential [8]. Consequently, the stability of a colloidal system may be improved or reduced based on whether the repulsive forces are enhanced or diminished. Zeta potential not only determines the colloidal stability but also affects the electrokinetic phenomena like electrophoresis and sedimentation.

5.5 Electrokinetic Properties of Colloids

Electrokinetic effects define the motion of charged colloidal particles in relation to the counterion cloud of the dispersion medium in the presence of an external electric field. The net effect relies on the ions' distribution within the electric double layer and the diffuse layer. This kind of behavior is important for the comprehension of colloidal stability and interactions between various charged particles.

5.5.1 Electrophoresis

Electrophoresis is an evident phenomenon in which the migrated charged colloidal particles are induced by the applied electric field [2]. The positively charged particles move toward the cathode and the negatively charged particles move toward the anode. The surface charge of the colloidal particles and the existing electric double layer in the dispersion medium are accountable for such particle movement. The charged colloidal particles are denoted by the yellow color dots in Figure 5.5, which are migrating toward the oppositely charged electrode in the presence of an electric field. The electrophoresis velocity is controlled by the zeta potential, which is one of the important parameters for quantifying the interaction between two layers of particles.

Figure 5.5: Electrophoresis in colloids. (a) Before passing the electric current and (b) after electrophoresis, when the charged particles move toward oppositely charged electrodes. Figure generated and edited using Canva AI Image Generator [Internet]. Canva Pty Ltd; Available from: https://www.canva.com/ai-image-generator/.

The electrophoretic mobility (μ_e) of a colloidal particle is related to the zeta potential (ζ) by the Smoluchowski equation (for large particles in aqueous solutions with low surface conductivity) and can be expressed as:

$$\mu_e = \frac{\varepsilon \times \zeta}{\eta} \tag{5.5}$$

where ζ represents zeta potential in volts, ε is the dielectric constant of the medium, and η is the viscosity of the medium in Pa s. Beyond fundamental studies, this process finds extensive applications in the analysis of nanoparticles, characterization of biomolecules such as DNA, RNA, and proteins, water treatment processes, and in studying the surface charge of soil colloids. Moreover, it finds practical applications in effective pollutant removal and purification by the removal of suspended impurities.

5.5.2 Electro-osmosis

When a colloidal liquid comes in contact with a charged surface, an electric double layer is created at the interface. This layer develops a region of potential difference by combining the charged surface with a mobile layer of ions of opposite charge. Electro-osmosis is a fundamental electrokinetic phenomenon wherein the dispersion phase (liquid) moves relative to a stationary charged surface when an external electric field is applied, as shown in Figure 5.6. The electric field along the direction of the capillary or porous material exerts a force on the mobile counterions in the diffuse part of the electric double layer. These ions, typically found in the dispersion medium, begin to migrate toward the electrode of opposite charge. Since these ions are solvated, they carry their associated surrounding solvent molecules along, leading to a bulk flow of the liquid. This process is called electro-osmotic flow. The direction and magnitude of this flow depend on the sign and density of the surface charge, the ionic strength of the medium, and the applied field strength. Electro-osmotic mobility (μ_o) is typically used to measure the flow, which relates the velocity of the liquid to the electric field strength, much like electrophoretic mobility for particles.

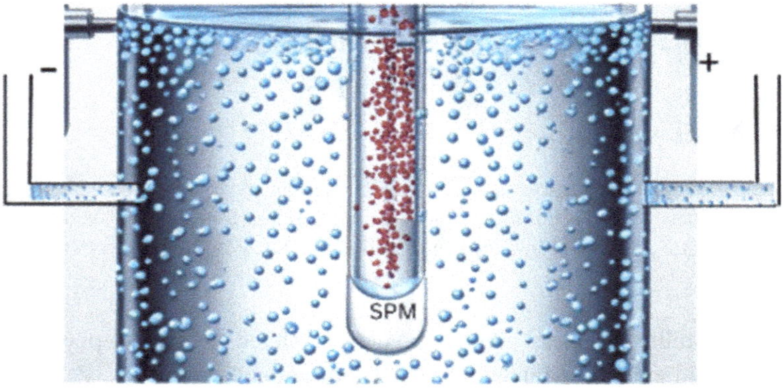

Figure 5.6: Electro-osmosis phenomena, where the dispersion medium crosses the barrier of a semipermeable membrane, in the presence of an applied electric field. Figure generated and edited using Canva AI Image Generator [Internet]. Canva Pty Ltd; Available from: https://www.canva.com/ai-image-generator/.

Electro-osmosis has significant practical applications, especially in systems where controlling the flow of liquid without mechanical pumps is desirable, such as in nanoparticle manipulations. In wastewater treatment, drug delivery, and soil analysis, dewatering and stabilization are maintained through membrane technologies. In microfluidics and lab-on-a-chip devices, electro-osmosis is utilized to transport fluids through microchannels, enabling precise control over the movement of reagents in analytical and diagnostic systems.

5.5.3 Sedimentation Potential (Dorn Effect)

Sedimentation potential is an electrokinetic phenomenon that arises when charged colloidal particles settle under the influence of gravity or centrifugation in the presence of an electric field. It is the reverse of electrophoresis, where the charged particles move instead of settling down. As these particles move downward, they carry their surface charge along with them, thereby leading to the temporary separation of charges. This, in turn, creates a dipole moment and disrupts the symmetry of the electric double layer that surrounds them. However, the counterions in the surrounding fluid do not move at the same rate. The charge separation creates an electric field and an electric potential difference between the top and bottom of the liquid column, which is called the sedimentation potential. In essence, a small electric field is generated in the direction opposite to sedimentation. Electrodes positioned at different places in suspension can be used to measure sedimentation potential. It highlights the intricate coupling between mechanical forces (like gravity) and electrical responses in colloidal systems. The magnitude of sedimentation potential depends upon certain characteristics of the dispersed phase and dispersion medium, such as size, concentration, and surface charge density of the colloidal particles and viscosity and dielectric constant of the dispersion medium. The sedimentation potential (E_s) can be described by the Dorn equation:

$$E_s = \frac{Z \times e \times v}{\varepsilon \times \varepsilon_0} \tag{5.6}$$

where v is the sedimentation velocity, ε is the relative permittivity of the medium, and ε_0 represents vacuum permittivity (8.854×10^{-12} C^2 N^{-1} m^{-2}).

5.5.4 Streaming Potential

When a colloidal solution is allowed to flow through a narrow channel, porous plug, or a membrane under an external pressure gradient, the generated potential is called the streaming potential. Unlike sedimentation potential, which results from the gravitational settling of colloidal particles, streaming potential is generated due to the movement of the liquid phase itself. When a liquid flows through a capillary or porous structure with charged walls, the surface attracts counterions, forming an electric double layer. The layer closest to the surface is immobile, while the outer diffuse layer contains mobile ions. As pressure-driven flow carries the liquid forward, it also drags these mobile counterions from the diffuse layer along with it. This leads to a net transport of charge in the direction of flow, creating a separation of charges and developing an electric potential across the length of the channel. Unlike electrophoresis and electro-osmosis, streaming potential is created as a consequence of flow pressure. Colloidal particles in contact with the surface and the mobile ions create an electric double layer in the system. This resulting potential, known as the streaming potential, gradually builds up until the saturation of charge accumulation occurs.

Streaming potential can be used to determine the surface interactions through adsorption kinetics of particles onto the colloidal surface using zeta potential, dielectric constant of the liquid, and the electrical conductivity of the solution.

This relationship is described by the Helmholtz-Smoluchowski equation:

$$E_{\text{sm}} = \frac{\zeta \times \eta}{\varepsilon \times \ \varepsilon_0 \times \kappa} \Delta P \tag{5.7}$$

Here, E_{sm} is the streaming potential, ζ is the zeta potential, η is the viscosity of the liquid, ε and ε_0 are the relative and absolute permittivity of the medium, κ is the conductivity, and ΔP is the applied pressure difference. The equation indicates that higher pressure gradients or greater zeta potentials will result in larger streaming potentials, while higher conductivity or viscosity can reduce the effect.

5.6 Comparison of Properties of Colloidal Solutions

In order to have a better understanding of colloidal behavior, different characteristics and properties can be compared. Table 5.1 highlights the applications and causes of optical, kinetic, and electrical characteristics of colloids.

Table 5.1: Comparison of optical, kinetic, and electrical properties of colloidal systems.

Property type	Key phenomena	Cause	Observation/effect	Examples/ applications
Optical	Tyndall effect, color, and ultramicroscopy	Scattering of light by colloidal particles	Visibility of the light path and apparent color of the sol	Distinction between true and colloidal solutions
Kinetic	Brownian motion, diffusion, and sedimentation	Random motion due to molecular collisions	Stability against settling, slow diffusion	Determination of particle size
Electrical	Electrophoresis, electro-osmosis, streaming, and sedimentation potentials	Presence of an electric double layer and surface charge	Movement of particles or liquid under external fields	Zeta potential and colloid stability studies

5.7 Suspension in Chemistry: Description, Characteristics, and Applications

A suspension is a heterogeneous mixture in which solid particles are dispersed in a liquid or gas medium, but the particles are large enough to settle over time due to gravity. Suspensions have a particle size greater than 1,000 nm (1 µm). They appear cloudy or opaque. Suspensions are heterogeneous (particles can be seen with the naked eye or under a low-power microscope). Examples include muddy water, sand in water, and flour in water [9].

Colloids: A colloid is a type of heterogeneous mixture in which tiny particles of one substance are dispersed uniformly throughout another substance. The dispersed particles are larger than molecules but too small to settle out or be filtered by ordinary methods [10]. Colloids are different from suspensions, as depicted in Table 5.2.

Table 5.2: Difference between suspension and colloids.

Feature	Suspension	Colloid
Particle size	>1,000 nm	1–1,000 nm
Visibility	Visible to the naked eye or with a microscope	Not visible to the naked eye, visible under an ultramicroscope
Settling	Particles settle on standing (unstable)	Do not settle on standing (relatively stable)
Filtration	Can be separated by filter paper	Cannot be separated by ordinary filter paper
Tyndall effect	May or may not show	Always show the Tyndall effect
Appearance	Cloudy or opaque	Translucent or sometimes milky
Stability	Unstable; particles settle without stirring	Relatively stable; particles remain dispersed

5.7.1 Sedimentation

Sedimentation is the process in which suspended particles settle down at the bottom of the container over time due to gravity. It happens when the particles are denser than the medium, and the suspension is left undisturbed [11]. The rate of sedimentation depends on:

1. **Particle size**: Larger particles settle faster.
2. **Density difference between the particle and medium**: The greater the difference in density between the particles and the medium, the faster the rate of sedimentation. Also, if the particles are much heavier than the liquid, they will settle quickly.
3. **Viscosity of the medium**: Viscosity is the thickness or resistance to flow of the liquid. Higher viscosity (thicker liquids like syrup) slows down sedimentation. Lower viscosity (like water) allows particles to settle more quickly.
4. **Temperature:** Higher temperature reduces viscosity, increasing sedimentation. For example, in muddy water, soil particles settle at the bottom when the mixture is allowed to stand.

5.7.2 Stability of Suspensions

Stability in a suspension refers to the ability of particles to remain evenly distributed throughout the medium without settling over time. Suspensions are inherently unstable, meaning the particles settle down over time, and require stirring or shaking before use (e.g., in medicines). The stability of suspensions can be improved by using suspending agents (e.g., gelatin, starch, and gum), reducing particle size to enhance Brownian motion, increasing the viscosity of the medium, and adding surfactants or emulsifiers to prevent particle clumping. Table 5.3 highlights the major distinctions between stable and unstable suspensions, providing a clearer understanding of the factors influencing their behavior.

Table 5.3: Difference between stable and unstable suspension.

Stable suspension	Unstable suspension
Particles stay suspended for longer	Particles settle quickly
Requires little or no shaking before use	Forms sediment at the bottom and clear liquid on top
Important in products like medicines, paints, and drinks	Needs to be shaken before each use
Less sedimentation	More sedimentation

5.7.3 Practical Applications of Suspensions

1. Medical field

(i) Antacids like milk of magnesia are suspensions of magnesium hydroxide that help neutralize excess stomach acid.
(ii) A large number of antibiotic syrups are prepared in the form of suspensions, which is why they need to be shaken before use to ensure proper mixing.
(iii) Vaccines are formulated as suspensions containing inactivated microbes that stimulate the body's immune response.

2. Agriculture

(i) Pesticides are mostly formulated as suspensions so they can be sprayed evenly on crops to protect them from pests and diseases.
(ii) Fertilizers are also available in suspension form, which allows nutrients to be distributed more uniformly in the soil.

3. Construction

(i) When cement or lime is mixed with water, it forms a suspension that is commonly used for plastering in construction work.

(ii) Paints and coatings are also suspensions, as they contain pigment particles dispersed in a liquid binder to provide color and protective finishing.

4. Environmental science

(i) In environmental studies, sediment present in rivers, lakes, or other water bodies is analyzed in suspension form to assess water quality and ecological conditions [4].

(ii) Wastewater treatment processes involve the removal of suspended solids from water to make it cleaner and safer for reuse or discharge.

5. Food industry

(i) Fruit juices often contain pulp, which remains suspended in the liquid and gives the drink its natural texture and flavor.

(ii) In dishes such as curries and soups, spices stay suspended in the liquid medium, enhancing both taste and aroma.

(iii) Chocolate milk is an example of a suspension where fine cocoa particles are dispersed throughout the milk to create a uniform and tasty beverage.

5.8 Emulsions

Emulsions are mixtures composed of two immiscible liquids, in which one liquid is dispersed within the other. In such systems, one liquid serves as the dispersed phase, while the other functions as the dispersion medium. In simple terms, emulsions are a type of colloid in which both the dispersed phase and the dispersion medium are liquids. An emulsion features a continuous phase (the dispersion medium) and a dispersed phase, separated by an interface. A common example is the mixture of oil and water. When shaken together, the oil forms tiny droplets that spread throughout the water, creating an emulsion. The term "emulsion" may also be broadly used to describe other mixed systems like solutions, gels, or suspensions.

A simple example of an emulsion is oil mixed with water. On its own, oil does not dissolve in water, but when the two are shaken together, the oil breaks into numerous small globules that remain suspended within the water. This creates a temporary emulsion. However, because oil and water are immiscible, the droplets eventually coalesce unless a stabilizing agent such as soap, detergent, or an emulsifier is added. These agents reduce the surface tension between the two liquids and prevent the droplets from merging back together, thereby making the emulsion more stable.

The boundary separating the dispersed phase from the continuous phase is called the interface. The properties of this interface strongly influence the behavior of emulsions, including their stability, texture, and appearance. When light is scattered evenly across all wavelengths, the emulsion appears white. However, in a dilute emulsion, shorter wavelengths and higher frequency light, such as blue light, are scattered more strongly. This results in a bluish appearance, a phenomenon known as the Tyndall effect.

Emulsions are of great importance in both natural and industrial contexts. They are found in everyday items such as milk (a natural emulsion of fat droplets in water), cosmetics (such as creams and lotions), pharmaceuticals (where drugs are delivered in emulsion form), and food products (such as mayonnaise, butter, and salad dressings) [7]. Their widespread applications highlight the need to understand how they form, how they can be stabilized, and how their properties influence their practical uses [12].

5.8.1 Types of Emulsions

Emulsions are broadly classified into two types, depending on the dispersion and continuous phases. Both are pictorially represented in Figure 5.7:

1. **Oil-in-water (O/W) emulsion**: In this type of emulsion, oil serves as the dispersed phase, while water acts as the continuous phase. These emulsions are typically nongreasy, water-washable, and widely used in the food and cosmetic industries. Common examples include milk, vanishing creams, shaving creams, and certain liquid pharmaceutical emulsions.
2. **Water-in-oil (W/O) emulsion:** In this type of emulsion, water is the dispersed phase and oil is the continuous phase. These emulsions tend to be greasy and are often used for their moisturizing properties, thanks to the oil-based medium. Examples include butter, cold creams, and certain topical ointments [1, 2].

The properties associated with both have been summarized in Table 5.4.

5.8.1.1 Multiple and Microemulsions (Advanced Types)

Multiple emulsions and microemulsions represent advanced colloidal systems with significant scientific and industrial importance. Multiple emulsions, often referred to as double emulsions, are complex structures formed in a two-step process where one emulsion (such as oil-in-water) is further emulsified into another of the opposite type (such as water-in-oil). The resulting systems, typically oil-in-water-in-oil or water-in-oil-in-water, are capable of encapsulating both hydrophilic and lipophilic substances simultaneously, making them highly useful in pharmaceuticals, cosmetics, and food

Figure 5.7: Emulsions of oil-in-water (O/W) and water-in-oil (W/O). Figure generated and edited using Canva AI Image Generator [Internet]. Canva Pty Ltd; Available from: https://www.canva.com/ai-image-generator/.

technology. They offer advantages such as controlled release, partitioning, and protection of sensitive ingredients, though their stability remains a major challenge.

Microemulsions, on the other hand, are thermodynamically stable, transparent or translucent dispersions of oil and water stabilized by surfactants, often in combination with cosurfactants. Unlike multiple emulsions, microemulsions form spontaneously under the right conditions and consist of droplet sizes in the nanometer range (10–100 nm), giving them unique optical clarity and high stability. They can exist in different structures such as O/W, W/O, or bicontinuous systems. Microemulsions are widely studied for their ability to solubilize poorly water-soluble drugs, enhance bioavailability, deliver active ingredients in cosmetics, and improve formulations in food, agrochemicals, and petroleum industries. Together, multiple emulsions and mi-

Table 5.4: Emulsion types and their characteristics.

Feature	Oil-in-water (O/W) emulsion	Water-in-oil (W/O) emulsion
Dispersed phase	Oil	Water
Continuous phase	Water	Oil
Texture	Nongreasy	Greasy
Washability	Easily washable with water	Difficult to wash with Water
Common uses	Foods and cosmetics	Moisturizing products
Examples	Milk, vanishing cream, shaving cream, and liquid pharmaceuticals	Butter, cold creams, and topical ointments

croemulsions expand the possibilities of modern formulation science, enabling multi-functional products in healthcare, personal care, and industrial applications, but also present challenges in terms of design, stability, etc. [1, 2]. The main difference between the two is tabulated in Table 5.5.

Table 5.5: Multiple emulsions versus microemulsions.

Feature	Multiple emulsions	Microemulsions
Definition	Emulsions within emulsions (double layer)	Thermodynamically stable, isotropic mixture of oil and water
Structure	Complex, with two emulsion phases (e.g., water-in-oil-in-water or oil-in-water-in-oil)	Simple, uniform dispersion (oil-in-water, water-in-oil, or discontinuous)
Stability	Kinetically stable (may separate over time)	Thermodynamically stable (spontaneously forms and is stable)
Droplet size	Micrometer range	Nanometer range (10–100 nm)
Appearance	Often opaque or milky	Transparent or translucent
Formation method	Requires stepwise emulsification	Formed spontaneously with surfactants and cosurfactants
Applications	Drug delivery (controlled release), cosmetics, food, and fuels	Drug delivery (enhanced solubility), oil recovery, cosmetics, and nanoparticle synthesis
Surfactant requirements	Lower concentration	Higher concentration often with cosurfactants
Examples	Vanishing cream (water-in-oil-in-water), sustained-release formulations	Nanodrug carriers, clear cleansing oils, and microemulsion fuels

5.8.2 Preparation of Emulsions and Their Stabilization

In the preparation of emulsions, the two immiscible liquids must be mixed so that droplets of one liquid are finely distributed in the other [14]. This can be done by various methods discussed below:

1. **Mechanical agitation (shaking or stirring)**: Manually shake or stir the two immiscible liquids together using a blender or stirrer.
2. **Homogenization**: The mixture is forced through a narrow nozzle or small opening at high pressure, breaking one liquid into very fine droplets.
3. **Ultrasonication:** High-frequency sound waves (ultrasound) are used to break droplets into smaller sizes. Ultrasonication produces very fine droplets and stable emulsions.

4. **Phase inversion method:** Changing the volume ratio of oil and water or temperature so that the emulsion switches from O/W to W/O, or vice versa.
5. **Using emulsifying agents (emulsifiers):** Adding substances that help to form and stabilize the emulsion.

5.8.3 Role of Emulsifying Agents

Emulsions have an unstable nature, and without stabilization, the dispersed droplets tend to coalesce or separate over time. To overcome this limitation, emulsifying agents, also known as surfactants, are used. Emulsifiers are surface-active agents (surfactants) that reduce the interfacial tension between oil and water. They form a protective film around dispersed droplets, preventing coalescence.

Beyond their primary role, emulsifying agents contribute to stability in several additional ways. They help maintain a uniform droplet size by resisting flocculation, they enhance the viscosity of the dispersion medium, which slows down droplet movement, and in many cases, they provide electrostatic or steric repulsion between droplets.

Approaches to stabilizing emulsions: Emulsions are thermodynamically unstable systems, and their stability is improved using various strategies:

1. **Emulsifier addition**: Emulsifiers are surface-active substances that lower interfacial tension between the continuous and dispersed phases, ensuring stability. Natural emulsifiers like lecithin (derived from egg yolk) and casein (derived from milk), and synthetic ones like soaps, detergents, and sodium stearate are common examples.
2. **Viscosity increase:** Thickening agents like gums or starches raise the viscosity of the continuous phase. This retards the movement of droplets and slows coalescence, thus improving the stability of the emulsion [6].
3. **Electrostatic stabilization:** When droplets become like charged, usually because of emulsifiers, they repel one another. This electrostatic repulsion inhibits droplets from getting close to each other and coalescing.
4. **Steric stabilization:** Polymers or proteins may adsorb on the surface of droplets, creating an insulating physical barrier against droplet approach. Such steric hindrance can effectively decrease coalescence and enhance stability [15].

Stabilizing emulsions is crucial, as without proper stabilization, they become unstable and eventually undergo processes as mentioned below:

1. **Creaming**: This occurs when dispersed droplets rise to the surface and form a concentrated layer at the top of the emulsion.
2. **Sedimentation:** This is the settling of droplets toward the bottom due to their higher density compared to the surrounding medium.
3. **Coalescence:** It is the process by which smaller droplets join to form larger droplets, reducing stability and causing the breakdown of the emulsion.

4. **Phase separation:** This is the last stage of emulsion instability, where the dispersed and continuous phases detach entirely and form distinct layers, resulting in the breakdown of the system.

5.8.4 Applications of Emulsions in Industry and Daily Life

Emulsions are essential to the pharmaceutical industry because they make it possible for oil-based medications, such as cod liver oil, to be delivered efficiently. Emulsions are used in cosmetics to improve skin absorption, such as in creams and lotions. To enhance texture and flavor, the food industry uses emulsions in products like milk, butter, and mayonnaise. Emulsion paints are used in construction to provide non-toxic, smooth, and fast-drying wall surfaces. Emulsions help ensure that oil-based insecticides are applied consistently to crops in agriculture [14].

5.8.4.1 Industrial Applications of Colloids

1. **Paints and inks:** Colloids are used to achieve uniform dispersion of pigments. Pigments, inks, or dye particles are suspended in a solvent like water or alcohol. Moreover, many modern inkjet printers rely on nanosized pigment colloids to deliver precise printing and consistent color stability.
2. **Food services industry**: Emulsions like milk, mayonnaise, and ice cream are colloidal systems.
3. **Wellness products**: Creams, lotions, and shampoos are colloids designed for smooth application and stability.
4. **Rubber and plastics**: Latex, a type of colloid, serves as a key material in the production of rubber-based products.
5. **Cement and ceramics**: Colloidal silica improves both strength and durability. It consists of a stable suspension of amorphous silica (SiO_2) nanoparticles in water or another liquid medium.
6. **Lubricating agents**: Colloidal suspensions reduce friction in machines. MoS_2 (molybdenum disulfide) particles are dispersed in grease or oil to create a colloidal lubricant.
7. **Detergent action of soap**: Dirt adheres to greasy or oily substances that accumulate on the fabric surface. Since grease is not easily wetted by water, washing the cloth with water alone is ineffective. When soap is added, it reduces the interfacial tension between grease and water, allowing the grease to emulsify in water. Rubbing the fabric then helps to loosen and remove the dirt along with the emulsified grease [13, 15].

5.9 Medical Applications of Colloids

1. **Drug delivery systems**: Nanocolloids enable precise and controlled drug delivery. An example is Doxil, a liposomal formulation of the chemotherapy drug doxorubicin, which is commonly used in cancer treatment.
2. **Blood substitutes**: Colloidal solutions like dextran are used to maintain blood volume.
3. **Diagnostic agents**: Colloidal gold (Au) is used as a diagnostic agent, particularly in nuclear medicine.
4. **Antiseptics**: Colloidal silver has antibacterial properties.
5. **Wound healing**: Colloidal silver is known for its antibacterial properties and is used in wound dressings and antiseptics.
6. **Radiology**: Barium sulfate colloidal suspensions are widely used in X-ray imaging, particularly for picturing the gastrointestinal (GI) tract. They improve the contrast of GI structures by increasing their opacity to X-rays.
7. **Gene delivery**: The genetic material can be released into the cell via liposomes, which fuse with cell membranes and encasing nucleic acids (DNA, siRNA, and mRNA) as colloids. mRNA vaccines, such as those created by Pfizer-BioNTech and Moderna, use lipid nanoparticles, a kind of colloidal technology, to efficiently transfer mRNA into human cells.

5.10 Environmental Applications of Colloids

1. **Water purification**: Colloids help in flocculation and sedimentation processes to remove impurities [11]. For example, when alum is added to water, it reacts to form $Al(OH)_3$, a gelatinous precipitate. This precipitate aids in destabilizing colloidal particles such as clay, organic matter, and microorganisms, which are suspended in the water.
2. **Pollution control:** Colloidal particles can capture and eliminate airborne or waterborne contaminants. Titanium dioxide (TiO_2), in nanosized colloidal form, is utilized for photocatalytic coatings or in air cleaners. TiO_2 is utilized to destroy dangerous air pollutants such as volatile organic compounds, nitrogen oxides (NO_x), and formaldehyde.
3. **Remediation of soil**: Treatment of heavy metal or chlorinated organic contaminated soils through the use of nanoscale zero-valent iron, which is a colloidal suspension of iron nanoparticles usually measuring between 1 and 100 nanometers in diameter. Due to their high reactivity and tendency to migrate through soil, these nanoparticles are very efficient for in-situ remediation, allowing them to directly degrade or immobilize contaminants beneath the ground surface.

4. **Wastewater treatment:** Sewage generally has colloidal substances like dirt and organic matter in it, which are easily removed by coagulation followed by sedimentation. Colloidal or reactive silica is commonly utilized in wastewater treatment processes for the removal of contaminants.
5. **Slow release of agrochemicals:** Nanoemulsion-based colloidal carriers are utilized for the delivery of pesticides, herbicides, and fertilizers in slow-release mode to reduce leaching. Example includes chitosan-based formulations as colloids for the delivery of eco-friendly pesticides.
6. **Oil spill clean-up**: Emulsifiers help disperse oil into fine droplets, creating emulsions that enhance microbial breakdown of the oil [13]. For example, sorbitan monooleate, a surfactant that is used to break the oil into tiny droplets, forms an emulsion. The formation of tiny droplets increases the surface area for oil-degrading microbes, thereby speeding up the biodegradation process in marine environments.
7. **Heavy metal removal:** Manganese dioxide and iron oxide colloids, in the form of nanoparticles, adsorb toxic heavy metals like lead, mercury, cadmium, and arsenic from water.

Practice Questions

1. Define the Tyndall effect and explain why it is observed in colloidal systems.
2. Compare and contrast true solutions, colloids, and suspensions in the context of the Tyndall effect.
3. List three practical applications of the Tyndall effect.
4. Describe an experiment to demonstrate the Tyndall effect in a classroom setting.
5. Why is the sky blue during the day and red during sunset? Relate your answer to the Tyndall effect.
6. What is Brownian motion, and how does it help in the stability of colloidal systems?
7. Explain how temperature and particle size affect the rate of Brownian motion in a colloidal solution.
8. Why does Brownian motion become less noticeable for larger particles in a suspension?

9. Compare and contrast Brownian motion and diffusion. Are they governed by the same physical principles?

10. Discuss the significance of Brownian motion in distinguishing true solutions from colloidal dispersions.

11. How does the presence of colloidal particles affect the viscosity of a dispersion medium?

12. Why does diffusion occur more slowly in colloidal systems compared to true solutions?

13. What are the two distinct layers of the electric double layer? How do they differ in terms of ion mobility?

14. How does the ionic strength of the medium affect the thickness of the electric double layer?

15. Explain the role of the electric double layer in the stability of a colloidal solution.

16. What is the significance of the slipping plane in electrokinetic phenomena?

17. How does the pH of the medium influence the structure of the electric double layer?

18. What are electrokinetic phenomena? Name the four main types.

19. Define zeta potential. Why is it important in electrokinetic processes?

20. Differentiate between electrophoresis and electro-osmosis.

21. What is the role of the electric double layer in electrokinetic phenomena?

22. Explain the concept of streaming potential in brief.

23. How does sedimentation potential develop in a colloidal solution?

24. Mention any two practical applications of electrokinetic effects.

25. What factors influence the magnitude of electro-osmosis in a capillary system?

26. What is electrophoresis, and how is it used to determine particle mobility?

27. How does pH affect the electrophoretic mobility of colloidal particles?

28. State any two differences between electrophoresis and sedimentation potential in terms of cause and effect.

29. Define suspension and give two common examples.

30. Why do particles in suspension settle over time while colloidal particles do not?

31. How does the appearance of a suspension differ from that of a colloid?

32. What happens when you let a suspension sit undisturbed for a long period of time, and why?

33. What is sedimentation? Describe the process with an example.

34. How does the particle size of particle affect the rate of sedimentation?

35. Give two ways by which sedimentation can be increased in a suspension.

36. What is the role of suspending agents in stabilizing suspensions?

37. A bottle of suspension medicine says, "Shake well before use." Explain why shaking is important.

38. A student adds starch to a suspension and notices the particles settle more slowly. What property of the medium has been changed, and how does this affect sedimentation?

39. Define an emulsion. What are its two main components?

40. Differentiate between the dispersed phase and dispersion medium in an emulsion.

41. Give examples of each of O/W and W/O emulsions.

42. Why are O/W emulsions preferred in pharmaceutical applications over W/O emulsions?

43. Define multiple emulsions and give one example.

44. What is a microemulsion? How is it different from a regular emulsion?

45. Give the differences between multiple emulsions and microemulsions in terms of parameters of appearance and stability?

46. Name and explain any two methods of preparing emulsions.

47. Why are emulsifying agents important in emulsions?

48. Explain the role of surfactants and cosurfactants in forming microemulsions.

49. How do microemulsions enhance drug stability and delivery in pharmaceutical applications?

50. What happens to an emulsion that is not stabilized properly? Explain concerning common instability phenomena.

51. Discuss the various strategies used to stabilize emulsions with suitable examples.

52 How do colloidal particles contribute to the uniformity of pigments in paints and inks?

53. How does soap function as a colloidal system to remove greasy dirt from fabrics?

54. What role does colloidal silver play in both antiseptics and wound healing?

55. Describe the use of colloidal barium sulfate in radiological imaging.

56. Explain how alum acts on colloidal impurities during water purification.

57. In oil spill clean-up, how does emulsification enhance microbial degradation of oil?

58. Discuss how the stability of colloids affects product shelf life in the food and cosmetics industry.

59. What is the role of latex as a colloid in rubber manufacturing?

60. Explain the industrial application of colloids.

References

[1] Hunter RJ. *Foundations of colloid science*. 2nd ed. Oxford: Oxford University Press; 2001.

[2] Shaw DJ. *Introduction to colloid and surface chemistry*. 4th ed. Oxford: Butterworth-Heinemann; 1992.

[3] Israelachvili JN. *Intermolecular and surface forces*. 3rd ed. London: Academic Press; 2011.

[4] Buffle J, Leppard GG. Characterization of aquatic colloids and macromolecules. 1. Structure and behavior of colloidal material. *Environ Sci Technol*. 1995;29(9):2176–2184. doi: 10.1021/es00009a008.

[5] Atkins P, De Paula J, Keeler J. *Atkins' physical chemistry*. 11th ed. Oxford: Oxford University Press; 2018.

[6] Spielman LA. Particle deposition in the presence of electrostatic double layer repulsion. *J Colloid Interface Sci*. 1977;60(3):497–514. doi: 10.1016/0021-9797(77)90134-1.

[7] Russel WB, Saville DA, Schowalter WR. *Colloidal dispersions*. Cambridge: Cambridge University Press; 1991.

[8] Hunter RJ. *Zeta potential in colloid science: Principles and applications*. London: Academic Press; 1981.

[9] Schramm LL. *Emulsions, foams, and suspensions: Fundamentals and applications*. Weinheim: Wiley-VCH; 2006.

[10] Pc J, Jain M. *Engineering chemistry*. 16th ed. New Delhi: Dhanpat Rai Publishing; 2017.

[11] Gregory J. *Particles in water: Properties and processes*. Boca Raton: CRC Press; 2005.

[12] Kuriacose JC, Rajaram J. *General and physical chemistry*. New Delhi: Tata McGraw Hill Education; 2015.

[13] Hiemenz PC, Rajagopalan R. *Principles of colloid and surface chemistry*. 3rd ed. New York: Marcel Dekker; 1997.

[14] Puri BR, Sharma LR, Pathania MS. *Principles of physical chemistry*. 49th ed. Jalandhar: Vishal Publishing Co.; 2022.

[15] Birkett JD, Lester JN. *Endocrine disrupters in wastewater and sludge treatment processes*. Boca Raton: CRC Press; 2003.

Pratibha Sharma, Deepak Yadav and Sujata Kumari*

6 Surface Chemistry of Colloids

Abstract: The surface chemistry of colloidal systems is fundamental to understanding their stability, interactions, and diverse applications. Colloids, consisting of microscopic particles dispersed within a continuous medium, exhibit unique properties due to surface interactions and interfacial phenomena. The surface potential, which arises from charge distribution at particle interfaces, governs interparticle interactions and significantly influences dispersion and stability. Monomolecular films, formed by the adsorption of a single molecular layer on surfaces, possess unique properties and are widely applied in coatings, pharmaceuticals, and biological systems. Langmuir-Blodgett films, a specialized type of monomolecular film, allow precise molecular organization and find applications in nanotechnology and materials science. Electrical aspects, including the electric double layer and zeta potential, play a crucial role in determining colloidal stability. Electrostatic repulsions prevent aggregation, while van der Waals forces influence dispersion. Balancing these attractive and repulsive forces is therefore essential for maintaining dispersion and suspension stability, which is critical in industries such as drug delivery, food emulsions, and ceramics. Recent advancements have further enhanced our understanding of colloidal surface chemistry, highlighting its pivotal role in ensuring the stability and functionality of colloidal systems across diverse scientific and technological fields.

6.1 Introduction

Colloidal science is an interdisciplinary field that lies at the interface of chemistry, physics, biology, and materials science. It deals with systems in which one substance, composed of microscopic particles, is uniformly dispersed throughout another. These dispersed particles are larger than simple molecules but small enough to resist rapid settling under gravity, typically ranging in size from 1 nm to 1,000 nm (1 μm) [1]. Colloids are commonly encountered in everyday life, for example, fog, smoke, milk, paint, and even biological fluids such as blood. The distinctive properties and wide-ranging applications of colloids largely arise from the very high surface area of dispersed particles relative to their volume, which makes surface and interfacial phenomena central to their behavior.

The stability of a colloidal system, its ability to remain dispersed without undergoing aggregation or separation, depends on a delicate balance of forces acting at the surfaces of particles. These forces are governed by the chemical and physical nature of the particle-medium interface. Understanding such surface interactions is not just of academic interest; it is also critical for the design and control of industrial pro-

https://doi.org/10.1515/9783112208236-006

cesses and advanced materials. For example, the shelf life of pharmaceutical formulations, the texture of food products, the efficiency of drug delivery systems, and the performance of catalysts all rely on effective regulation of colloidal stability. Figure 6.1 illustrates a typical laboratory setup for the synthesis of surface-capped nanocrystals, in which capping agents help the nanocrystals remain dispersed in solution by preventing aggregation. This example demonstrates how surface modification at the nanoscale plays a vital role in controlling colloidal stability. By tailoring the surface chemistry through the choice of suitable capping molecules, researchers can tune the size, shape, solubility, and functional properties of the nanocrystals, which are crucial for their applications in catalysis, drug delivery, optoelectronics, and biomedical imaging.

This chapter presents a comprehensive overview of the surface chemistry of colloidal systems. It begins by defining colloids and differentiating them from true solutions and coarse suspensions. The discussion then turns to the concept of surface potential, its origin, and its influence on interparticle interactions. Monomolecular films, including the specialized Langmuir-Blodgett (LB) technique, will be examined in terms of their formation, properties, and applications. The electrical aspects of colloidal stability, particularly the electric double layer (EDL) and zeta potential, will be discussed in detail to explain how electrostatic repulsions counterbalance attractive forces. The chapter also highlights the role of van der Waals and other intermolecular forces in maintaining stability. Finally, recent advancements in colloidal surface chemistry, including new characterization tools and emerging applications, are presented to emphasize the continuing importance of this field in science and technology.

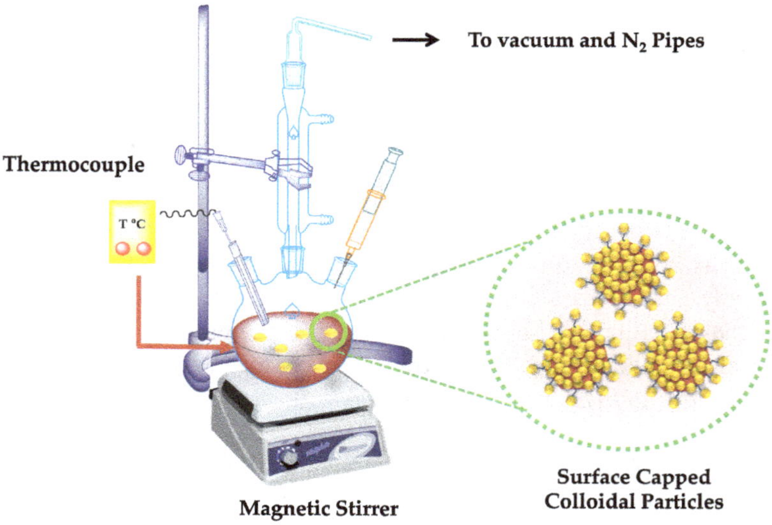

Figure 6.1: Schematic representation of a typical glassware setup used in the colloidal solution growth of nanocrystals (NCs), yielding surfactant-capped NCs dispersible in various solvents.

6.2 Fundamentals of Colloidal Systems and Surface Chemistry

6.2.1 Colloids

Colloids represent a distinct state of matter defined primarily by the size of the dispersed particles. Unlike true solutions, in which solute particles are molecularly dispersed, or coarse suspensions, in which particles settle readily under gravity, colloids occupy an intermediate domain. Their defining feature is that at least one dimension of the dispersed particles lies between approximately 1 nm and 1,000 nm (1 µm) [2]. Within this size range, the particles are large enough to establish well-defined interfaces with the surrounding medium, yet small enough to remain suspended due to the Brownian motion and/or interparticle repulsive forces.

A colloidal system is heterogeneous in nature and always consists of two phases:

1. A dispersed phase (the particles)
2. A continuous phase (the dispersion medium)

Depending on the physical states of these two phases, colloids can be classified into several types, as shown in Figure 6.2 [3].

(i) **Sols**: These are colloidal systems where solid particles are uniformly distributed within a liquid medium. The particles are typically stabilized by electrostatic repulsion or by adsorption of stabilizing agents that prevent aggregation. Sols are common in both natural and industrial systems. Examples include paint, where pigment particles are dispersed in a liquid binder, ink, which contains finely divided carbon or dye particles, and milk of magnesia, in which magnesium hydroxide particles are dispersed in water. Sols often exhibit the Tyndall effect (scattering of light by colloidal particles), which distinguishes them from true solutions.

(ii) **Solid sols**: These are colloidal systems in which solid particles are dispersed in a solid medium. Unlike ordinary solid mixtures, the dispersed particles remain uniformly distributed at the colloidal scale without settling or separating. Their stability often depends on strong interactions at the particle interfaces. Examples include colored gemstones like ruby (where chromium oxide particles are dispersed in aluminum oxide), opal (hydrated silica dispersed in a solid matrix), and certain alloys that exhibit colloidal-scale distribution of components. Solid sols hold great importance in materials science, as they frequently display distinctive optical, electrical, and mechanical properties that set them apart from bulk solids.

(iii) **Emulsions**: These consist of liquid droplets dispersed in another immiscible liquid. Since the two liquids do not naturally mix, emulsions require stabilizing agents, known as emulsifiers or surfactants, which reduce interfacial tension and form protective layers around droplets. Common examples include milk (fat droplets dispersed in water), mayonnaise (oil droplets stabilized by egg yolk proteins), and vinaigrette (oil dispersed in vinegar when shaken). Emulsions are crucial in food products, cosmetics, pharmaceuticals, and paints.

(iv) **Foams**: Foams are colloidal systems where gas bubbles are dispersed within a liquid or solid medium. The stability of foams depends on the viscosity of the medium and the presence of surfactants or stabilizers that prevent bubble coalescence. Everyday examples include whipped cream, where air is dispersed in cream; shaving foam, where gas is dispersed in a liquid soap solution; and pumice stone, where gas bubbles are trapped within a solid volcanic rock. Foams are widely used in fire-fighting, insulation, packaging, and food processing.

(v) **Solid foams**: These are colloidal systems characterized by the dispersion of gas bubbles within a solid matrix. The gas phase is trapped within the solid network, resulting in a lightweight, porous material. Solid foams can be natural or synthetic. Examples include pumice stone (formed when volcanic lava cools quickly, trapping gas bubbles), sponges, bread (where carbon dioxide bubbles are trapped in baked dough), and Styrofoam (used in packaging and insulation).

(vi) **Aerosols**: Aerosols are formed when either liquid droplets or solid particles are dispersed in a gas. They can be naturally occurring, like fog (water droplets in air), mist, and dust particles, or man-made, such as smoke from combustion processes and aerosol sprays used in deodorants or paints. Aerosols play an important role in atmospheric science, climate studies, and medicine (e.g., inhalers for asthma deliver drugs in aerosol form).

(vii) **Solid aerosols**: These are colloidal systems where fine solid particles are dispersed throughout a gaseous medium. Unlike ordinary dust clouds that settle rapidly, solid aerosols remain suspended due to their very small particle size and interaction with air currents. Natural examples include smoke (carbon or ash particles dispersed in the air) and atmospheric dust. Artificial examples include industrial emissions, aerosol sprays containing powdered solids, and airborne pollutants. They are also important in health science because inhalation of particulate aerosols can have significant biological effects, both beneficial (e.g., drug delivery via dry powder inhalers) and harmful (e.g., respiratory diseases caused by air pollution).

(viii) **Gels**: Gels are semirigid colloidal systems where a liquid is trapped within a three-dimensional solid network, giving them both liquid- and solid-like properties. This network prevents the liquid from flowing freely, resulting in a soft but stable structure. Examples include gelatin (a protein-based gel used in desserts), jelly, and agar gels commonly used in microbiology. Gels find applications in the food industry, pharmaceuticals (as drug carriers), cosmetics (lotions, creams), and biomedical materials (hydrogels for tissue engineering).

The stability and properties of different colloidal systems mainly depend on the nature of the interface that exists between the dispersed phase and the surrounding medium. Factors such as surface charge, adsorption of stabilizing agents, and the balance of attractive and repulsive forces at this interface govern whether a colloid remains stable or undergoes aggregation.

6.2.2 Introduction to Surface Chemistry

Surface chemistry, also referred to as interfacial science, is the branch of chemistry that studies the phenomena occurring at the boundary between two phases. An interface is the boundary that exists between two distinct bulk phases, such as solid-liquid, liquid-gas, or liquid-liquid. In colloidal systems, the interfaces are extremely large due to the very high surface-to-volume ratio of the dispersed particles. For example, just 1 cm^3 of material divided into particles of 10 nm diameter would expose a total surface area comparable to several tennis courts [4]. This vast surface area results in a large fraction of atoms or molecules residing at the surface rather than in the bulk, giving colloidal systems unique chemical and physical properties. Some of the key surface chemistry concepts relevant to colloids include:

(i) **Surface tension (for liquid-gas/liquid-liquid interfaces) and surface energy (for solid-liquid/solid-solid interfaces):** These are measures of the excess energy at the interface due to the imbalance of intermolecular forces experienced by molecules at the surface compared to those in the bulk. Since systems tend to minimize surface energy, colloidal particles may aggregate if repulsive forces are not sufficient to counteract this tendency [5].

(ii) **Adsorption:** The accumulation of molecules from a gas or liquid phase onto a solid or liquid surface. This process is critical for establishing surface charge, forming protective layers, and mediating interparticle interactions.

(iii) **Wetting:** The tendency of a liquid to spread over a solid surface is governed by the balance between adhesive forces (liquid-solid interactions) and cohesive forces (liquid-liquid interactions). Wetting is quantified by the contact angle and is of great importance in practical applications such as detergency, painting, coating, and mineral flotation.

In colloidal systems, the surface is far from being a passive boundary. Instead, it serves as an active site where interactions with the dispersion medium, other colloidal particles, and dissolved species take place. These interfacial interactions regulate both the stability of the dispersion and its optical, electrical, and rheological characteristics, ultimately defining the diverse applications of colloidal materials.

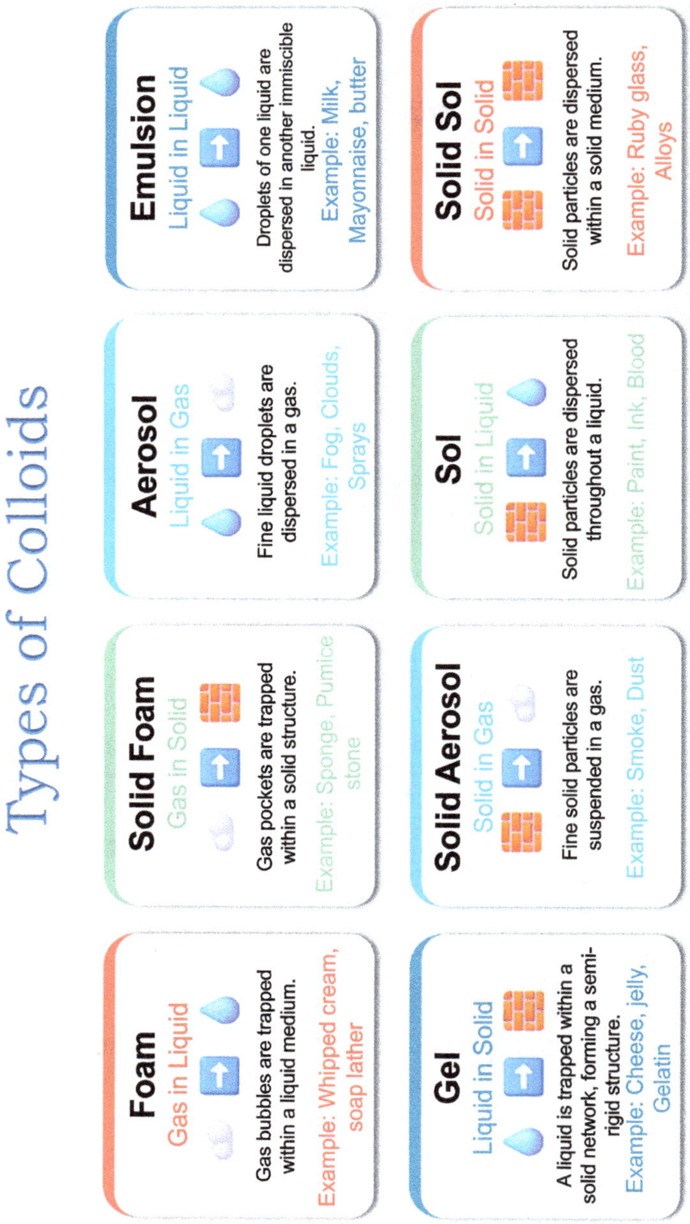

Types of Colloids
Foam
Gas in Liquid
Gas bubbles are trapped within a liquid medium.
Example: Whipped cream, soap lather
Solid Foam
Gas in Solid
Gas pockets are trapped within a solid structure.
Example: Sponge, Pumice stone
Aerosol
Liquid in Gas
Fine liquid droplets are dispersed in a gas.
Example: Fog, Clouds, Sprays
Emulsion
Liquid in Liquid
Droplets of one liquid are dispersed in another immiscible liquid.
Example: Milk, Mayonnaise, butter
Gel
Liquid in Solid
A liquid is trapped within a solid network, forming a semi-rigid structure.
Example: Cheese, jelly, Gelatin
Solid Aerosol
Solid in Gas
Fine solid particles are suspended in a gas.
Example: Smoke, Dust
Sol
Solid in Liquid
Solid particles are dispersed throughout a liquid.
Example: Paint, Ink, Blood
Solid Sol
Solid in Solid
Solid particles are dispersed within a solid medium.
Example: Ruby glass, Alloys

Figure 6.2: Classification of colloids based on physical state.

6.3 Surface Potential and Interparticle Interactions

The surface potential is a critical parameter in colloidal science. It represents the electrical potential at the interface of a colloidal particle relative to the bulk solution. Its origin is multifaceted and arises from several mechanisms, leading to the development of a net electrical charge on the particle surface. This surface charge, in turn, dictates the nature and strength of interparticle interactions, profoundly influencing the stability, dispersion, and aggregation characteristics of the entire colloidal system.

6.3.1 Origin of Surface Potential

The generation of surface charge on colloidal particles can occur through several primary mechanisms [6]:

1. **Ionization of surface functional groups:** Many inorganic and organic materials possess surface groups that can ionize when immersed in an aqueous medium. For example, metal oxides (such as silica, SiO_2) have surface silanol groups (Si-OH) that can deprotonate in basic solutions to form Si–O^- and H^+ or protonate in acidic solutions to form Si–OH^{2+}. Similarly, proteins have carboxyl (–COOH) and amino (–NH_2) groups that can ionize depending on the pH, leading to a pH-dependent surface charge. The isoelectric point (IEP) is the pH at which the overall surface charge of a particle becomes zero.
2. **Differential adsorption of ions:** Colloidal particles can selectively adsorb ions from the surrounding electrolyte solution. This selective adsorption can occur due to specific chemical affinity between the ion and the surface, or simply due to size and polarizability preferences. For instance, common ions like H^+ or OH^- are often preferentially adsorbed or desorbed, contributing to the surface charge.
3. **Isomorphic substitution within the crystal lattice:** In some crystalline materials, particularly clays, atoms of one valence can be substituted by atoms of another valence within the crystal lattice (e.g., Al^{3+} replacing Si^{4+} in a tetrahedral site). This results in a permanent structural charge imbalance within the particle, which is neutralized by the adsorption of counterions from the surrounding solution onto its surface. Such a charge is generally independent of pH.
4. **Adsorption of charged surfactants or polyelectrolytes:** The intentional addition of charged macromolecules, such as surfactants or polyelectrolytes, can significantly modify the surface charge of colloidal particles. These molecules adsorb onto the surface, exposing their charged groups to the solution, thereby imparting a desired charge. This is a common strategy for stabilizing or destabilizing dispersions.
5. **Preferential dissolution of ions from the lattice:** For ionic crystals, one type of ion may preferentially dissolve from the crystal lattice, leaving behind an excess of ions of the opposite charge on the particle surface. For example, in a silver

iodide (AgI) precipitate, I^- ions might preferentially dissolve, leaving a positive surface charge, or Ag^+ ions might dissolve, leaving a negative surface charge, depending on the conditions.

Regardless of the mechanism, the development of a surface charge creates an electric field that extends into the surrounding solution. This electric field influences the distribution of nearby ions, giving rise to an EDL and profoundly affecting the interactions, aggregation tendencies, and overall stability of colloidal particles.

6.3.2 Role in Interparticle Interactions

The surface potential, and the resulting charge on colloidal particles, plays a pivotal role in governing particle-particle interactions, which directly determine the stability, dispersion, and aggregation behavior of colloidal systems. Particles with similar surface charges experience electrostatic repulsion, which acts to keep them apart and prevent aggregation.

When two similarly charged colloidal particles come close, the EDLs surrounding them start to overlap. This overlap alters the local ion distribution, increasing the counterion concentration between the particles. As a result, there is a rise in local osmotic pressure, which generates a repulsive force that pushes the particles apart. The magnitude and effectiveness of this electrostatic repulsion depend on several factors:

1. **Magnitude of surface potential/surface charge:** The electrostatic repulsion between particles increases with the magnitude of the surface potential. Particles with higher surface potentials (or equivalently, higher surface charge densities) create stronger electric fields, which in turn produce more substantial repulsive forces when their double layers overlap. If the surface potential falls below a critical threshold, the repulsion is insufficient to counterbalance attractive forces (mainly van der Waals attractions), and the particles aggregate.
2. **Concentration and valence of electrolytes:** The surrounding electrolyte concentration has a profound influence on colloidal stability. The presence of electrolytes (ions in the solution) can screen the surface charge, reducing the effective range and strength of the electrostatic repulsion and hence the effective thickness of the EDL. Higher concentrations or higher valence counterions (ions opposite in charge to the particle) are more effective at screening [7]. This phenomenon is central to the concept of ionic strength and its effect on colloidal stability, often leading to coagulation at high salt concentrations. This principle is summarized by the Schulze-Hardy rule, which states that the ability of electrolytes to cause coagulation increases dramatically with the valency of the counterion. For example, trivalent ions (Al^{3+} and Fe^{3+}) are far more effective in destabilizing a negatively charged colloid than monovalent ions (Na^+, K^+).

3. **Distance between particles:** The strength of the repulsive force is highly dependent on the distance between the particles. As two particles move closer, their EDLs increasingly overlap, leading to stronger repulsion. Conversely, at larger separations, the interaction is negligible. This steep distance dependence is why colloidal systems can appear stable for longer periods but may rapidly destabilize if particles are forced into close contact by stirring, drying, or external fields.

The stability of a colloidal system is governed by the balance between electrostatic repulsive forces and attractive interactions (such as ever-present van der Waals forces). This balance dictates whether the particles stay dispersed or undergo aggregation through coagulation or flocculation. A sufficiently high surface potential (and consequently, a high zeta potential, which will be discussed later) ensures that the electrostatic repulsion barrier is strong enough to overcome the attractive forces, thus maintaining dispersion. Without adequate repulsion, particles would inevitably come into contact and aggregate due to the omnipresent van der Waals attractions, leading to sedimentation, creaming, or phase separation. This delicate balance forms the cornerstone of understanding and controlling colloidal stability, a concept further elaborated by the DLVO (Derjaguin-Landau-Verwey-Overbeek) theory [8].

6.4 Monomolecular Films and LB Technology

The study of monomolecular films is closely related to colloid science because both deal with phenomena occurring at interfaces. Just as the stability of colloidal dispersions depends on interparticle forces, such as electrostatic repulsion and van der Waals attraction, the stability and organization of monolayers are governed by the balance of molecular interactions at the interface. In fact, monolayers can be viewed as two-dimensional analogs of colloidal systems, where intermolecular forces dictate packing, phase behavior, and functionality. Understanding these relationships helps explain how surface chemistry controls colloidal stability, self-assembly, and the design of advanced nanostructured materials.

6.4.1 Monomolecular Films

A monomolecular film, or monolayer, is a single layer of molecules adsorbed or spread at an interface. Such films are usually formed by amphiphilic molecules that possess a hydrophilic (water-attracting) head group and a hydrophobic (water-repelling) tail. At the air-water interface, these molecules spontaneously arrange themselves so that the hydrophobic tails extend into the air, while the hydrophilic heads stay in contact with the water. This spontaneous self-assembly results in a highly ordered, two-dimensional molecular layer.

Key aspects and properties of monomolecular films include the following:

1. **Formation mechanisms:** Monolayers can form spontaneously via adsorption from solution onto a solid surface (e.g., self-assembled monolayers [SAMs]) or by spreading an insoluble amphiphilic substance onto a liquid surface, most commonly water, in a device called a Langmuir trough [9]. In SAMs, molecules dissolved in a solution spontaneously adsorb onto a solid surface and arrange themselves into a highly ordered monolayer. The process is driven by strong interactions between the functional group of the molecule and the substrate, and is widely used in nanotechnology, biosensors, and corrosion protection. In the second one, amphiphilic molecules (e.g., fatty acids and phospholipids) are dissolved in a volatile organic solvent and then carefully spread onto the surface of water. A Langmuir trough, equipped with movable barriers, is used to compress the molecules and study their packing behavior. These are used in biological membranes.
2. **Surface pressure:** The surface pressure (Π) of a monolayer refers to the decrease in the surface tension of the subphase caused by the formation of the film and is expressed as:

$$\Pi = \gamma_0 - \gamma \tag{6.1}$$

where γ_0 is the surface tension of the pure subphase and γ is the surface tension in the presence of the monolayer. Measuring surface pressure as a function of area per molecule (a π-A isotherm) provides crucial information about the molecular packing and phase transitions within the monolayer:

(i) **Molecular orientation and packing:** The orientation of molecules at the interface is governed by their amphiphilic nature. As the molecules are compressed, they undergo phase transitions analogous to 3D bulk phases (gas, liquid-expanded, liquid-condensed, and solid), exhibiting different packing densities and molecular arrangements [10]. The area of a close-packed monolayer is defined as the minimum surface area occupied by a single amphiphilic molecule in a tightly packed monolayer.

(ii) **Properties:** Monolayers exhibit unique properties that make them distinct from their bulk counterparts. Because they are confined to a two-dimensional interface, even slight molecular changes can drastically alter the physical and chemical behavior of the system. For example, a closely packed monolayer can reduce the rate of water evaporation by covering the surface like a molecular shield. Monolayers alter the interaction of a surface with liquids. By carefully choosing the molecular head and tail groups, one can make a surface more hydrophilic (water-attracting) or hydrophobic (water-repelling). They can act as ultrathin lubricating layers that reduce friction between solid surfaces.

6.4.1.1 Applications of Monomolecular Films

Monomolecular films find diverse applications across various fields:

(i) **Lubrication and corrosion inhibition:** Ultrathin films can significantly reduce friction and wear on surfaces or act as barriers against corrosive agents [11]. Their ability to form ordered, smooth, and tightly packed films makes them useful in microelectronics, coatings, and surface treatments where low friction is desired.

(ii) **Coatings:** They can be used to modify surface properties, such as hydrophobicity/hydrophilicity, adhesion, and optical characteristics. A monolayer of hydrophobic molecules can turn glass (normally hydrophilic) into a water-repellent surface.

(iii) **Biological systems:** Monolayers serve as excellent model systems for studying biological membranes, protein-lipid interactions, and drug permeability [12]. Because many biological membranes are composed of amphiphilic molecules like phospholipids, Langmuir monolayers are often used as simplified models for studying membrane structure and dynamics. They allow researchers to examine how proteins, drugs, or nanoparticles interact with membranes under controlled conditions.

(iv) **Drug delivery**: As ultrathin coatings on nanoparticles or drug carriers, they can control release rates and improve biocompatibility.

6.4.2 Langmuir-Blodgett (LB) Films

LB films are a specialized type of monomolecular film that is transferred layer by layer from a liquid surface (typically water) onto a solid substrate [13]. This technique, pioneered by Irving Langmuir and Katharine Blodgett, allows for exquisite control over the thickness and molecular architecture of the deposited film, enabling the fabrication of highly ordered, multilayered structures with precise molecular orientation. In this method, a solid substrate (such as glass, silicon, or metal oxide) is vertically dipped through a compressed monolayer in a Langmuir trough. During immersion and withdrawal, the organized molecular layer adheres to the surface, creating a uniform coating. By repeating this process, multilayer films with controlled thickness and architecture can be fabricated.

The process of creating LB films involves:

1. **Monolayer formation on a Langmuir trough:** An insoluble amphiphilic substance is spread onto the surface of a subphase (usually water) in a Langmuir trough. Barriers compress the molecules, forming a stable, well-packed monolayer at a desired surface pressure.
2. **Substrate immersion and transfer:** A solid substrate (e.g., silicon wafer and glass slide) is then vertically dipped through the monolayer. As the substrate is withdrawn or reimmersed, the monolayer is transferred onto its surface due to the balance of surface tension and adhesive forces.

3. **Layer-by-layer deposition:** By repeatedly dipping and withdrawing the substrate through the monolayer, multiple layers can be deposited, forming a multilayered LB film. The orientation of each layer (Y-type for head-to-head or tail-to-tail, X-type, and Z-type) can be controlled by the dipping protocol.

6.4.2.1 Advantages of LB Films

1. **Precise control over thickness:** Each layer is typically a few nanometers thick, allowing for atomic-scale precision in film thickness.
2. **Molecular organization:** The molecules within each layer are highly ordered and oriented, imparting anisotropic properties to the film.
3. **Layer-by-layer architecture:** The ability to deposit multiple layers allows for the construction of complex, multifunctional thin films with controlled composition and structure.
4. **Versatility:** A wide range of materials, including fatty acids, polymers, proteins, and even nanoparticles, can be incorporated into LB films.

6.4.2.2 Applications of LB Films

The fine control offered by the LB technique makes these films invaluable for advanced applications:

1. **Sensors:** The ordered molecular arrangement enhances sensitivity and selectivity in chemical and biosensors [14]. For example, gas sensors, pH sensors, and immune sensors have been developed using LB films.
2. **Nanotechnology:** LB films are used as templates for nanofabrication, to create ordered nanoparticle arrays, or as components in molecular electronic devices.
3. **Optical devices:** Their controlled thickness and refractive index make them suitable for antireflection coatings, optical filters, and waveguides.
4. **Biomaterials and drug delivery:** LB films can mimic biological membranes, facilitate cell adhesion, or encapsulate active pharmaceutical ingredients for controlled release.
5. **Surface wetting and friction control:** Modifying the surface energy with precise LB coatings can tailor wetting properties or create ultralow friction surfaces.

The manipulation of matter at the molecular scale using monomolecular and LB film techniques underscores the pivotal role of surface chemistry in advancing materials science and engineering.

6.5 Electrical Aspects of Colloidal Stability

In the previous sections, we discussed how interparticle interactions and the organization of molecules at interfaces strongly influence the stability of colloidal systems. In lyophobic (solvent-repelling) colloids, stability is not due to any inherent attraction between the dispersed phase and the medium; instead, it is governed by the interfacial forces acting at the particle-medium boundary. These include short-range attractive forces, such as van der Waals interactions, and longer-range repulsive forces, which are primarily of electrostatic origin.

The delicate balance between these opposing forces determines whether a colloidal system will remain stable (well-dispersed) or become unstable (aggregate and precipitate). In practice, the repulsion arising from the electrical properties of the particle surface is especially important, as it keeps particles from coming close enough for van der Waals attractions to take over. Thus, understanding the electrical aspects of colloidal stability is crucial. Two key concepts form the foundation of this understanding: the EDL, which describes the distribution of ions around a charged colloidal particle, and the zeta potential, which quantifies the effective electrostatic potential at the slipping plane. Together, these parameters allow us to predict, interpret, and manipulate the stability of colloidal dispersions in scientific and industrial applications.

6.5.1 Electric Double Layer (EDL)

When a colloidal particle is dispersed in an electrolyte solution, it acquires a surface charge. To preserve electroneutrality, counterions (oppositely charged ions) from the bulk solution are drawn toward the charged surface, whereas coions (similarly charged ions) are repelled away. This arrangement of charges around the particle forms what is known as the EDL (Figure 6.3) [15].

The EDL is a dynamic interfacial region that extends outward from the particle surface into the surrounding bulk solution, rather than being a fixed structure.

It is typically conceptualized as having two main parts:

1. **Stern layer (or Helmholtz layer):** This is the inner region immediately adjacent to the charged particle surface. It consists of a layer of strongly adsorbed counterions that are strongly adsorbed to the surface, and some may even be specifically adsorbed (chemisorbed). Within this layer, the electrical potential decreases almost linearly with distance from the surface. The inner Helmholtz plane represents the location of specifically adsorbed ions directly bound to the surface, while the outer Helmholtz plane marks the closest position that hydrated ions can approach [16].
2. **Diffuse layer (or Gouy-Chapman layer):** This is the outer region of the EDL, where the counterions are less tightly bound and are distributed diffusely in the solution, extending into the bulk phase. The distribution of ions in this layer is

governed by the combined effects of electrostatic attraction to the charged surface and thermal agitation (Brownian motion). The interplay of these effects produces a diffuse layer of ions whose concentration gradually decreases with increasing distance from the particle surface. Mathematically, the electrical potential within this diffuse region decays roughly exponentially and approaches zero at sufficiently large distances into the bulk solution [17].

The concept of the EDL is central to explaining colloidal stability. As two charged colloidal particles move closer, their diffuse layers overlap, increasing the ion concentration in the intervening region. This results in an osmotic pressure gradient that draws solvent into the gap, pushing the particles apart and producing electrostatic repulsion. The extent of the diffuse layer is described by the Debye length (κ^{-1}), which decreases with increasing ionic strength of the solution [18]:

(i) In dilute solutions, the Debye length is relatively large, meaning the diffuse layer extends further into the solution.
(ii) In concentrated electrolyte solutions, the Debye length becomes shorter, causing the potential to decay more rapidly.

6.5.2 Zeta Potential (ζ-Potential)

The surface potential corresponds to the electrical potential at the particle's surface, but direct measurement of this quantity is often challenging. A more practically measurable and highly significant parameter related to colloidal stability is the zeta potential (ζ) [19].

Zeta potential is the electrical potential measured at the slipping (shear) plane, which separates the compact Stern layer, where ions are firmly associated with the particle surface from the diffuse layer, where ions and solvent molecules remain mobile relative to the particle. When a charged particle moves through a liquid (e.g., under an applied electric field), the ions and solvent molecules within the diffuse layer beyond the slipping plane are carried along with the particle. The zeta potential, therefore, reflects the effective charge of the particle as it moves through the medium.

6.5.2.1 Significance of Zeta Potential

The magnitude of the zeta potential directly reflects the strength of electrostatic repulsion between like-charged particles in a dispersion, making it a crucial parameter for predicting colloidal stability:

(i) **High zeta potential (typically >±30 mV):** Indicates a strong electrostatic repulsion between particles. This repulsion is sufficient to overcome the attractive

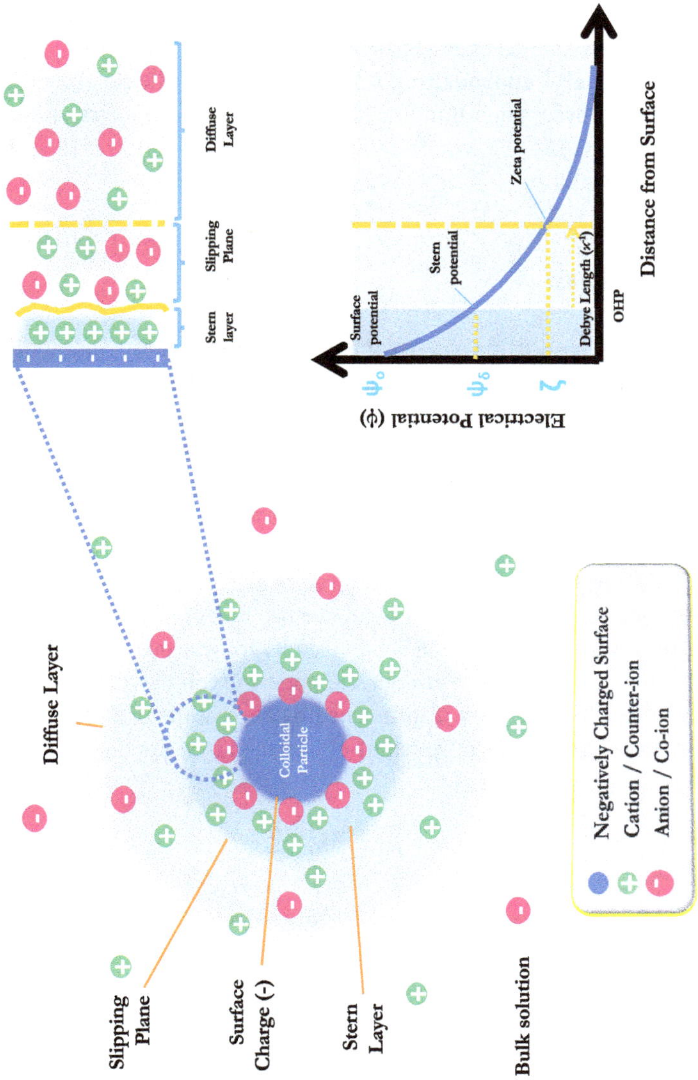

Figure 6.3: Schematic illustration of the electrical double layer (EDL) developed over negatively charged colloidal particles.

forces (like van der Waals forces), leading to a stable dispersion where particles remain individually dispersed and resist aggregation.

(ii) **Low zeta potential (typically approaching 0 mV):** Suggests weak electrostatic repulsion. The attractive forces then dominate, leading to particle aggregation, flocculation, or coagulation. The IEP, where the zeta potential is zero, represents the point of maximum instability.

6.5.2.2 Measurement Techniques

Zeta potential is typically measured using electrokinetic phenomena, most commonly electrophoresis. In this technique, an electric field is applied to a colloidal dispersion, prompting charged particles to move toward the oppositely charged electrode. The migration velocity is proportional to the zeta potential, and by measuring this velocity (electrophoretic mobility), the zeta potential can be determined using models such as the Henry equation [20] [21]. Other techniques include electro-osmosis, streaming potential, and sedimentation potential.

The freely suspended charged particles of an electrolyte are attracted toward the oppositely charged electrodes on the application of an electric field. The movement of these suspended particles toward the electrode is opposed by the viscous force of the medium. When equilibrium is established between the opposing forces, the particles migrate at a constant velocity. This velocity is influenced by several factors, including the zeta potential, viscosity, and dielectric constant of the medium, applied voltage gradient, and the strength of the electric field:

$$\text{Electrophoretic mobility, } U_e = \frac{2\varepsilon\zeta f(\kappa a)}{3\eta} \tag{6.2}$$

The velocity of a particle per unit electric field is known as electrophoretic mobility (U_e). Henry's equation provides the relation between electrophoretic mobility and zeta potential as shown above, where ε is the dielectric constant, η is the viscosity of the medium, and $f(\kappa a)$ is the Henry function. The parameter κ is the Debye length, which means the measure of the thickness of the double layer. In other words, the Debye length is defined as the range around a charged particle within which its electric field influences the arrangement of free ions in the surrounding solution. Its unit is the reciprocal of length. It is affected by the concentration of monovalent salts, e.g. NaCl, as with the increase in concentration of the salt, which leads to more ionic shielding, and as a result, the Debye length or the thickness of the electrical double layer decreases. The parameter "a" denotes the radius of the particle, and hence κa is the ratio of the particle radius and the Debye length.

Moreover, there are two limiting cases regarding the size of Henry's function (κa), i.e. for large κa (>200) and for very small κa (<0.1) (Figure 6.4). Under the conditions when the particle radius is smaller than the Debye length, then Henry's function is reduced to Huckle limit of 1, whereas when the particle radius is larger than the Debye length, then Henry's function follows Smoluchowski limit of 1.5 for aqueous solution, and Henry's equation reduces to Smoluchowski equation as follows:

$$U_e = \frac{\varepsilon \zeta}{\eta} \tag{6.3}$$

This equation is valid for the nanosized particles (≪100 nm) or for dilute electrolytes with concentration <10 to 5 M. This equation has less application for normal aqueous suspensions, whereas the Huckle equation is useful for the nonaqueous solutions.

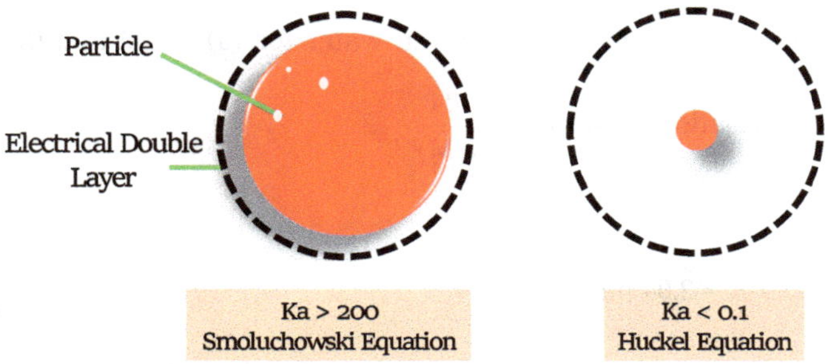

Figure 6.4: Schematic representation of the limiting case of Henry's function, where κ^{-1} is the Debye length and "*a*" is the particle radius.

The Debye length can be measured using the mentioned relation, where ε_0 is the permittivity of free space (8.854×10^{-12} Fm^{-1}), ε_r is the dielectric constant, k is the Boltzmann constant (1.6022×10^{-19} C), I is the ionic strength in terms of molarity, and N_A is the Avogadro's number (6.022×10^{23} mol^{-1}):

$$\kappa^{-1} = \left(\frac{\varepsilon_0 \varepsilon_r kT}{2{,}000\, e^2 I N_A} \right)^{\frac{1}{2}} \tag{6.4}$$

where the ionic strength (I) can be measured from the following equation in terms of molarity. In this equation, C_i is the measure of the concentration of ionic species, and Z_i is the valency:

$$I = \frac{1}{2} \sum C_i Z_i^2 \tag{6.5}$$

The Debye length can be reduced to the following form for water at 298 K:

$$\kappa^{-1} = \frac{0.304}{\sqrt{I(M)}} \tag{6.6}$$

6.5.2.3 Numerical Problems

Problem 1: A latex in 1 mM KCl has electrophoretic mobility, $\mu_e = -3.0 \times 10^{-8}$ m^{2}V^{-1} s^{-1}; using viscosity $\eta = 1.0 \times 10^{-3}$ Pa s permittivity $\varepsilon_r = 78.5$, and $\varepsilon_0 = 8.854 \times 10^{-12}$ Fm^{-1}. Calculate zeta potential (ζ).

Solution: Water permittivity:

$$\varepsilon = \varepsilon_r \times \varepsilon_0 = 78.5 \times 8.854 \times 10^{-12} = 6.95 \times 10^{-10}\ \text{Fm}^{-1}$$

Zeta potential using Smoluchowski formula:

$$\zeta = \frac{\mu_e \eta}{\varepsilon} = (-3.0 \times 10^{-8} \times 1.0 \times 10^{-3})/(6.95 \times 10^{-10}) = -0.0432\ \text{V} = -43.2\ \text{mV}$$

Problem 2: At 25 °C, calculate the Debye length κ^{-1} in water for (a) 1 mM and (b) 10 mM 1:1 electrolyte.

Solution: Using the practical relation $\kappa^{-1}\ (\text{nm}) = \dfrac{0.304}{\sqrt{I(\text{M})}}$

(a) $I = 0.001$

$$\kappa^{-1} = 0.304 / \sqrt{0.001} = 9.61\ \text{nm}$$

(b) I = 0.010

$$\kappa^{-1} = 0.304 / \sqrt{0.010} = 3.04\ \text{nm}$$

Problem 3: At 25 °C, a nanoparticle has $\mu_e = -2.5 \times 10^{-8}$ m^{2}V^{-1} s^{-1} in 10 mM KCl, viscosity 1.00×10^{-3} Pa s, and permittivity 6.95×10^{-10} Fm^{-1}. Compute the zeta potential and verify that the Debye length is of the order of a few nanometers for 10 mM.

Solution: 1. Zeta potential:

$$\zeta = \frac{\mu_e \eta}{\varepsilon} = \frac{-2.5 \times 10^{-8} \times 1.0 \times 10^{-3}}{6.95 \times 10^{-10}} = -3.60 \times 10^{-2}\ \text{V}$$

$$= -36\ \text{mV}$$

2. Debye length (10 nM):

$$\kappa^{-1}\ (\text{nm}) = \frac{0.304}{\sqrt{I(\text{M})}}$$

$$= 3.04\ \text{nm}$$

check: $|\zeta| \sim 25$–50 mV with κ^{-1} few nm is a typical number and is self-consistent.

6.5.3 Electrostatic Repulsion and DLVO Theory

Electrostatic repulsion is the primary mechanism by which similarly charged colloidal particles prevent aggregation. The presence of overlapping EDLs creates a repulsive force barrier that must be overcome for particles to come into sufficiently close contact to aggregate.

The widely accepted DLVO theory provides a quantitative framework for understanding the stability of lyophobic colloids by considering the total interaction energy

(V_T) between two approaching particles as the sum of two primary components [8, 21]:

$$V_T = V_A + V_R \tag{6.7}$$

where V_A is the attractive energy, primarily due to van der Waals forces. These are always attractive and act over relatively short distances. V_R is the repulsive energy, primarily due to electrostatic repulsion arising from the overlap of the EDLs. This force is repulsive for similarly charged particles.

A typical DLVO energy profile plots the total interaction energy as a function of the distance between particles. This plot usually reveals:

(i) **A primary minimum:** A deep, attractive minimum at very short distances, where particles can aggregate irreversibly if the repulsive barrier is overcome.
(ii) **A primary maximum (energy barrier):** A repulsive maximum (the "energy barrier") at intermediate distances. If this barrier is sufficiently high (>15–20 kT, where k is Boltzmann's constant and T is temperature), it prevents particles from reaching the primary minimum, thus providing colloidal stability.
(iii) **A secondary minimum:** A shallower attractive minimum at larger distances, which can lead to reversible flocculation.

The magnitude of the repulsive energy barrier is strongly influenced by both the zeta potential and the ionic strength of the solution. A higher zeta potential leads to a taller barrier, promoting stability. Conversely, increasing ionic strength compresses the diffuse layer, lowers the barrier, and can lead to coagulation (critical coagulation concentration). The DLVO theory, therefore, elegantly explains the "salt effect" on colloidal stability.

In summary, the electrical aspects, particularly the EDL and zeta potential, are fundamental to understanding and controlling the electrostatic repulsive forces that are crucial in maintaining the stability of many colloidal dispersions.

6.6 Balancing Attractive and Repulsive Forces for Colloidal Stability

The persistent stability of a colloidal dispersion, meaning the dispersed particles remain separate and evenly distributed within the continuous medium, is not a passive state but rather a dynamic equilibrium resulting from the delicate balance between various attractive and repulsive forces. While electrostatic repulsion is a major player, other forces also significantly influence interparticle interactions and overall colloidal behavior.

6.6.1 Attractive Forces

Several types of attractive forces act between colloidal particles, tending to draw them together and promote aggregation:

1. **Van der Waals forces:** These are universal, short-range attractive forces that originate from transient fluctuations in electron distribution, which generate instantaneous dipoles, and, in turn, induce corresponding dipoles in adjacent molecules or particles [22]. They are always attractive and are present between any two atoms or molecules. In the context of colloids, these forces are additive over the entire volume of the particles, meaning they can become significant for larger particles. The magnitude of van der Waals attraction is influenced by the Hamaker constant, which depends on the materials involved and the intervening medium.
2. **Hydrophobic interactions:** In aqueous systems, hydrophobic particles or regions of molecules tend to minimize their contact with water. This "desire" to exclude water drives hydrophobic surfaces to associate, effectively reducing the total surface area exposed to water. This is a significant attractive force for nonpolar particles or regions in an aqueous dispersion [23].
3. **Capillary forces:** In the presence of a liquid bridge between particles (e.g., due to incomplete wetting or evaporation), capillary forces can exert a strong attractive pull, leading to aggregation. This is particularly relevant in drying processes or at high particle concentrations.
4. **Bridging flocculation:** If polymers or macromolecules are present in the dispersion, they can adsorb onto multiple particles simultaneously, forming bridges that link the particles together and lead to flocculation. This mechanism is often intentionally employed for particle removal in wastewater treatment [24].

6.6.2 Repulsive Forces

To counteract these attractive forces and maintain stability, several types of repulsive forces can be engineered or naturally arise in colloidal systems:

1. **Electrostatic repulsion:** As extensively discussed in Section 6.5, this force arises from the overlap of EDLs around similarly charged particles. It provides a significant energy barrier that prevents particles from approaching closely enough for the strong primary minimum (due to van der Waals forces) to dominate.
2. **Steric repulsion:** This stabilization mechanism relies on the adsorption or grafting of nonadsorbing polymer chains onto colloidal particle surfaces [25]. As two polymer-coated particles come close, their polymer layers start to overlap. This overlap results in two main repulsive effects:

(i) **Osmotic (mixing) effect:** As the polymer segments interpenetrate, the local concentration of polymer increases in the overlap region, leading to a higher osmotic pressure and a net influx of solvent, which pushes the particles apart.
(ii) **Volume restriction (elastic) effect:** Further compression of the polymer layers restricts the conformational freedom of the polymer chains. This reduction in entropy leads to a repulsive elastic force that resists further compression. Steric stabilization is generally less sensitive to ionic strength and pH changes, compared to electrostatic stabilization, making it a robust stabilization mechanism.

3. **Electrosteric stabilization:** This combines aspects of both electrostatic and steric stabilization. It typically involves the adsorption of charged polyelectrolytes onto the particle surface. The polyelectrolytes provide both electrostatic repulsion (due to their charge) and steric repulsion (due to their polymeric chain extending into the solvent) [26]. This dual mechanism often results in highly stable dispersions, even in high ionic strength environments.
4. **Depletion flocculation:** In contrast to steric stabilization, if nonadsorbing polymers are present in the bulk solution at high enough concentrations, they can cause an attractive force between particles. This occurs because the polymers are excluded from the narrow gap between closely approaching particles, leading to a region of lower polymer concentration and higher solvent chemical potential. This results in an osmotic pressure imbalance that pushes the particles together. While a form of attraction, understanding depletion forces is important for controlling colloidal stability.

6.6.3 Mechanisms of Stabilization and Destabilization

Controlling the balance of these forces is paramount for either stabilizing a colloidal dispersion or intentionally destabilizing it for separation processes.

6.6.3.1 Stabilization Mechanisms

1. **Electrostatic stabilization:** This is accomplished by imparting a sufficiently high surface charge and zeta potential to the particles, thereby generating a strong electrostatic repulsive barrier. Higher magnitudes ($|\zeta| > 30$ mV) favor stability, while values near zero promote coagulation. This often involves adjusting the pH or adding specific electrolytes that adsorb onto the surface.
2. **Steric stabilization:** It is achieved by adsorbing or grafting suitable polymer chains onto the particle surface, creating a physical barrier to aggregation. Surfactants and polymers may enhance stability by providing additional electrostatic or steric repulsion.

3. **Electrosteric stabilization:** A combination of both, utilizing charged polymers for robust stability.

6.6.3.2 Destabilization Phenomena

When repulsive forces are insufficient to overcome attractive forces, colloidal systems can undergo various destabilization processes. These processes determine how particles associate and separate from the dispersion medium, leading to visible changes such as sedimentation, creaming, or phase separation. The major destabilization pathways include:

(i) **Coagulation:** This refers to the irreversible aggregation of colloidal particles into larger, compact masses. It generally occurs when the electrostatic double layer is suppressed or collapsed, thereby eliminating the repulsive forces that normally keep particles apart. A common cause is the addition of high concentrations of electrolytes or multivalent ions, which neutralize surface charges and lower the energy barrier for aggregation. Once coagulated, the system often loses its ability to be redispersed, leading to permanent phase separation.

(ii) **Flocculation:** This is the reversible aggregation of colloidal particles into loosely bound, open-structured clusters known as flocs. Unlike coagulation, the interparticle attractions in flocculation are relatively weak, typically arising from polymer bridging, van der Waals attractions in the secondary minimum of the DLVO potential, or depletion forces. Because the forces holding flocs together are not strong enough to cause permanent fusion, they can often be redispersed by gentle agitation or dilution. Flocculation is a crucial process in water treatment, food stabilization, and pharmaceuticals, where controlled aggregation is desirable for efficient separation or product stability. The difference between these two processes is tabulated in Table 6.1.

(iii) **Creaming:** This phenomenon refers to the upward movement of dispersed phase droplets or particles within a dispersion, leading to the formation of a concentrated layer at the top of the continuous phase. Creaming occurs primarily due to density differences between the dispersed and continuous phases (e.g., oil droplets rising in oil-in-water emulsions) and is further influenced by particle size, viscosity of the medium, and gravitational forces. While creaming does not necessarily result in irreversible aggregation, it can significantly impact the appearance, uniformity, and stability of colloidal systems. Importantly, creamed systems can sometimes be redispersed upon shaking or homogenization, but prolonged creaming may eventually progress to coalescence or phase separation.

(iv) **Sedimentation:** This process refers to the settling of dispersed particles or droplets under the influence of gravity, resulting in their accumulation at the bottom of the container. Sedimentation generally takes place when the dispersed phase is denser than the surrounding medium (e.g., clay particles in water). The rate of sedimentation depends on factors such as particle size, density difference between the phases, and viscosity of the medium, often described by Stokes' law for spherical particles. Similar to creaming, sedimentation does not necessarily imply permanent destabilization, as particles may be redispersed by agitation. However, if particles undergo aggregation or coalescence during sedimentation, it can lead to irreversible phase separation and loss of colloidal stability.

(v) **Ostwald ripening:** This is a gradual destabilization process in which larger particles or droplets grow at the expense of smaller ones. The driving force is the difference in solubility arising from surface curvature: smaller particles, having a higher surface energy, are more soluble than larger ones. As a result, molecules diffuse from the smaller particles into the continuous phase and redeposit onto larger particles, leading to an overall increase in the average particle size with time. This phenomenon is particularly common in emulsions, foams, and nanoparticle dispersions, and ultimately promotes phase separation, if left unchecked.

Table 6.1: Difference between coagulation and flocculation.

Aspect	Coagulation	Flocculation
Definition	Coagulation involves destabilizing suspended or colloidal particles by neutralizing the electrical charges that facilitates to maintain their spacing	Flocculation involves lightly mixing the destabilized particles to foster their agglomeration into larger settleable flocs
Mechanism	Charge neutralization and particle destabilization through the addition of chemicals called coagulants (alum, ferric salts, etc.)	Bridging and aggregation of destabilized particles through slow mixing, often aided by polymers (flocculants)
Treatment stage	First stage, during which chemical reactions take place to overcome particle repulsion	The following stage, in which physical contacts under controlled agitation develop large flocs
Intensity of mixing	Vigorous and quick mixing to distribute the coagulant evenly	Slow and gentle mixing to permit flocs to develop without disintegration
Time required	Very short (seconds to a few minutes)	Relatively longer (15–30 min, depending on the system)
Outcome	Microfloc or destabilized particle formation	Visible, larger, settleable, or filterable floc formation

6.6.4 Industrial and Biological Applications of Stability Control

The ability to manipulate colloidal stability is critical across a vast array of industries and biological systems:

1. **Drug delivery:** Nanoparticle-based drug delivery systems rely heavily on controlled stability to ensure drugs remain dispersed in physiological fluids, reach target sites without premature aggregation, and release their cargo effectively. Surface modification with polymers (e.g., PEGylation) is often used for steric stabilization and stealth properties [27].
2. **Food emulsions:** Products like milk, mayonnaise, and salad dressings are complex oil-in-water or water-in-oil emulsions. Their stability, texture, and shelf-life depend on the careful balance of attractive and repulsive forces mediated by proteins, polysaccharides, and emulsifiers that adsorb at the oil-water interface [28].
3. **Paints and coatings:** Paints are dispersions of pigment particles in a liquid medium. Achieving long-term stability against sedimentation or flocculation is essential for uniform color and consistency. Surfactants and polymers are used to provide electrostatic or steric stabilization.
4. **Ceramics processing:** In the manufacture of advanced ceramics, suspensions of ceramic powders must be highly stable and homogeneous to ensure uniform green body formation and defect-free final products. Controlling the surface charge and rheology of ceramic slurries is paramount [29].
5. **Cosmetics and personal care:** Lotions, creams, and shampoos are often emulsions or suspensions whose stability determines their appearance, feel, and efficacy.
6. **Wastewater treatment:** Conversely, in wastewater treatment, the goal is often to destabilize colloidal impurities (e.g., suspended solids and oil droplets) to promote their aggregation and subsequent removal by sedimentation or filtration. This is achieved by adding coagulants (e.g., aluminum sulfate and ferric chloride) or flocculants (e.g., polyelectrolytes) that neutralize surface charge or bridge particles [30].

The mastery of balancing attractive and repulsive forces allows for the rational design of colloidal systems with tailored stability profiles, from long-term dispersions to controlled aggregation, enabling their diverse functionalities.

6.7 Recent Advancements and Future Directions in Colloidal Surface Chemistry

Colloidal surface chemistry has witnessed remarkable progress in recent years, largely due to the advent of advanced characterization methods, the integration of computational modeling, and the increasing demand for tailored colloidal systems across diverse technologies. These advancements have not only deepened the understanding of interfacial phenomena but also expanded the scope of colloid-based applications in medicine, energy, environment, and materials science. Colloidal surface chemistry plays a pivotal role in enabling innovative applications across frontier technologies:

1. **Nanomedicine and theranostics:** Colloidal nanoparticles are indispensable for drug delivery, bioimaging, and diagnostics. Their performance strongly depends on surface properties such as charge, functionalization, and biocompatibility. Smart coatings and targeting ligands improve circulation time, cellular uptake, and site-specific action, making them critical for next-generation therapeutics [31].
2. **Advanced materials and metamaterials:** Colloidal self-assembly is a bottom-up approach to creating ordered architectures with tailored optical, electronic, or mechanical properties. Examples include colloidal crystals that serve as photonic bandgap materials and building blocks for metamaterials [32].
3. **Environmental remediation:** Functionalized colloids and nanoparticles are being designed for pollutant removal from water and air. Applications include iron oxide nanoparticles for heavy metal removal and titanium dioxide photocatalysts for the degradation of organic contaminants [33].
4. **Energy applications:** Colloids are integral in solar energy harvesting (quantum dot solar cells and dye-sensitized cells), energy storage (colloidal electrolytes in batteries), and catalysis (nanocatalysts with enhanced surface activity). Surface functionalization directly influences their efficiency and longevity.
5. **Soft robotics and microfluidics:** Colloidal assemblies are being explored for constructing dynamic, reconfigurable materials essential in soft robotics. In microfluidics, tailored colloids enhance fluid manipulation, separation processes, and sensing capabilities [34].

The future of colloidal surface chemistry lies in integrating multiscale characterization, predictive computational modeling, and advanced synthetic strategies [35]. Key directions include:

(i) Designing multifunctional, sustainable, and biocompatible colloids.
(ii) Developing hybrid systems that combine responsiveness to multiple stimuli.
(iii) Applying machine learning and simulation tools to predict colloidal behavior under complex environments.
(iv) Expanding applications in precision medicine, green technologies, and adaptive materials.

With these advances, colloidal systems are expected to play an increasingly central role in next-generation technologies, bridging fundamental science with practical innovation.

Practice Questions

1. Define a colloid and give one example.
2. What is the typical size range of colloidal particles?
3. Name two types of monomolecular films.
4. What is zeta potential?
5. Mention one application of LB films.
6. What are van der Waals forces?
7. Define the Stern layer in the electric double layer.
8. What is an isoelectric point?
9. List any two methods of destabilizing a colloidal system.
10. What are the various industrial applications of colloidal surface chemistry?
11. Why do colloidal particles not settle under gravity?
12. Explain why increasing ionic strength reduces colloidal stability.
13. Why does pH affect the zeta potential of a colloidal system?
14. Why is steric stabilization less sensitive to pH changes than electrostatic stabilization?
15. Why does temperature affect the behavior of PNIPAM-based smart colloids?
16. Why is the LB technique preferred for molecular-level control over thin film deposition?
17. Differentiate between true solution, colloid, and suspension with examples.
18. Describe the formation mechanism and structure of monomolecular films.
19. Explain the structure and significance of the electric double layer.

20. Discuss the role of zeta potential in predicting colloidal stability.

21. Elaborate on the DLVO theory with an appropriate energy-distance diagram. Explain electrostatic and steric stabilization mechanisms with examples.

22. Describe how bridging flocculation occurs in colloidal systems.

23. Explain electrostatic and steric stabilization mechanisms with examples.

24. Explain the principle and process of forming LB films.

25. What are the key forces involved in the stability of colloids? Explain their balance.

26. Write a short note on the application of surface chemistry in drug delivery or environmental remediation.

27. A colloidal particle has an electrophoretic mobility of 4.5×10^{-8} $m^2\ V^{-1}\ s^{-1}$ in a medium of viscosity 0.001 Pa s and dielectric constant 78.5. Calculate the zeta potential using Henry's equation (assume Smoluchowski approximation).

28. Given: Hamaker constant for a system is 10^{-20} J. Calculate the van der Waals attraction energy between two particles of radius 50 nm at a distance of 2 nm.

29. Calculate the total surface area of 1 cm^3 of solid divided into particles of diameter 10 nm (assume spherical particles).

30. A colloidal dispersion has a particle number concentration of 10^{12} particles per mL. If each particle is coated with a polymer layer of thickness 2 nm, estimate the total volume occupied by the polymer in 1 mL of dispersion.

Hints for Practice Questions

Question No. 27:
Given: Electrophoretic mobility, $U_e = 4.5 \times 10^{-8}\ m^2(Vs)^{-1}$
Viscosity, $\eta = 0.001$ Pa s
Dielectric constant, $\varepsilon_r = 78.5$
The Smoluchowski equation is as follows:

$$U_e = \frac{\varepsilon\,\zeta}{\eta}$$

where $\varepsilon = \varepsilon_r \varepsilon_0$ and $\varepsilon_0 = 8.854 \times 10^{-12}$ F m^{-1}

1. Calculate the permittivity of the medium

$$\varepsilon = \varepsilon_r \varepsilon_0 = 78.5 \times 8.854 \times 10^{-12} = 6.95039 \times 10^{-10}\ \text{F m}^{-1}$$

2. Calculation of zeta potential:

$$\zeta = \frac{U_e \eta}{\varepsilon}$$

$$\zeta = \frac{(4.5 \times 10^{-8}) * 0.001}{6.95039 * 10^{-10}}$$

$$\zeta = 0.0647446\ \text{V}$$

Answer: ≈0.0647 V = 64.7 mV (positive, since mobility given positive).

Question No. 28:

Solution: Given: Hamaker constant $A_H = 1.0 \times 10^{-20}$ J

Particle radius, $a = 50$ nm $= 5.0 \times 10^{-9}$ m

Surface-surface separation, $h = 2$ nm $= 2.0 \times 10^{-9}$ m

Using (Derjaguin/short-range approximation for two equal spheres, $h \ll a$)

$$U \approx -\frac{A_H a}{12h}$$

$$U = -\frac{(1.0 \times 10^{-20}) \times (5.0 \times 10^{-9})}{12 \times (2.0 \times 10^{-9})}$$

$$U = -2.0833333 \times 10^{-21}$$

Answer: $U = -2.08 \times 10^{-21}$

Question No. 29:

Given: Total solid volume $V_{tot} = 1\ \text{cm}^3 = 1.0 \times 10^{-6}\ \text{m}^3$

Particle diameter, $d = 10$ nm

$r = 5.0 \times 10^{-9}$ m.

Number of particles, $N = \frac{V_{tot}}{V_{particle}} = \frac{V_{tot}}{\frac{4}{3}\pi r^3}$

Total surface area, $A_{tot} = N \times 4\pi r^2$

Therefore, $A_{tot} = \frac{V_{tot}.4\pi r^2}{\frac{4}{3}\pi r^3} = \frac{3V_{tot}}{r}$

$$A_{tot} = \frac{3 \times (1.0 \times 10^{-6})}{5.0 \times 10^{-9}}$$

$$A_{tot} = 0.6 \times 10^3 = 600\ \text{m}^2$$

Question No. 30:

Given: Particle number concentration, $N = 10^{12}$ particles per mL,
Polymer thickness, $t = 2$ nm $= 2.0 \times 10^{-9}$ m
Particle core diameter = 10 nm; radius, $r = 5.0 \times 10^{-9}$ m

Volume of polymer shell per particle: $V_{\text{shell}} = \frac{4}{3}\pi\left[(r+t)^3 - r^3\right]$

$$V_{\text{shell}} = \frac{4}{3} \times 3.14\left[\left(5.0\times10^{-9} + 2.0\times10^{-9}\right)^3 - \left(5.0\times10^{-9}\right)^3\right]$$

$$V_{\text{shell}} = \frac{4}{3} \times 3.14\left[\left(7.0\times10^{-9}\right)^3 - \left(5.0\times10^{-9}\right)^3\right]$$

$$V_{\text{shell}} = \frac{4}{3} \times 3.14\left[\left(3.430\times10^{-25}\right) - \left(1.25\times10^{-25}\right)\right]$$

$$V_{\text{shell}} = \frac{4}{3} \times 3.14\left[2.18\times10^{-25}\right]$$

$$V_{\text{shell}} = 9.13\times10^{-25}\ \text{m}^3 \text{ per particle}$$

Total polymer volume in 1 mL: $V_{\text{poly, tot}} = N \times V_{\text{shell}} = 10^{12} \times 9.13 \times 10^{-25} = 9.13 \times 10^{-13}$ m^3
1 mL = 1.0×10^{-6} m^3
So, in mL:

$$V_{\text{poly, tot}} = \frac{9.13\times10^{-13}}{1\times10^{-6}} = 9.13 \times 10^{-7}\ \text{mL}$$

Answer: Total polymer volume ≈ 9.13×10^{-13} m^3 = 9.13×10^{-7} mL

References

[1] Hunter RJ. *Zeta potential in colloid science: Principles and applications*. Academic Press; 2013 Sep 3.
[2] Hiemenz PC. *Principles of colloid and surface chemistry. Electrophoresis and other electrokinetic phenomena*. 1977.
[3] Everett DH. *Basic principles of colloid science*. Royal Society of Chemistry; 2007 Oct 31.
[4] Adamson AW, Gast AP. *Physical chemistry of surfaces*. New York: Interscience Publishers; 1967.
[5] Israelachvili JN. *Intermolecular and surface forces*. Academic Press; 2011 Jul 22.
[6] Lyklema J. *Fundamentals of interface and colloid science. Vol. II: Solid-liquid interfaces*. Academic Press; 1995.
[7] Verwey EJ, Overbeek JT. Theory of the stability of lyophobic colloids. *J Colloid Sci*. 1955;10(2):224–225.
[8] Derjaguin B, Landau LJ. The theory of stability of highly charged lyophobic sols and coalescence of highly charged particles in electrolyte solutions. *Acta Physicochim URSS*. 1941 Jan;14(633–52):58.
[9] GL G Jr. *Insoluble monolayers at liquid-gas interfaces*. Interscience Publishers; 1966.
[10] Petty MC. *Langmuir-Blodgett films: An introduction*. Cambridge University Press; 1996. p. 252.
[11] Bhushan B. *Principles and applications of tribology*. John Wiley & Sons; 2013 Feb 15.
[12] Tredgold RH. *Order in thin organic films*. Cambridge University Press; 1994 Mar 3.

[13] Blodgett KB. Films built by depositing successive monomolecular layers on a solid surface. *J Am Chem Soc*. 1935 Jun;57(6):1007–1022.

[14] Matharu Z, Bandodkar AJ, Gupta V, Malhotra BD. Fundamentals and application of ordered molecular assemblies to affinity biosensing. *Chem Soc Rev*. 2012 Jan 17;41(3):1363–1402.

[15] Butt HJ, Graf K, Kappl M. *Physics and chemistry of interfaces*. John Wiley & Sons; 2023 Feb 7.

[16] Stern O. The theory of the electrolytic double-layer. *Z Elektrochem*. 1924;30(508):1014–1020.

[17] Gouy MJ. Sur la constitution de la charge électrique à la surface d'un électrolyte. *J Phys Theor Appl*. 1910;9(1):457–468.

[18] Debye P, Hückel E. The theory of electrolytes. I. Freezing point depression and related phenomena [Zur Theorie der Elektrolyte. I. Gefrierpunktserniedrigung und verwandte Erscheinungen. *Phys Z*. 1923 Jul;24:185–206.

[19] Smoluchowski MV *Handbuch der Elektrizität und des Magnetismus. Band II*. Barth-Verlag: Leipzig; 1921. pp. 366–427.

[20] Henry DC. The cataphoresis of suspended particles. Part I. – The equation of cataphoresis. *Proc R Soc Lond A Math Phys Sci*. 1931 Sep 1;133(821):106–129.

[21] Overbeek JT. Thermodynamics of electrokinetic phenomena. *J Colloid Sci*. 1953 Aug 1;8(4):420–427.

[22] Hamaker HC. The London – Van der Waals attraction between spherical particles. *Physica*. 1937 Oct 1;4(10):1058–1072.

[23] Tanford C. *The hydrophobic effect: Formation of micelles and biological membranes*. John Wiley & Sons; 1980.

[24] Gregory J. The role of colloid interactions in solid-liquid separation. *Water Sci Technol*. 1993 May 1;27 (10):1–7.

[25] Napper DH. *Polymeric stabilization of colloidal dispersions*. CRC Press; 1983.

[26] Fleer G, Stuart MC, Scheutjens JM, Cosgrove T, Vincent B. *Polymers at interfaces*. Springer Science & Business Media; 1993 Sep 30.

[27] Gref R, Minamitake Y, Peracchia MT, Langer R. Poly(ethylene glycol)-coated biodegradable nanospheres for intravenous drug administration. In: *Microparticulate systems for the delivery of proteins and vaccines*. CRC Press; 2020 Jul 24. pp.279–306.

[28] Dickinson E. Hydrocolloids at interfaces and the influence on the properties of dispersed systems. *Food Hydrocoll*. 2003 Jan 1;17(1):25–39.

[29] JO C III, Aksay IA. Processing of highly concentrated aqueous α-alumina suspensions stabilized with polyelectrolytes. *J Am Ceram Soc*. 1988 Dec;71(12):1062–1067.

[30] Gregory J, O'Melia CR. Fundamentals of flocculation. *Crit Rev Environ Sci Technol*. 1989 Jan 1;19 (3):185–230.

[31] Lu AH, Salabas EE, Schüth F. Magnetic nanoparticles: Synthesis, protection, functionalization, and application. *Angew Chem Int Ed*. 2007 Feb 12;46(8):1222–1244.

[32] Yin Y, Lu Y, Gates B, Xia Y. Template-assisted self-assembly: A practical route to complex aggregates of monodispersed colloids with well-defined sizes, shapes, and structures. *J Am Chem Soc*. 2001 Sep 12;123(36):8718–8729.

[33] Qu X, Alvarez PJ, Li Q. Photochemical transformation of carboxylated multiwalled carbon nanotubes: Role of reactive oxygen species. *Environ Sci Technol*. 2013 Dec 17;47(24):14080–14088.

[34] Whitesides GM. Soft robotics. *Angew Chem Int Ed*. 2018 Apr 9;57(16):4258–4273.

[35] Johnson E, Koh A. Recent advances in smart emulsion materials: From synthesis to applications. *Adv Eng Mater*. 2024 Dec;26(24):2400995.

Harsh Kumar Rai, Nilesh Singh, Rashika, Tannavi Badhan, Deepak Yadav and Sushmita*

7 Application of Colloidal Chemistry

Abstract: Colloids are special types of heterogeneous mixtures made up of two components: the dispersed phase (the particles) and the dispersion medium (the substance in which the particles are distributed). Depending on the physical states of these two phases, colloids can appear in different forms such as aerosols, emulsions, foams, sols, and suspensions. Hydrocolloids, which are mainly derived from polysaccharides and proteins, are widely recognized for their capacity to alter the texture and flow behavior of aqueous systems. They can increase viscosity or promote gel formation, thereby functioning as versatile additives across different industries. Their applications range from stabilizing and emulsifying mixtures to regulating crystallization and improving rheological properties, making them indispensable in sectors such as food technology, pharmaceuticals, cosmetics, and other industrial formulations.

This chapter highlights the broad spectrum of applications of colloidal chemistry in modern science and technology. Specific focus is given to its role in petroleum recovery, drug delivery and formulation, municipal water purification, detergency and cleaning action, coatings, and photographic processes. Collectively, these applications underscore the significance of colloidal systems in bridging fundamental chemistry with practical utility across multiple sectors.

7.1 Introduction

Colloidal systems consist of very small particles, usually between 1 nm and 1 µm in size, dispersed in another medium. Because these particles have a very large surface area compared to their volume, they possess high surface energy. This gives them properties that are quite different from the same material in bulk form or as individual molecules.

This special behavior explains why colloidal chemistry is important in science, technology, and industry. Among colloidal systems, hydrocolloids are especially significant. These are long-chain polymers, commonly polysaccharides or proteins that easily disperse in water. When mixed with water, hydrocolloids can greatly change the flow and texture of the system. They can increase viscosity, form gels, stabilize emulsions, and regulate crystal formation [1]. Due to these functions, hydrocolloids are widely applied in food products, medicines, cosmetics, and even in water treatment.

https://doi.org/10.1515/9783112208236-007

7.1.1 Functional Roles of Colloidal Systems

1. Helping in the stabilization or breakdown of emulsions.
2. Forming gels and adjusting viscosity.
3. Controlling crystallization and flow properties.
4. Assisting in separation and purification processes.

7.1.2 Industrial Relevance

The versatility of colloidal chemistry comes from the fact that its properties can be adjusted depending on the dispersed phase and the dispersion medium. Because of this, colloids are useful in a wide range of fields:

1. **Petroleum industry**: Specially designed colloidal systems help to stabilize heavy oil emulsions. This reduces viscosity and allows oil to flow more easily through rock formations. Polymeric surfactants are often used to improve emulsion stability and process efficiency.
2. **Pharmaceuticals**: Colloids are used in emulsions, suspensions, and combined formulations to increase drug solubility, improve bioavailability, and even enable targeted drug delivery.
3. **Water and wastewater treatment**: Coagulants and flocculants are applied to remove suspended colloidal particles, making water cleaner and safer.
4. **Materials science**: Products such as latex coatings, detergents, fire-fighting foams, and photographic films depend on colloidal formulations for better performance.
5. **Textiles and leather**: Colloidal solutions are applied in processes like dyeing, tanning, and finishing to improve product quality.
6. **Environmental applications**: Electrostatic precipitators (ESPs) help remove colloidal particles from industrial exhaust gases. Natural processes like river delta formation and cloud seeding are also explained through colloidal behavior.

Beyond these examples, colloidal chemistry plays an important role in sustainable technologies, healthcare advancements, and environmental protection. The ability to design and control colloidal systems with specific properties is also opening new possibilities in nanotechnology, advanced materials, and green industry.

7.2 Petroleum Industry

The petroleum industry is one of the most important sectors where colloidal chemistry has wide-ranging applications. From crude oil extraction in deep reservoirs to its processing, transport, and refining, colloidal systems, especially emulsions and dispersions, play a key role in efficiency and productivity.

Heavy oils and bitumen, because of their high density and viscosity, readily form stable emulsions with water. These emulsions can have two opposite effects: under controlled conditions, they can assist in enhanced oil recovery (EOR), but if uncontrolled, they create major operational problems. For this reason, the ability to understand, control, and manipulate colloidal behavior is crucial for improving oil recovery, lowering processing costs, ensuring smooth transport, and designing eco-friendly remediation methods.

7.2.1 Heavy Oil and Emulsion Formation

Heavy oil is defined as crude oil with an American Petroleum Institute (API) gravity of 20° or less, making it naturally dense and viscous. In different stages of petroleum recovery and processing, heavy oil frequently forms emulsions with water. These emulsions can be beneficial, for example, in aiding oil recovery, or problematic, as they may hinder separation, transportation, and refining processes.

Beneficial roles in enhanced oil recovery (EOR) are:

1. Emulsification reduces oil viscosity, allowing easier mobilization in reservoir rock.
2. It improves flow dynamics and displacement efficiency, thereby boosting recovery yields.

Operational challenges:

(i) Uncontrolled emulsions hinder separation, foul equipment, and raise chemical treatment costs.

(ii) They disrupt pipeline transport, reduce refining efficiency, and complicate wastewater treatment.

7.2.2 Types and Methods of Emulsion Generation

In petroleum systems, emulsions are generally classified according to the nature of the dispersed phase and the continuous phase. The two most common types are as follows:

1. Oil-in-water (O/W) emulsions: Oil droplets are dispersed within a continuous water medium.
2. Water-in-oil (W/O) emulsions: Water droplets are dispersed in an oil medium.

The stability, properties, and applications of these emulsions depend on the type formed as well as the conditions under which they are generated.

7.2.2.1 Surfactant-Based Emulsions

One of the most effective ways of forming emulsions in petroleum systems is through the use of surfactants (surface-active agents). Surfactants lower the interfacial tension (IFT) between oil and water, which promotes droplet formation and enhances emulsion stability [2]. Depending on their chemical nature, surfactants are classified as follows:

(i) **Anionic surfactants** (e.g., alkyl sulfates and sulfonates): provide strong repulsion between droplets and are effective under alkaline conditions.
(ii) **Cationic surfactants** (e.g., quaternary ammonium salts): adsorb onto negatively charged surfaces and are useful where charge reversal is required in rock-fluid interactions.
(iii) **Nonionic surfactants** (e.g., ethoxylated alcohols and sorbitan esters): highly tolerant to changes in salinity and temperature, making them suitable for harsh reservoir conditions.
(iv) **Zwitterionic surfactants**: contain both positive and negative charges, offering excellent stability and adaptability over a wide pH range.

The main mechanism of surfactants is the reduction of IFT, which not only stabilizes emulsions but also improves the efficiency of EOR processes.

7.2.2.2 Polymeric Emulsions

Polymers are introduced into petroleum systems to modify rheology and control mobility. They anchor surfactant molecules at the oil-water interface, enhance viscosity, and prevent coalescence of droplets. Depending on the compatibility with reservoir conditions, both natural polymers (e.g., xanthan gum and guar gum) and synthetic polymers (e.g., polyacrylamides and block copolymers) are utilized:

(i) They are particularly important in EOR, where emulsion stability under high temperature, pressure, and salinity is required.
(ii) By increasing solution viscosity, polymers improve sweep efficiency and control the channeling of the displacing fluid.

7.2.2.3 Alkaline Flooding and Emulsion Generation

Alkaline chemicals such as sodium carbonate or sodium hydroxide are injected into reservoirs to react with naturally occurring acidic components of crude oil (e.g., naphthenic acids). This reaction generates in situ surfactants, which reduce IFT and stabilize emulsions:

(i) The method is strongly dependent on the chemical compatibility of the alkaline agent with reservoir brine and crude composition.
(ii) It is frequently used in EOR applications, where the in situ formation of surfactants reduces oil viscosity and enhances recovery.

7.2.2.4 Solvent-Induced Emulsions

Certain solvents are introduced to adjust the polarity and miscibility of the petroleum system. Solvents such as ethanol, acetone, and ketones can promote emulsion formation by altering the interfacial properties between oil and water:
(i) Solvent-based emulsification is often used in pipeline transportation, where reduced viscosity improves flow assurance.
(ii) However, the economic feasibility and environmental impact of solvent usage must be carefully evaluated.

7.2.2.5 Solid Particle-Stabilized Emulsions (Pickering Emulsions)

Solid nanoparticles adsorb at the oil-water interface, creating a physical barrier that prevents droplet coalescence. These Pickering emulsions are highly stable and exhibit long-term resistance to breakdown:
(i) Carbon-based nanoparticles (graphene, carbon nanotubes, and fullerenes) provide mechanical stability and conductivity.
(ii) Clay minerals and hydroxides are naturally abundant and inexpensive stabilizers.
(iii) Metal oxides (e.g., silica, alumina, ZnO, and TiO_2) offer tunable surface chemistry and robustness under extreme conditions.
(iv) Polymeric and lipid nanoparticles enable controlled drug delivery and remediation applications, in addition to petroleum use.

This method is gaining importance due to its environmental sustainability, as solid particles can often replace chemical surfactants.

7.2.2.6 Microemulsions

Microemulsions are thermodynamically stable systems formed spontaneously under certain conditions without requiring high shear forces. They consist of nanometer-sized droplets stabilized by surfactants and cosurfactants:
(i) In petroleum recovery, microemulsions provide superior solubilization of crude components and enhanced sweep efficiency.

(ii) Their ability to remain stable under varying pressure and temperature conditions makes them highly effective for EOR.

In certain scenarios, emulsions can form spontaneously without the addition of surfactants. This occurs due to the natural composition of crude oil and the physicochemical conditions of the reservoir, which favor self-assembly. Although less common, such systems are of interest in reducing chemical costs and improving eco-friendliness. The choice of method to be used to generate emulsion is determined by factors such as reservoir conditions, crude oil composition, brine chemistry, temperature, and economic feasibility. Surfactants and polymers are most widely used, while nanoparticle and microemulsion-based techniques represent emerging, sustainable frontiers in petroleum colloid technology. Table 7.1 provides a classification based on stabilizing mechanism and medium.

Table 7.1: Methods of heavy oil emulsion generation.

Methods	Emulsion nature	Emulsion types
Surfactant	Anionic, cationic, nonionic, and zwitterionic	Reduction of interfacial tension (IFT)
Polymeric	–	Based on chemical compatibility with the surrounding fluids and environment
Alkaline	Ionic	Reacts with acidic components of crude oil to form natural surfactants
Solvent	–	Alters polarity and fluid conditions
Solid particles (e.g., nanoparticles)	Metal oxides, clays, and carbon materials	Adsorption of solid particles at oil-water interface
Microemulsion	–	In situ emulsification enhancing sweep efficiency
Others	–	Spontaneous surfactant-free emulsification

7.2.3 Emulsions Across Petroleum Streams

Emulsions are encountered at different stages of petroleum operations, and their role varies significantly depending on whether the system is at the upstream, midstream, or downstream level. The type of emulsion, its mechanism, and the ultimate objective all determine how effectively petroleum production, transport, and processing are carried out:

1. **Upstream (reservoir level):** At the reservoir level, emulsions are deliberately generated to aid in crude oil recovery. Both O/W and W/O emulsions may be employed depending on the geological and fluid conditions [3]. The emulsified sys-

tems reduce crude viscosity, improve sweep efficiency, and assist in trapping or mobilizing oil droplets within the porous rock matrix. The main aim is to enhance the recovery of heavy oil and improve the overall efficiency of EOR techniques.
2. **Midstream (production and transportation):** During production and pipeline transport, emulsions are often managed to improve fluid handling. Here, O/W emulsions are preferred. Formation of O/W emulsions lowers the apparent viscosity of crude oil, thereby facilitating smoother flow through pipelines. This minimizes pumping costs, reduces the risk of blockages, and ensures consistent production rates. The main objective is to simplify petroleum production and guarantee reliable long-distance transportation through pipelines.
3. **Downstream (refining and treatment):** In downstream operations, emulsions are usually undesirable, as crude must be separated from water before refining. The process of demulsification, breaking stable emulsions to recover oil and remove water, is critical at this stage. Efficient demulsification improves crude quality, protects refining equipment from corrosion and fouling, and reduces operational costs.

Thus, emulsions play a dual role across petroleum streams: they are beneficial in upstream and midstream processes for recovery and transport, but require careful elimination in downstream refining. Additionally, their engineered use in environmental applications highlights their versatility beyond conventional petroleum operations. The practical use of heavy oil emulsions in the petroleum industry is governed by the following:
(i) Identification of emulsion type (O/W or W/O) and its stabilization mechanism.
(ii) Control of emulsion generation for efficient oil recovery and channeling.
(iii) Prevention of undesirable inversion (O/W ↔ W/O) due to changes in oil-water ratio, chemical composition, pH, salinity, or temperature.

7.3 Cosmetic Industry

In the cosmetics industry, polymeric surfactants are widely used because of their ability to stabilize modern personal care formulations. Their unique function comes from the way they adsorb and arrange themselves at interfaces (liquid or solid-liquid-liquid). This interfacial behavior helps to:
1. Control droplet size
2. Prevent coalescence (droplets merging)
3. Improve product texture and feel

Such properties are essential for producing high-performance and long-lasting cosmetic products.

Block copolymers: A major class of polymeric surfactants is block copolymers, often with A-B or A-B-A structures. In these systems:

(i) The B block usually anchors the molecule to oil or water regions.
(ii) The A block provides steric stabilization and improves compatibility with the surrounding medium.

This arrangement makes block copolymers highly efficient stabilizers.

Polymeric surfactants are particularly useful [4] in the following:

1. **Nanoemulsions**: These are mainly used where they reduce Ostwald ripening and extend stability. Ostwald ripening (It is a physicochemical process in which larger particles or droplets grow at the expense of smaller ones in a dispersed system over time).
2. **Multiple emulsions:** These emulsions such as W/O/W (water-in-oil-in-water) and O/W/O (oil-in-water-in-oil), helps maintaining their structural integrity.
3. **Vesicular systems:** These (e.g., liposomes) are used in both cosmetics and pharmaceuticals, where they enhance robustness and performance.

In addition to stability, these surfactants are generally mild and non-irritating to the skin, making them especially valuable for sensitive formulations such as baby products, sunscreens, and dermatological creams.

7.3.1 Common Cosmetic Formulations and Their Functional Structures

1. **Lotions**: Most lotions are formulated as O/W emulsions, where oil droplets are dispersed in a continuous manner [5]. Their flow behavior is usually shear thinning, which means that the viscosity decreases when stress or force is applied. At rest or low shear rates ($\approx 0.1\ s^{-1}$), the lotion remains relatively thick, which helps maintain stability and gives a smooth texture. During application or high shear, the viscosity decreases and allowing the lotion to spread easily across the skin, offering light, nongreasy absorption.
2. **Hand creams**: These may be structured as either O/W or W/O emulsions. To achieve a thicker, more luxurious texture (often referred to as having "body"), specialized surfactant blends and thickening agents are used. These can organize into liquid crystalline phases, forming gel-like networks that improve product feel, enhance water retention, and provide prolonged moisturization.
3. **Nail polishes**: These are colloidal suspensions of pigments in nonaqueous, thixotropic media. Thixotropy ensures that the formulation flows easily during application (low viscosity under shear) and gradually regains its viscosity upon standing, preventing pigment settling. Careful regulation of relaxation time ensures optimal leveling, glossy finish, and long-lasting adhesion.

4. **Foundations:** Foundations are hybrid systems, combining emulsions and suspensions to deliver both coverage and consistency. Pigments and fillers are dispersed within either O/W or W/O emulsions, depending on the desired finish (lightweight vs. long-lasting). Controlled thixotropic behavior ensures uniform application, smooth film formation, and even distribution of colorants across skin with different textures.

Together, these examples highlight how colloid and interface science underpins the performance of cosmetic products. By controlling rheology, phase behavior, and stability through polymeric surfactants and structured emulsions, the cosmetic industry can design formulations that are both functional and appealing to consumers.

7.4 Pharmaceuticals

Pharmaceuticals and pesticides are frequently formulated as colloidal dispersions, where their interfaces exhibit unique properties distinct from those of their bulk phases. A comprehensive understanding of interfacial interactions and behaviors is therefore essential in designing stable, effective formulations. The performance, stability, and bioavailability of these systems are governed by the interplay of various particles present within the dispersion [6].

7.4.1 Types of Disperse Systems in Pharmaceuticals

Pharmaceutical formulations can exist in several colloidal forms:

1. **Suspensions (solid dispersed in liquid):** Solid particles, which may be of organic or inorganic nature, are dispersed in a liquid medium. Depending on the solvent, these may be:
   (i) Aqueous suspensions (water as dispersion medium)
   (ii) Nonaqueous suspensions (organic liquids as dispersion medium)
2. **Emulsions (liquid dispersed in liquid):** When the continuous phase is aqueous, the formulation is termed an **aqueous emulsion** (e.g., O/W creams and parenteral emulsions). If the continuous phase is nonaqueous (e.g., polar oil dispersed in a nonpolar oil), the formulation is known as a nonaqueous emulsion.
3. **Suspoemulsions (mixed systems):** These are complex systems where both droplets and solid particles coexist within a liquid medium, providing multifunctional delivery platforms.

7.4.2 Thermodynamic and Kinetic Considerations

Disperse systems are usually thermodynamically unstable, since their formation requires input of external energy (nonspontaneous). However, once formed, they can exhibit kinetic stability, enabling practical use in pharmaceutical and agricultural products. This instability can be explained through the processes of emulsification (oil phase dispersed into droplets) and comminution (solid phase broken into particles). Figure 7.1 shows the emulsification process as well as the process of comminution in a bulk solid, which is depicted in a cuboidal form [7].

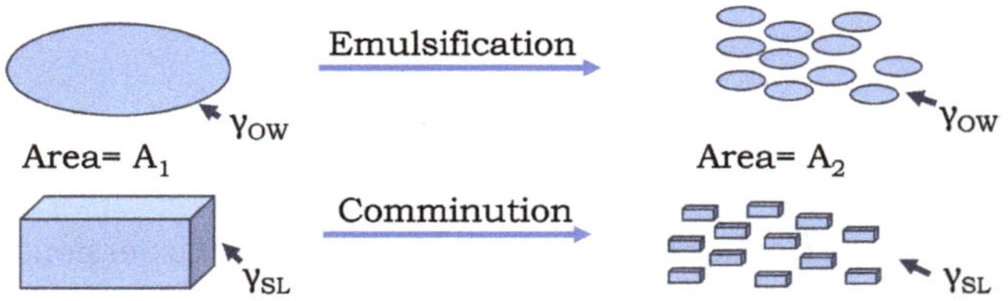

Figure 7.1: The process of emulsification and comminution.

Both processes involve an increase in the interfacial area (ΔA) given as:

$$\Delta A = A_2 - A_1 \tag{7.1}$$

A_2 represents the cumulative surface area of the dispersed droplets or particles, while A_1 refers to the original surface area of the undispersed bulk phase, such as oil or solid. If no adsorption of molecules takes place at the interface then the IFT γ_{OW} (IFT of oil) or γ_{SL} (IFT of solid) is comparatively sizable and as a result the interfacial free energy required to create the interface $\Delta A\gamma_{OW}$ or $\Delta A\gamma_{SL}$ is also large, because of this comminution/emulsification is opposed by the interfacial free energy. It is useful to take into consideration that in emulsification/comminution, a large number of droplets or particles are formed, which causes an increase in total entropy of the system. This promotes the process of emulsification, though the increase in value is small relative to the value of interfacial free energy. Using the second law of thermodynamics, we can give the following expression for the free energy of emulsification/comminution (dispersion):

$$\Delta G_{\mathrm{form}} = \Delta A\gamma_{OW} - T\Delta S_{\mathrm{conf}} = \Delta A\gamma_{SL} - T\Delta S_{\mathrm{conf}} \tag{7.2}$$

In the majority of dispersion processes, ΔG_{form} is large and positive. Hence, emulsification/comminution processes show nonspontaneity, as a result, overtime the particles/droplets aggregate in order to reduce the total energy of the system. For preventing aggregation or coalescence, an energy barrier is created between the particles/droplets, which prevents them from coming too close. This barrier is caused by the repulsive forces present between particles/droplets, which counter the van der Waals forces of attraction between the particles.

As discussed, emulsification and comminution processes inherently generate thermodynamically unstable systems with high interfacial free energy [8]. To prevent these dispersions from collapsing through aggregation or coalescence, stabilization strategies are essential. These strategies operate by introducing repulsive interactions that create an energy barrier large enough to counteract the natural van der Waals attractive forces.

In pharmaceutical colloidal formulations, the stability of the system depends on a fine balance between the attractive and repulsive forces acting at the particle interface. The van der Waals forces, which naturally occur, tend to draw particles or droplets together, leading to aggregation or coalescence. To prevent this, stabilizing mechanisms are applied. One approach is electrostatic stabilization, in which surface charges on the particles create repulsive double-layer forces that keep them apart. Another is steric stabilization, where polymers or surfactants adsorb on the particle surface and act as a physical barrier, blocking close contact. In some cases, polymeric chains extend into the surrounding medium, providing both entropic and enthalpic resistance to particle aggregation. The combination of these mechanisms builds an energy barrier that maintains particle separation. Although colloidal systems are not thermodynamically stable, these stabilization strategies give them kinetic stability, allowing for long shelf life, consistent drug release, and reliable therapeutic performance.

Pharmaceutical significance of stabilization:

(i) Ensures long shelf life by preventing aggregation, sedimentation, or creaming
(ii) Enables controlled release by maintaining uniform particle/droplet size distribution
(iii) Improves the bioavailability of poorly soluble drugs by keeping them in dispersed form
(iv) Enhances reproducibility in therapeutic action, as instability often leads to dose variation

7.4.3 Suspoemulsions

A suspoemulsion (SE) refers to a formulation that combines both solid and liquid (or low-melting-point solid) active ingredients dispersed within a nonaqueous base [9]. For successful SE development, it's essential that both solid and liquid components are compatible with the nonaqueous medium and capable of remaining uniformly suspended.

One effective technique involves separately preparing:

(i) A suspension concentrate containing the solid active component
(ii) A solution or melt containing the liquid or low-melting-point active ingredient

These two preconcentrates are then blended under moderate agitation, which enables precise control over the final composition and structural characteristics of the formulation.

7.4.3.1 Effect of Wet Milling

Wet milling involves grinding materials while they are suspended in a liquid, which helps control particle size, prevent overheating, and improve dispersion. Wet milling of combined solid-liquid actives is typically discouraged due to:
(i) The lubricating effect of the liquid phase, which can interfere with the grinding efficiency.
(ii) Insufficient solid content concentration, making milling less effective.
(iii) Potential disruption of surfactant performance and particle size uniformity, resulting in broader size distributions and reduced stability.

7.4.3.2 Role of Surfactants in Suspoemulsion (SE) Formulations

Surfactants, including emulsifiers, dispersing agents, and wetting agents, play a critical role in water-based SEs. Their functions include:
(i) Stabilizing the interface between dispersed solids and liquid droplets
(ii) Preventing aggregation or flocculation that could lead to phase separation
(iii) Serving dual roles in promoting both dispersion of solids and emulsification of liquids

Their presence ensures structural integrity and long-term stability of these hybrid systems, which are increasingly relevant in advanced cosmetic and pharmaceutical applications.

7.5 Drug Formulation

In pharmaceutical sciences, emulsions and microemulsions are widely applied as advanced drug delivery systems for both hydrophobic and hydrophilic drugs. These systems are particularly advantageous for drugs with poor solubility or limited membrane permeability, where conventional formulations often fail to achieve adequate therapeutic levels. In recent years, micro- and nanoemulsion technologies have attracted significant attention, as they provide an effective strategy to overcome solubility challenges and enhance the bioavailability of active pharmaceutical ingredients [10].

A commonly employed method in pharmaceutical colloidal formulations is the use of O/W emulsions, in which the drug is solubilized within the dispersed oil phase. This approach not only improves the solubility of hydrophobic drugs but also avoids the risk of drug precipitation, which is frequently observed with cosolvent-based formulations at the administration site. Furthermore, submicron emulsions, where droplet sizes fall in the nano to submicron range, have demonstrated superior performance in terms of both stability and therapeutic action. One of the key advantages of such colloidal delivery systems is the enhancement of solubility and bioavailability. The small droplet sizes provide large interfacial areas, which facilitate better drug dispersion and absorption. Another advantage is their low-energy input and thermodynamic stability, since microemulsions often form spontaneously with minimal energy requirements, thereby ensuring long-term stability. Additionally, these systems allow for controlled and targeted release, as their colloidal nature enables sustained release and, in some cases, site-specific drug delivery. A further benefit is the protection of sensitive drugs, because encapsulation within the nonpolar domains of O/W emulsions shields labile drugs from hydrolysis, oxidation, and other degradation pathways. By integrating all of these benefits, emulsion-based formulations represent an innovative platform for next-generation pharmaceuticals, particularly in addressing the formulation challenges posed by modern drugs with poor water or lipid solubility.

7.5.1 Emulsions in Drug Formulation

One of the most widely utilized colloidal systems in pharmaceutical formulations is the emulsion. In O/W emulsions, hydrophobic drugs are incorporated into the dispersed oil phase, which enhances their apparent solubility and facilitates both parenteral and oral administration. These systems are particularly advantageous for drugs with poor water solubility, as the oil droplets act as carriers that enable effective dispersion in aqueous environments [11]. Advanced forms of emulsions, such as microemulsions and nanoemulsions, provide further benefits because of their very small droplet sizes and high kinetic stability. Such characteristics not only reduce the energy required for their preparation but also contribute to extended shelf life and resistance to phase separation. In addition, the fine droplet distribution significantly improves drug absorption across biological membranes, thereby increasing bioavailability and therapeutic efficiency. For example, the immunosuppressant cyclosporine is formulated as a microemulsion to enhance its oral bioavailability [12], while propofol, a widely used intravenous anesthetic, is administered in the form of an O/W emulsion [13].

In the case of submicron emulsions, even though substantial amounts of both oil and water phases are present, the system may appear transparent due to the ultrafine dispersion of oleic and aqueous domains, invisible to the naked eye. Its pictorial representation is shown in Figure 7.2. These systems can exist as micellar emulsions, where the interfacial boundary continuously shifts through spontaneous molecular motion, making it highly dynamic. Structurally, microemulsions are often categorized into three primary types as follows:

(i) Water-in-oil (W/O): Water droplets dispersed in a continuous oil phase
(ii) Oil-in-water (O/W): Oil globules suspended in a continuous aqueous phase
(iii) Bicontinuous structures: Interwoven networks of oil and water with comparable proportions

Typically, the inherent solubility between oil and water is minimal. However, when a surfactant or amphiphilic agent is added, it promotes interfacial blending. At a critical surfactant concentration, the system can transition into a homogeneous and thermodynamically stable mixture.

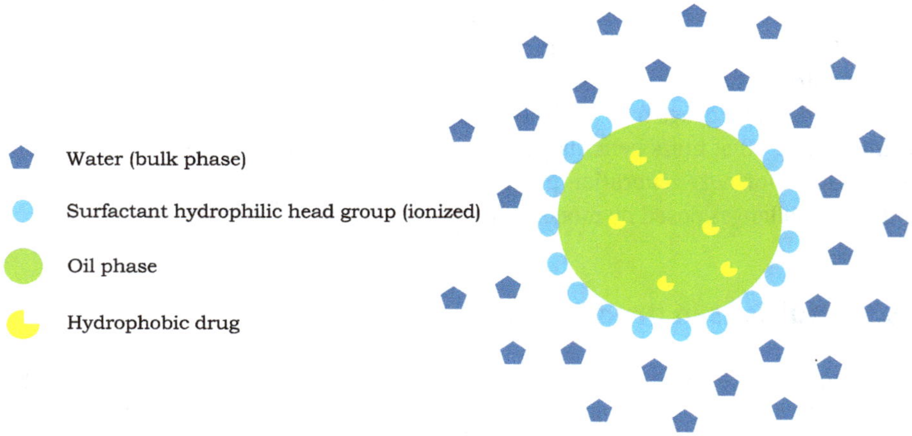

Figure 7.2: The structure of the drug-loaded emulsion system.

In submicron emulsion systems, the oil phase can penetrate and expand the lipophilic region of the surfactant's monolayer. This interaction influences the system's interfacial curvature, which in turn affects droplet formation and stability. Specifically:

(i) Short-chain oils tend to infiltrate the surfactant's hydrophobic tail region more easily than longer chains.

(ii) This leads to negative interfacial curvature, promoting the generation of smaller droplets, an essential trait for stable microemulsions.

To achieve the desired structural and functional characteristics, the chosen surface-active agent should:

(i) Lower IFT for ease of droplet formation

(ii) Provide a flexible interfacial film to accommodate curvature variations

(iii) Offer hydrophobicity tailored to the emulsion type

7.5.2 Classification of Surfactants for Submicron Emulsion Systems

Surfactants employed in microemulsion formulation are generally classified according to the nature of their hydrophilic (polar) head groups. The major categories, along with representative examples, are summarized in Table 7.2.

Table 7.2: Classification of surfactants.

Types	Charge nature	Examples
Anionic	Negative	Sodium lauryl sulfate and potassium laurate
Cationic	Positive	Quaternary ammonium and halides
Non-ionic	Neutral	Brij®35 and sorbitan monooleate
Zwitterionic/ampholytic	Both positive and negative	Phospholipids and sulfobetaines

Combining ionic and nonionic surfactants can enhance system performance, improving droplet size distribution, interfacial flexibility, and long-term stability.

7.5.2.1 Role of HLB in Surfactant Selection

The hydrophilic-lipophilic balance (HLB) is a critical parameter guiding surfactant choice:
(i) Low HLB (3–6): Supports W/O submicron emulsions.
(ii) High HLB (8–18): Promotes O/W microemulsion systems.

HLB quantifies the relative dominance of hydrophilic versus lipophilic moieties [14], helping formulators align surfactant characteristics with desired emulsion architecture. Surfactant-based micelles can encapsulate poorly soluble drugs in their hydrophobic cores, whereas polymeric nanoparticles provide opportunities for controlled release and site-specific targeting. For example, paclitaxel, a poorly soluble anticancer agent, has been successfully formulated as albumin-bound nanoparticles, offering improved delivery to tumor sites [15].

7.5.3 Advantages and Challenges

The stability of colloidal drug formulations depends on a delicate balance between attractive and repulsive forces. The van der Waals forces promote aggregation of particles or droplets, while electrostatic, steric, or polymeric stabilization mechanisms provide repulsive barriers that prevent coalescence. Surfactants, emulsifiers, and polymers are commonly employed to achieve this stability, ensuring that the formulations remain kinetically stable for extended periods. This stability is critical for maintaining drug uniformity, prolonging shelf life, and achieving reproducible therapeutic action.

The application of colloid chemistry in drug formulation offers several distinct advantages. These include enhanced solubility and dissolution rate of poorly soluble drugs, improved oral and parenteral bioavailability, protection of labile drugs against hydrolysis and oxidation, and the ability to design controlled or sustained-release dosage forms [16]. Importantly, colloidal delivery systems also enable targeted drug delivery, thereby reducing systemic side effects and improving patient compliance.

Despite their potential, colloidal formulations also face challenges. Issues such as physical instability (e.g., Ostwald ripening, phase separation, or flocculation), toxicity of certain surfactants, difficulties in large-scale manufacturing, and stringent regulatory requirements need to be carefully addressed. Nonetheless, advances in nanotechnology and the development of novel biocompatible colloidal carriers are expected to overcome many of these limitations. In the future, smart colloidal systems capable of responding to physiological triggers, such as pH or temperature changes, are likely to revolutionize drug delivery by providing precision and personalization.

7.6 Coatings and Paints

Coating materials are integral to everyday life, serving purposes that range from enhancing the appearance of household walls to ensuring visibility and durability in road markings. Beyond aesthetics, coatings play a vital role in protecting surfaces from environmental damage, chemical corrosion, and mechanical wear. In response to growing demands for improved performance and cost efficiency, the coatings industry is simultaneously adapting to stricter environmental regulations and sustainability goals. One of the most significant advances has been the development of polymer latex binders, designed with diverse particle morphologies to meet functional and environmental requirements.

7.6.1 Composition of Waterborne Latex Coatings

Modern waterborne latex coatings are colloidal systems that rely on the stable dispersion of multiple ingredients shown in Figure 7.3. Typical formulations include the following:

(i) **Water and organic solvents**: Serve as the dispersion medium and evaporation agents.
(ii) **Polymer binders**: Colloidal polymer particles that coalesce into a continuous film after drying.
(iii) **Dispersants and surfactants**: Maintain pigment stability and prevent flocculation.
(iv) **Rheology modifiers**: Impart shear-thinning or thixotropic behavior, essential for smooth application.
(v) **Pigments and extenders**: Provide color, opacity, and texture while influencing gloss and reflectivity.
(vi) **Additives**: Enhance durability, UV resistance, antimicrobial activity, or flow properties.

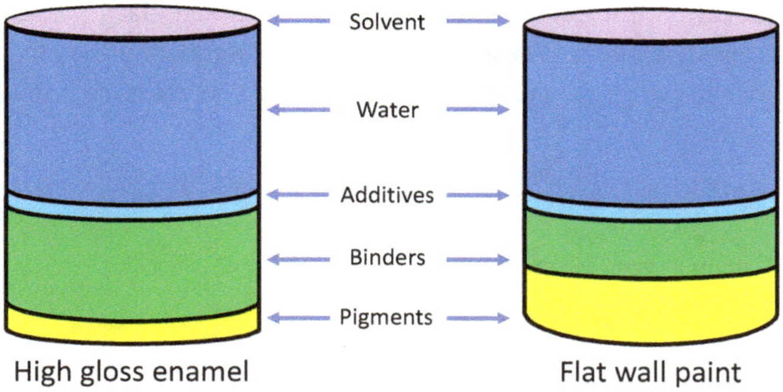

Figure 7.3: Major components in a can of paint.

7.6.2 Role of the Polymer Binder

The polymer binder is the central component that governs coating performance. Its function extends beyond merely binding pigment particles; it determines adhesion to surfaces, film durability, mechanical strength, and final surface appearance [17]. The binder typically exists as colloidal particles in suspension, with sizes ranging from 20 to 600 nm and concentrations of 20–60% by weight, depending on the formulation.

A critical property influenced by the binder is gloss. By varying the ratio of polymer binder to inorganic components (pigments and extenders), formulators can precisely control light reflection and surface smoothness. Higher binder levels usually produce glossy, smooth films, while higher pigment concentrations yield matte finishes with greater opacity.

7.6.2.1 Film Formation and Colloid Principle

The transformation of a latex coating into a continuous protective film is a direct consequence of colloid science. This film-formation process is strongly dependent on particle size, particle softness (glass transition temperature of the polymer), and stabilizing surfactants [18]. The process of drying involves the following three stages, as shown in Figure 7.4:

Stage I – concentration phase: As water gradually evaporates from the latex dispersion, the polymer binder particles become increasingly concentrated within the remaining liquid medium.

Stage II – particle packing: Once the concentration reaches a critical threshold, the binder particles begin to approach one another closely. If the particle size distribution is sufficiently uniform, this leads to the formation of a tightly packed and orderly arrangement.

Stage III – film formation: Continued evaporation intensifies capillary forces, which draw the particles even closer. These forces eventually overcome the electrostatic repulsion between particles, causing them to deform and coalesce into a seamless, continuous film.

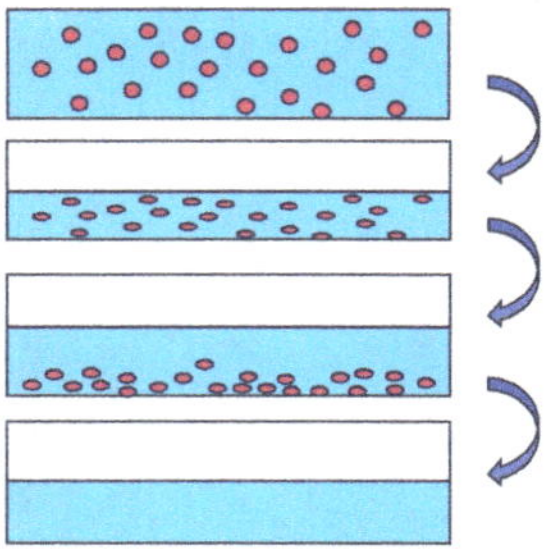
Stage I: Water evaporates concentrating latex particles
Stage II: Particles pack and come into irreversible contact
Stage III: Particles coalesce and form a continuous film

Figure 7.4: Different stages of coating deposition.

7.6.3 Applications of Colloid Chemistry in Paints and Coatings

Colloid science underpins the formulation, performance, and durability of paints. Paints are essentially multicomponent colloidal systems, where pigments, extenders, and additives are finely dispersed in a continuous medium stabilized by surfactants and polymeric binders [19]. The principles of colloid chemistry are applied in the following ways:

(i) Colloid chemistry ensures that pigment particles remain uniformly dispersed in the medium, preventing aggregation and sedimentation.
(ii) Surfactants and dispersants adsorb on pigment surfaces, imparting electrostatic or steric stabilization to keep the suspension stable during storage.
(iii) Polymer latex binders are colloidal particles that coalesce during drying to form a continuous protective film.
(iv) Hydrodynamic and associative thickeners modify colloidal networks, allowing paints to spread easily during application but resist sagging afterwards [20].
(v) Gloss, opacity, and color strength depend on how colloidal pigments and fillers scatter and reflect light. Proper control of particle dispersion allows for tailored matte or glossy finishes.
(vi) Colloid-based formulations enhance resistance to UV radiation, moisture, and chemical attack by embedding protective agents within the colloidal matrix.
(vii) Smart colloidal additives (e.g., silica nanoparticles) can impart self-cleaning or antifouling properties.
(viii) Modern waterborne paints rely on colloidal dispersions (latex systems) to reduce volatile organic compounds.

(ix) Colloid chemistry enables eco-friendly coatings with comparable or superior performance to solvent-based systems.

7.7 Purification of Municipal Water

Surface waters contain suspended colloidal particles, such as soil, clay, dirt, and organic matter, which give the water a muddy and cloudy appearance. These particles, known as total suspended solids, are of colloidal size and mass, preventing them from settling under the influence of gravity. Moreover, microorganisms and bacteria often adhere to these colloidal particles, making the water unsafe for human use. Municipal water treatment, therefore, makes extensive use of colloidal chemistry, particularly the principles of surface charge regulation and aggregation. The sol particles present in untreated water generally carry a negative surface charge that leads to mutual electrostatic repulsion and prevents them from aggregating, thereby keeping them dispersed in suspension [21].

To render water suitable for consumption, raw water from lakes, rivers, or reservoirs is directed to treatment plants, where the initial and most critical step is the removal of suspended colloidal matter. This is accomplished through coagulation and flocculation processes.

In the coagulation stage, positively charged chemical substances known as coagulants are added to the water. These neutralize the surface charge of colloidal particles, enabling them to aggregate into larger clusters called *microflocs*. Common coagulants include aluminum salts such as aluminum sulfate and aluminum chloride. These compounds hydrolyze in water to form aluminum hydroxide, which enmeshes colloidal impurities and promotes their aggregation. Among them, aluminum sulfate is more widely used due to its lower cost [22], while aluminum chloride is preferred only in specific applications. Iron-based coagulants such as ferric chloride and ferric sulfate operate in a similar manner, producing ferric hydroxide flocs that capture impurities. Optimal dosing of coagulants is critical for efficient removal. Following coagulation, the process of flocculation is carried out. This involves gentle mixing of the water to encourage the collision and binding of microflocs into larger, heavier flocs. This process is represented in Figure 7.5 for better understanding. Polymers are often added to enhance bridging between particles, resulting in more stable aggregates. The resulting flocs are separated by sedimentation in specially designed tanks or clarifiers, where the heavier aggregates gradually settle to the bottom and are removed through sludge outlets.

After sedimentation, the water undergoes filtration to remove residual suspended particles and microorganisms. Filters are typically composed of sand, gravel, or activated carbon. Advanced multimedia filters, which incorporate multiple layers of materials with varying particle sizes, allow efficient capture of a wider range of impurities. Two common filtration methods used in water treatment plants are gravity

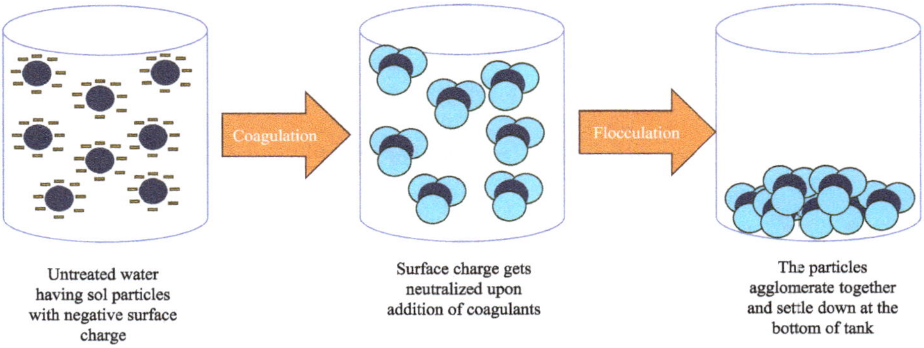
Coagulation
Flocculation
Untreated water having sol particles with negative surface charge
Surface charge gets neutralized upon addition of coagulants
The particles agglomerate together and settle down at the bottom of tank

Figure 7.5: The process of coagulation and settling down of flocculants.

filters and pressure filters [23]. Finally, the filtered water is subjected to disinfection, typically using chlorine or other disinfectants, to eliminate pathogenic organisms. This final step ensures that the water is microbiologically safe and suitable for human consumption.

Thus, the entire purification process illustrates how colloidal stability is first disrupted (via coagulation) and then manipulated (via flocculation) to achieve aggregation and removal of colloidal particles. Understanding particle charge, double-layer interactions, and aggregation kinetics from colloid chemistry is essential for designing efficient water treatment systems.

7.8 The Cleansing Action of Soaps and Detergents

Soaps are amphiphilic surfactants, meaning they possess affinity for both water and oil. Structurally, each soap molecule consists of two distinct parts, as shown in Figure 7.6:

(i) A polar, water-loving (hydrophilic) ionic head, which is insoluble in oils and grease.
(ii) A nonpolar, water-hating (hydrophobic) hydrocarbon tail, which readily dissolves in oils and grease.

This amphiphilic nature is the basis of their cleansing ability [24]. Surfactant molecules have the unique tendency to self-assemble in aqueous solutions into supramolecular structures known as micelles, a concept central to colloidal chemistry.

During washing, when soap comes into contact with grease or dirt, the hydrophobic tails embed themselves into the grease particle, while the hydrophilic heads remain in contact with the surrounding water. As more soap molecules accumulate, the grease particle becomes surrounded by surfactant molecules. This leads to the forma-

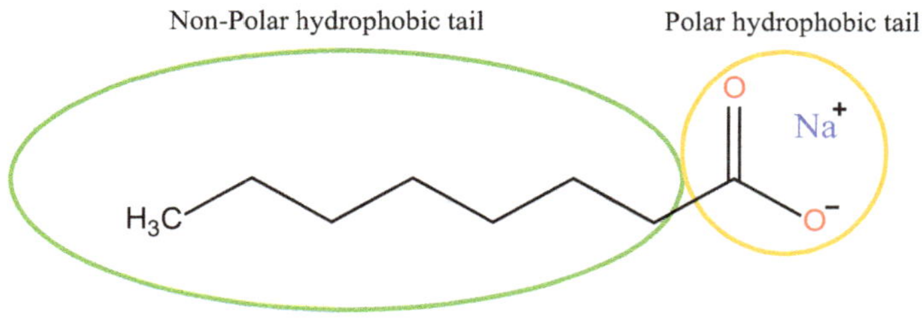

Figure 7.6: Structure of a soap molecule.

tion of micelles (shown in Figure 7.7), where the grease is trapped in the hydrophobic core and stabilized by hydrophilic heads facing outward toward the water [25]. The micelle formation reduces the surface tension and disperses large grease or oil droplets into smaller emulsified particles. These suspended micelles behave as a colloidal system in water, which can be easily removed by rinsing. Thus, the cleansing action of soaps is essentially an application of colloidal chemistry and micelle formation.

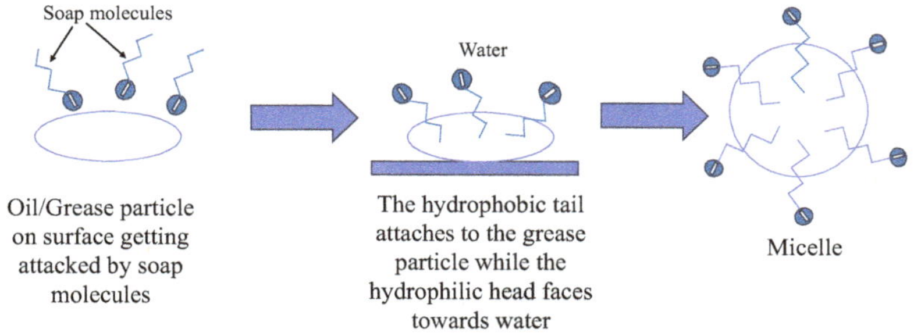

Figure 7.7: Micelle formation.

Detergents, which are synthetic surfactants with structures similar to soaps, function through the same mechanism of micelle formation [26]. However, unlike soaps, detergents remain effective even in hard water and acidic conditions, since they do not form insoluble precipitates with calcium or magnesium ions.

7.8.1 Types of Detergents

1. **Anionic detergents**: Detergent molecules having a negatively charged head are classified as anionic detergents, e.g., sulfate ion (RSO_4^-). They are used in dishwashing liquids, laundry detergents, etc.
2. **Cationic detergents**:: Detergent molecules having positively charged heads are classified as cationic detergents, e.g., tertiary ammonium ion ($RN(CH_3)_3^+$). They

are used in antiseptics and disinfectants, hair conditioners, plastic cleaners, and fabric softeners.

3. **Nonionic detergents**: Detergents having partially charged heads are known as nonionic detergents, e.g., alcohol (hydroxyl) and/or ether (ethoxy) functional groups. They are used in dishwashers and front-loading washing machines.

7.9 Delta Formation

Deltas are formed at the meeting point of rivers and seas. River water carries suspended colloidal particles of sand, clay, and organic matter, which are generally negatively charged. On the contrary, seawater is saline and contains a high concentration of positively charged ions such as Na^+, Mg^{2+}, and Ca^{2+}. When river water mixes with seawater, these positively charged ions neutralize the surface charge of the colloidal particles [27]. Once neutralized, the particles no longer repel each other and begin to aggregate, forming larger flocs. In the calmer coastal waters, these aggregates gradually settle and get deposited (Figure 7.8).

Over time, the accumulated sediments obstruct the main river channel, forcing the river to split into smaller branches known as distributaries. These distributaries spread sediments across the coastal plain, and as deposition continues, a network of interconnected channels develops, giving rise to a delta. The maximum sedimentation generally occurs near the mouth of the delta, where the flow velocity decreases significantly.

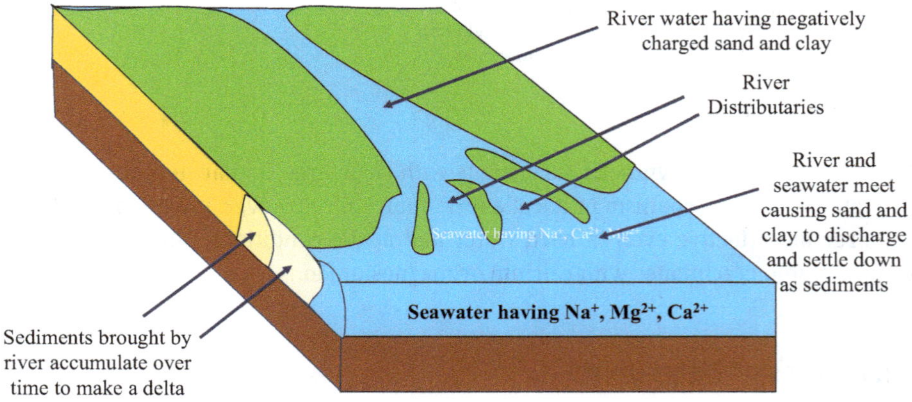

Figure 7.8: The formation of a delta.

7.9.1 Role of Colloidal Chemistry

The formation of deltas is a direct application of colloidal chemistry. The stability of colloidal particles in river water is disrupted when electrolytes from seawater are introduced. Coagulation and flocculation processes, driven by charge neutralization and aggregation, cause suspended colloids to settle. Thus, the principles of colloidal stability and coagulation underpin the natural phenomenon of delta formation [28].

7.10 Colloids in Photography

Colloids play a crucial role in the photographic industry, especially in the preparation of light-sensitive materials. Photographic films, papers, and slides are essentially colloidal systems in which finely dispersed particles are embedded in a continuous medium.

In most cases, the colloidal particles have a radius of 100 nm or less, which provides a very high surface area, minimizes light scattering, and ensures the mechanical stability of films during swelling and processing. In some specialized applications, however, larger colloidal particles are employed. For instance, high-speed (low-light) films use larger silver halide grains to enhance sensitivity, while certain photographic coatings incorporate O/W emulsions as lubricating layers [29].

7.10.1 Colloidal Materials in Photography

Some colloidal materials used in the photographic process are as follows:

1. **Silver halide emulsion:** The most significant colloidal system in photography is the silver halide-gelatin emulsion, where silver halide crystals are dispersed in gelatin and coated on a substrate such as glass, film, or paper [30]. Upon exposure to light, these crystals undergo a photochemical change, which is later amplified during development to form a visible image.

   The size of silver halide particles is critical for both light sensitivity and image resolution. Larger grains increase the likelihood of absorbing enough photons to form silver atom clusters that create the latent image. In X-ray films, grains measure several micrometers to ensure high coverage, while color films use grains of about 500 nm and graphic arts films around 250 nm. Daylight films and holographic plates, however, contain true nanoparticulate dispersions with sizes ranging from 20 to 30 nm. To further boost sensitivity, most grains undergo chemical ripening, where their surfaces are recrystallized and doped with trace amounts of gold, sulfur, and silver. This modified surface is essential for achieving the high light sensitivity required in modern photography.

2. **Color couplers:** The colors in prints, slides, and negatives are created by reactions between oxidized developers and leuco dyes called color couplers. To keep these couplers confined to the correct image-forming layer, they are incorporated in colloidal form within the film [31]. Because small molecules can easily diffuse through the coating layers, couplers are usually either dispersed in oil-water emulsions or chemically bonded to a polymeric backbone. Upon formulation as emulsions, couplers include a hydrophobic ballast to prevent them from dissolving in water and migrating to neighboring layers. Since couplers possess high melting points, there is a need for a cosolvent to help disperse them and optimize the film's sensitometric properties. The interlayers between image-forming layers are generally filled with oil-water emulsions that have UV-absorbing compounds/substances that undergo reaction with the oxidized developer and prevent cross-contamination.
3. **Gelatin:** Gelatin performs several important functions in photographic coatings and, other than water, makes up the bulk of the coated material. It plays a vital role in forming and stabilizing the various colloidal substances that capture light and create the image. These colloids include inorganic particles like silver halide crystals, where gelatin acts as a peptizer and helps regulate their growth, as well as organic components such as color couplers, stabilizers, and UV absorbers [32]. Ensuring the stability of these colloids is particularly critical in thin-film, multilayer coatings:
   (i) Solution of gelatin offers many valuable physical properties. It increases viscosity, causing a slowing down of sedimentation and ensuring a stable coating flow, and it exhibits shear-thinning behavior at high shear rates, allowing for faster coating speeds.
   (ii) Gelatin solutions can be gelled by lowering the temperature, forming physical gels in a thermoreversible process known as "chill-setting." This process is essential for storing the melt and is the first crucial step after coating, as it stabilizes the layer structure and provides mechanical resistance to external forces during drying. After drying, gelatin acts as a clear, flexible protective layer for the imaging chemicals. Subsequently, gelatin molecules are covalently crosslinked through reaction with a hardener, producing a chemical gel that creates a supportive network to maintain mechanical integrity during the reswelling stage, which is vital for photo-processing.

7.10.2 Role of Colloidal Chemistry

1. The dispersed state of silver halides and couplers ensures photosensitivity and color formation.
2. Gelatin acts as a protective colloid, regulating particle growth and preventing aggregation.

3. O/W emulsions provide localized distribution of couplers and additives. Together, these colloidal systems are fundamental for achieving image sharpness, stability, and the mechanical durability of photographic materials.

7.11 Colloids in Artificial Rain

Clouds are essentially colloidal dispersions of water droplets in air. Each droplet carries an electric charge, which helps keep it suspended. When finely divided dust or sand particles with opposite charge are introduced into such clouds, the droplets get neutralized. This neutralization promotes coagulation, leading to the formation of larger droplets that eventually fall as rain. This principle forms the basis of artificial rainmaking techniques, also known as cloud seeding. A pictorial representation is shown in Figure 7.9.

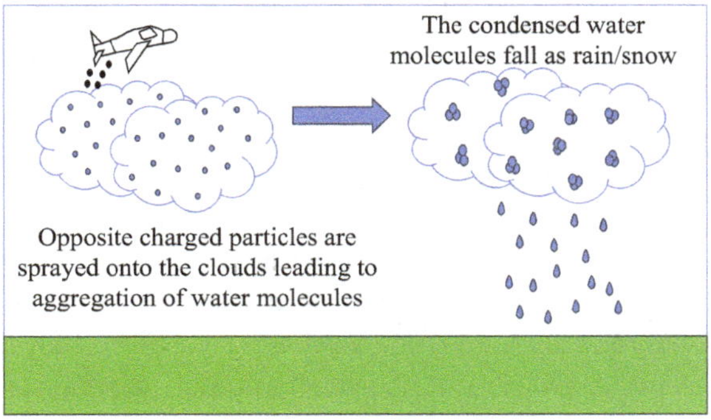

Figure 7.9: Artificial rain caused by cloud seeding.

Cloud seeding involves the addition of condensation nuclei (substances that act as centers for droplet formation) into clouds, thereby accelerating the natural process of raindrop or snowflake formation. It is widely studied for applications in weather modification, drought relief, and agricultural enhancement [33].

7.11.1 Cloud Seeding Methods

1. **Static cloud seeding:** In this method, chemicals such as potassium iodide, silver iodide, or dry ice (solid CO_2) are dispersed into clouds [34]. These chemicals provide nuclei around which moisture condenses more efficiently, thereby increasing the likelihood of rainfall.
2. **Dynamic cloud seeding:** This method is more complex. It uses ice crystals to enhance vertical air currents inside the cloud system. The stronger currents cause

more water to rise, condense, and accumulate within the clouds. Eventually, the increased water mass results in rainfall. Since this involves a sequence of meteorological events, it is technically more challenging than static seeding.

3. **Hygroscopic cloud seeding:** In this approach, flares or explosive devices are used to disperse hygroscopic salts (e.g., sodium chloride and calcium chloride) into the lower regions of clouds [35]. These salts act as moisture-attracting particles, encouraging water to accumulate, grow in size, and eventually fall as rain. Artificial rain exploits the principle of coagulation of colloidal particles. The charged colloidal water droplets in clouds are induced to aggregate by introducing condensation nuclei, thus forming larger droplets that can no longer remain suspended and fall to the ground as precipitation.

7.12 Electrical Precipitation of Smoke

Industrial smoke is a colloidal dispersion of solid particles such as carbon, arsenic compounds, metal oxides, and cement dust in the air. Since these suspended particles contribute to severe air pollution, they must be removed before releasing the gases into the atmosphere. This is achieved using devices called Electrostatic Precipitators (ESPs). ESPs work by charging the smoke particles and then attracting them to oppositely charged collection plates, where they accumulate. Once a substantial layer of particles builds up on the plates, mechanical rappers are used to dislodge and collect them. The cleaned air, free of particulate matter, is then released into the atmosphere. Figure 7.10 shows the schematic representation of the working of an ESP, showing ionization, charging of smoke particles, their deposition on oppositely charged plates, and release of clean gas.

The first ESP, known as the Cottrell precipitator, was invented in 1907 by Frederick Gardner Cottrell, a professor of chemistry at the University of California, Berkeley [36]. The device functions by passing untreated smoke through a series of sharp points charged to a high potential (20,000–70,000 V). These points release electrons at high velocity, which collide with gas molecules in the air, causing ionization. The resulting ions adsorb onto smoke particles, making them charged. Since the electrodes in the precipitator carry the opposite charge, the smoke particles are attracted to and deposited on them, leaving behind clean air. ESPs not only help in pollution control but also allow recovery of useful by-products. For example, arsenic oxide can be recovered from smelter smoke through this process [37].

7.12.1 Types of Electrostatic Precipitators

1. **Dry electrostatic precipitator**: It is used to collect dry pollutants like ash or cement. In a dry ESP, ionized particles are passed through electrodes where they

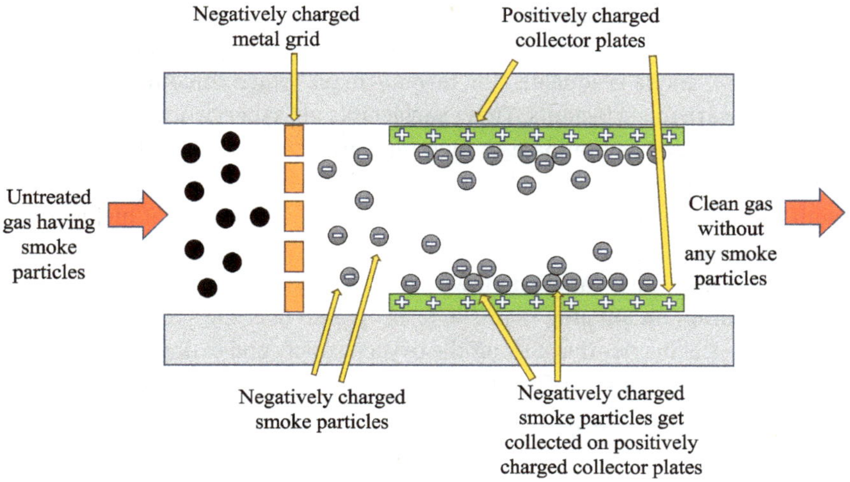

Figure 7.10: The working of the electrostatic precipitator.

get separated from the air and are deposited at collection devices [38]. A hopper is used to remove the collected particles from the collection devices.

2. **Wet electrostatic precipitator**: It operates with water-vapor-saturated air streams and is utilized in the removal of wet substances like oil, tar, and paint.
3. **Tubular precipitator**: It is a single-stage unit made up of tubes that have electrodes arranged in parallel, operating at very high voltages. It is suitable for cleaning industrial gases in confined spaces.
4. **Plate precipitator**: This is the most common type and it consists of large flat metal plates arranged vertically with rows of thin vertical wires between them. Plates are spaced 1–18 cm apart, and charged particles are attracted to the plates where they accumulate.

The working of an ESP is based on the precipitation (coagulation) of charged colloidal particles. By charging the colloidal smoke particles and providing oppositely charged electrodes, the suspended particles are neutralized and removed efficiently.

7.13 Other Applications of Colloids

7.13.1 Fire Foam

When sodium bicarbonate solution is mixed with alum, a carbon dioxide froth known as fire foam is produced. This froth is stabilized by the addition of colloids, such as dextrin, glue, or saponin, which protect the bubbles and maintain their finely dispersed form. The stabilization prevents rapid collapse of the foam, allowing it to degrade slowly upon exposure to burning material and thereby increasing its fire-extinguishing efficiency [39].

A well-known example is aqueous film forming foam, which contains both fluorocarbon and hydrocarbon surfactants. These surfactants act synergistically: the fluorocarbon surfactant stabilizes the foam at high temperatures, while the hydrocarbon surfactant enhances spreading on hydrocarbon fuel surfaces. Together, they produce a foam with superior resistance to heat and slow degradation, making it highly effective against liquid fuel fires [40].

Fire-fighting foams can be applied in two primary ways as follows:

1. **Direct application on burning fuel:** The foam spreads over the surface of the fuel, forming a physical barrier that cuts off the oxygen supply and extinguishes the fire.
2. **Pre-ignition application on fuel vapors:** Foam is sprayed over fuel vapors to suppress them, preventing ignition and thereby reducing the risk of fire.

7.13.2 Textile Dyeing

Chemicals that are used in the textile industry to increase the binding of pigments with fibers are known as mordants that are used alongside adjective dyes (dyes that require a mordant to bind with the fiber of fabric). The mordant creates pores in the fibers through which the colorant enters and adsorbs onto the substrate surface. In another method, mordants are applied along with dyes directly on the fiber, where they react with the dyestuff present inside the fiber and form insoluble compounds. These insoluble dye-mordant complexes improve the fastness and durability of the dyes. Mordants can be applied onto substrates before (pre-mordanting), during (meta-mordanting), or after (post-mordanting) the coloration step [41].

The process of dyeing is essentially a colloidal phenomenon. Most dyes exist as colloidal solutions when dissolved in water. Mordants assist in stabilizing these dye colloids and enhance their adsorption onto fibers. The fibers themselves act as porous substrates with high surface area, favoring adhesion of colloidal dye particles. Moreover, the insoluble dye mordant complexes formed inside the fibers behave like colloidal precipitates, which remain strongly bound and impart color fastness to the fabric. Thus, colloidal chemistry governs the dispersion of dye molecules in solution, their adsorption on fibers, and the precipitation/complexation of dye mordant systems within the fabric.

7.13.3 Leather Tannin

Leather originates from the structural framework of animal hides, predominantly composed of collagen, a fibrous protein organized into microcrystalline helices. These helical arrangements are stabilized by extensive hydrogen bonding between peptide nitrogen atoms and adjacent carbonyl groups. Surrounding the crystalline domains is an amorphous matrix, where polar and nonpolar side chains interact, and covalent crosslinks contribute to overall mechanical integrity.

In its dry form, native collagen exhibits high chemical and biological stability but tends to be stiff and brittle, limiting its practical utility. Upon hydration, however, collagen becomes susceptible to microbial degradation and loses thermal stability. In pure water, its crystalline regions begin to denature around 58°°C, while in saline environments, this transition can occur at temperatures as low as 37°°C [42].

7.13.3.1 Mechanism of Tanning

The process of tanning introduces crosslinks between collagen fibers, thereby stabilizing the fibril structure and improving thermal, chemical, and microbial resistance. Different tanning methods operate via distinct mechanisms:

(i) **Mineral tanning**: Coordination occurs between metal ions (e.g., chromium, aluminum, and zirconium) and the abundant carboxyl groups on collagen. This process produces some of the most hydrothermally stable and commercially valuable leathers, with chromium tanning being the most prevalent in the industry [43].

(ii) **Aldehyde tanning**: Involves the formation of covalent crosslinks between aldehydes and the limited number of amine groups present in collagen, resulting in a chemically stable material [21].

(iii) **Vegetable tanning**: Utilizes natural tannins (polyphenolic compounds of plant origin) that form hydrogen-bonded and hydrophobic interactions with collagen, imparting rigidity and color.

Tanning agents such as vegetable tannins and metal salts like chromium or aluminum frequently exist in colloidal form when dissolved in water. Because of their colloidal nature, they are able to diffuse evenly into the porous collagen matrix and interact with reactive functional groups. The adsorption of these colloidal agents onto collagen fibers promotes effective crosslinking, which imparts greater durability, flexibility, and resistance to the leather. Hence, leather tanning can be understood as a colloid-substrate interaction governed by adsorption, diffusion, and precipitation processes, all of which determine the final qualities of the leather.

Colloidal chemistry holds vital importance across a wide range of industries due to its capacity to manipulate matter at the microscopic level. Its applications are highly diverse: in the petroleum industry, it aids EOR by influencing viscosity and mobility; in cosmetics, polymeric surfactants stabilize advanced formulations; in pharmaceuticals, colloidal systems improve the delivery of drugs with poor solubility [44]. In materials science, colloids are fundamental to the development of paints and coatings, providing uniformity, strength, and texture. On a societal scale, colloidal principles are applied in water treatment through coagulation and flocculation, in the cleaning action of soaps and detergents, and in specialized processes such as photography, textile dyeing, and leather tanning. Additionally, colloidal systems contribute to environmental applications, including artificial rainmaking and air purification with ESPs [45].

Altogether, the extensive applications of colloidal chemistry demonstrate not only its scientific value but also its industrial and environmental importance. This broad scope emphasizes the need for ongoing research and innovation to fully exploit colloidal systems in solving current and future challenges.

Practice Questions

1. How are heavy oil emulsions formed in different streams of the petroleum industry?
2. What are emulsion-stabilizing agents? Explain their various types.
3. Discuss the challenges associated with breaking petroleum emulsions and the methods used for demulsification.
4. Explain the role of surfactants in EOR.
5. Why are polymeric surfactants used in the cosmetic industry?
6. Describe some commonly used formulations in the cosmetic industry.
7. Explain the significance of emulsions in creams and lotions.
8. What are the advantages of using nanoemulsions in cosmetic formulations?
9. What is an SE? Explain.
10. What factors influence emulsification and comminution in the pharmaceutical industry?
11. What are the two essential factors on which an SE system depends?
12. What are the advantages of submicron emulsions as drug carriers?
13. Explain lipid-based biphasic colloidal systems.

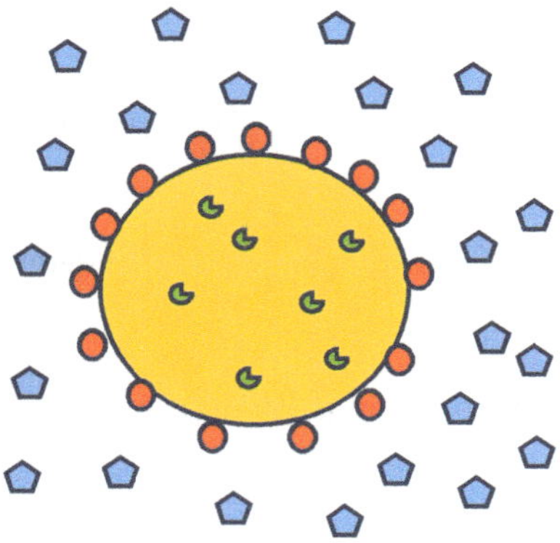

Figure 7.11: Loaded drug emulsion.

14. Identify the components in the given structure (shown in Figure 7.11) of a loaded drug emulsion system.
15. Mention the groups of surfactants used in the formulation of submicron emulsions.
16. Differentiate between microemulsions and nanoemulsions with examples.
17. Discuss the role of emulsions in controlled drug delivery systems.
18. What are the major components of modern waterborne latex coatings? Why is the polymer binder important?
19. What is the usual size range and concentration of polymer binder particles in paints?
20. Describe the formation of latex paint with the help of a diagram.
21. What are the principal thickening mechanisms in paints? Explain.
22. Discuss the role of surfactants in stabilizing latex paints.
23. Explain the role of coagulation in the municipal water treatment process.
24. Name some common coagulants used in water treatment and explain their mechanism of action.
25. Differentiate between coagulation and flocculation with suitable examples.

26. What are micelles, and how are they formed?

27. What are the different types of detergents? How do their structures differ from one another?

28. Explain the concept of critical micelle concentration and its importance in detergent action.

29. Why do colloidal particles used in the photography industry have a weight-average radius of ≤100 nm?

30. How does silver halide paper function in photography?

31. What effect does the size of silver halide particles have on the quality of a photograph?

32. Explain the role of gelatin in silver halide photography.

33. Describe the process of image development in silver halide photography.

34. What is cloud seeding? Discuss the various methods used for cloud seeding.

35. How does a Cottrell precipitator function?

36. Explain the different types of ESPs.

37. What is fire foam? How does it help in firefighting?

38. Discuss the role of colloids in atmospheric pollution control.

39. What are mordants? How do they increase the affinity between fiber and pigment?

40. What are the various types of mordants?

41. What is the purpose of leather tanning?

42. Explain the role of colloids in the dyeing process of textiles.

References

[1] Lee J, Babadagli TC. Comprehensive review on heavy-oil emulsion: Colloidal science and practical applications. Eng Sci. 2020;228:115962.

[2] Jain V, Demond AH. Conductivity reduction due to emulsification during surfactant-enhanced Aquifer Remediation. 1. Emulsion transport. *Environ Sci Technol*. 2002;36(24):5426–5433.

[3] Karsa DR, Davies S. Overview of Surfactant Use in Industry. In: Karsa DR (ed). *Industrial Applications of Surfactants IV*. Wiley-VCH Verlag GmbH & Co. KGaA: Weinheim, Germany; 2003. pp. 1–14.

[4] Yang Q, Xin X, Wang L, Lu H, Ren H, Tan Y, Xu G. Modification of the stability of oil-in-water nano-emulsions by polymers with different structures. *Colloid Polym Sci*. 2014 Jun;292(6):1297–1306.

[5] Baki G. *Introduction to cosmetic formulation and technology*. John Wiley & Sons; 2022 Dec 12.

[6] Lieberman H, Rieger M, Banker GS (eds). *Pharmaceutical dosage forms: Disperse systems*. CRC Press; 2020 Aug 26.

[7] Angardi V, Ettehadi A, Ö Y. Critical review of emulsion stability and characterization techniques in oil processing. *J Energy Resour Technol*. 2022 Apr 1;144(4):040801.

[8] Tadros TF. *Emulsions: Formation, stability, industrial applications*. Walter de Gruyter GmbH & Co KG; 2016.

[9] Kayea R, Secrest M, Lindner G. Novel Nonionic Star Polymeric Stabilizer in Aqueous Dispersion Formulations. In: *36th Symposium on Pesticide Formulation and Delivery Systems: Emerging Trends Building on a Solid Foundation*. ASTM International; 2016 Nov 30. pp. 76–89.

[10] Kolluru LP, Atre P, Rizvi SAA. Applications of colloidal systems as versatile drug delivery carriers for parenteral formulations. *Pharmaceuticals*. 2021;14(2):108.

[11] Khan BA, Akhtar N, Khan HM, Waseem K, Mahmood T, Rasul A, Iqbal M, Khan H. Basics of pharmaceutical emulsions: A review. *Afr J Pharm Pharmacol*. 2011 Dec 30;5(25):2715–2725.

[12] Dunn CJ, Wagstaff AJ, Perry CM, Plosker GL, Goa KL. Cyclosporin: An updated review of the pharmacokinetic properties, clinical efficacy and tolerability of a microemulsion-based formulation (Neoral®) su1 in organ transplantation. *Drugs*. 2001 Nov;61(13):1957–2016.

[13] Damitz R, Chauhan A. Kinetically stable propofol emulsions with reduced free drug concentration for intravenous delivery. *Int J Pharm*. 2015 May 30;486(1–2):232–241.

[14] Pant A, Jha K, Singh M. Role of Excipient's HLB values in microemulsion system. *J Pharm Biol Sci*. 2019 Apr;14:1–6.

[15] Desai N. Nanoparticle albumin-bound paclitaxel (Abraxane®). In: *Albumin in medicine: Pathological and clinical applications*. Singapore: Springer Singapore; 2016 Nov 2. pp. 101–119.

[16] Beija M, Salvayre R, Lauth-de Viguerie N, Marty JD. Colloidal systems for drug delivery: From design to therapy. *Trends Biotechnol*. 2012 Sep 1;30(9):485–496.

[17] Guy A, Marrion A. Coatings components beyond binders. *Chem Phys Coatings*. 2004;267–316.

[18] Koleske JV, Springate R, Brezinski D. Additives handbook. *Paint Coat Ind*. 2011;6:1–64.

[19] Kwaambwa H. A review of current and future challenges in paints and coatings chemistry. *Prog Multidiscip Res J*. 2013;3:75–101.

[20] Jiang S, Van Dyk A, Maurice A, Bohling J, Fasano D, Brownell S. Design of colloidal particle morphology and self-assembly for coating applications. *Chem Soc Rev*. 2017;46:3792–3807.

[21] Berger PS, Clark RM, Reasoner DJ, Rice EW, Domingo JWS. *Water drinking*. Academic Press (Elsevier); 2009.

[22] Stricklin T. *Filtration Techniques. SpringWell Water: How Do Cities Treat Drinking Water – Explained*. 2022.

[23] Pronk W, Ding A, Morgenroth E, Derlon N, Desmond P, Burkhardt M, Wu B, Fane AG. Gravity-driven membrane filtration for water and wastewater treatment: A review. *Water Res*. 2019 Feb 1;149:553–565.

[24] Saatchi A, Shiller PJ, Eghtesadi SA, Liu T, Doll GL. A fundamental study of oil release mechanism in soap and non-soap thickened greases. *Tribol, Int*. 2017;110:264–271.

[25] Soaps and Detergents. ScienceReady (2021). https://scienceready.com.au/pages/soaps-and-detergents
[26] Iwuozor KO. Properties and Uses of Colloids: A Review. *Sci Publ Group*. 2019.
[27] Mosley LM, Hunter KA, Ducker WA. Forces between colloid particles in natural waters. *Environ Sci Technol*. 2003 Aug 1;37(15):3303–3308.
[28] Smart JS, Moruzzi UL. Quantitative properties of delta channel networks. *Z Geomorph*. 1972;16 (3):268–282.
[29] Howe AM. Some aspects of colloids in photography. *Curr Opin Colloid Interface Sci*. 2000;5 (4):287–293.
[30] Maternaghan T. Halide photographic emulsions. *Technol Appl Dispersions*. 1994 Mar 30;52:373.
[31] Howe AM. Some aspects of colloids in photography. *Curr Opin Colloid Interface Sci*. 2000 Nov 1;5(5–6):288–300.
[32] Clayton W. Colloid chemistry. Annual Reports on the Progress of chemistry. Royal Society of Chemistry. 1925;22:281–332
[33] Sachan R, Tiwari A, Cloud seeding: Artificial rain making process. ResearchGate. 2022:127–128
[34] Azeez HM, Ibraheem NT, Hussain HH. Alternate chemical compounds as a condensation nucleus in cloud seeding. *Nat Environ Pollut Technol*. 2024 Sep 1;23(3):1795–1799.
[35] Bruintjes RT, Salazar V, Semeniuk TA, Buseck P, Breed DW, Gunkelman J. Evaluation of hygroscopic cloud seeding flares. *J Weather Modif*. 2012 Apr 1;44(1).
[36] Costa AB. A matter of life and breath: Frederick Gardner cottrell and the research corporation. *J Chem Educ*. 1985 Feb;62(2):135.
[37] Zhou H, Liu G, Zhang L, Zhou C. Formation mechanism of arsenic-containing dust in the flue gas cleaning process of flash copper pyrometallurgy: A quantitative identification of arsenic speciation. *Chem Eng J*. 2021 Nov 1;423:130193.
[38] Khattak Z, Ahmad J, Ali HM, Shah S. Contemporary dust control techniques in cement industry, Electrostatic Precipitator-a case study. *World Appl Sci J*. 2013;22(2):202–209.
[39] Yu X, Jiang N, Miao X, Li F, Wang J, Zong R, Lu S. Comparative studies on foam stability, oil-film interaction and fire extinguishing performance for fluorine-free and fluorinated foams. *Process Saf Environ Prot*. 2020;134:1–11.
[40] Malik P, Nandini D, Tripathi BP. Firefighting aqueous film forming foam composition, properties and toxicity: A review. *Environ Chem Lett*. 2024 Aug;22(4):2013–2033.
[41] Clark M (ed). *Handbook of Textile and Industrial Dyeing: Principles, Processes and Types of Dyes; Woodhead Publishing Series in Textiles, Vol. 1*. Cambridge, UK: Woodhead Publishing (Elsevier); 2011. ISBN 978-1-84569-695-5.
[42] Hassan MM, Harris J, Busfield JJC, Bilotti E. A review of the green chemistry approaches to leather tanning in imparting sustainable leather manufacturing. *Green Chem*. 2023;25(21):7384–7410.
[43] Harlan JW, Feairheller SH. Chemistry of the Crosslinking of Collagen during Tanning. In: Crosslinking P, Friedman M (eds). *Advances in Experimental Medicine and Biology*. Vol.86A. Springer: Boston, MA; 1977. pp. 365–378.
[44] Hiemenz PC, Rajagopalan R. *Principles of Colloid and Surface Chemistry*. revised and expanded. CRC press; 2016 Oct 4.
[45] Pashley RM, Karaman ME. *Applied colloid and surface chemistry*. John Wiley & Sons; 2021 Aug 4.

Namita Johar and Neeta Azad*

8 Surface Chemistry

Abstract: Surface chemistry is a branch of chemistry that focuses on the study of physical and chemical processes occurring at surfaces or interfaces when two or more phases, whether similar or different, come in contact with each other. These interacting phases can include solid-solid, solid-gas, liquid-gas, or liquid-liquid systems. Various phenomena such as adsorption, corrosion, colloidal behavior, and catalysis fall under the domain of surface chemistry. Unlike absorption, which involves the penetration of a substance into the bulk of a material, adsorption is a surface phenomenon that affects only the surface layer, leaving the bulk material largely unchanged. However, adsorption is often accompanied by other chemical processes, such as absorption and chemical reactions.

Surface chemistry has extensive applications in both industry and daily life. This chapter explores adsorption and related phenomena in detail. By the end of this chapter, learners will gain an understanding of physisorption and chemisorption, adsorption of gases, corrosion and its occurrence, adsorption isotherms, catalysis and its role in chemical reactions, and formation of colloids.

Adsorption is particularly significant in catalysis, where it facilitates chemical reactions and simplifies catalyst recovery. It also plays a key role in chromatography, a technique used for material separation. Additionally, adsorption is crucial for pollution control, as it helps remove contaminants from liquids and gases. In biological systems, adsorption influences critical processes such as enzyme activity, drug delivery, and toxin removal. This chapter comprehensively covers these topics, highlighting the significance of surface chemistry in scientific and practical applications.

8.1 Introduction

Surface chemistry is the branch of chemistry that deals with the study of chemical and physical phenomena occurring at the surfaces and/or interfaces of two phases when they come into contact with each other. When two or more different phases come into contact, multiple phenomena occur, such as chemical reactions, adsorption, and absorption. When the phenomenon is confined to the surface only and not the bulk, it is called adsorption. Since the interactions are confined only to the surface, it is often possible to separate the phases once the desired result has been achieved. Owing to this property, surface chemistry plays a central role in various applications, particularly in catalysis. Among the different surface phenomena, adsorption is the most significant, as it serves as the basis for a wide range of catalytic processes. Adsorption takes place due to the interaction between the surface molecules of two phases.

https://doi.org/10.1515/9783112208236-008

8.2 Adsorption

Adsorption is a surface phenomenon in which molecules accumulate on the boundary between two phases without penetrating the bulk. It occurs in systems such as solid-solid, solid-liquid, and liquid-gas interfaces. There is no bulk interaction. Adsorption is the adhesion of gas molecules or solute molecules from a solution onto the surface of a solid. In this process, the substance being adsorbed (gas or solute) is called the adsorbate, while the solid surface on which adsorption takes place is called the adsorbent (Figure 8.1). The plane that separates the adsorbate from the adsorbent is called as interface. Adsorption is classified into two types: physical adsorption (physisorption) and chemical adsorption (chemisorption). The other important associated terms are desorption, sorption, and occlusion.

8.2.1 Important Terms in Surface Chemistry

1. **Desorption:** The reverse of adsorption is called desorption, i.e., the removal of adsorbed molecules from the surface of the adsorbent.
2. **Sorption:** Sorption refers to the simultaneous occurrence of adsorption and absorption in a process.
3. **Occlusion:** Adsorption of gases on the surface of metal is known as occlusion.

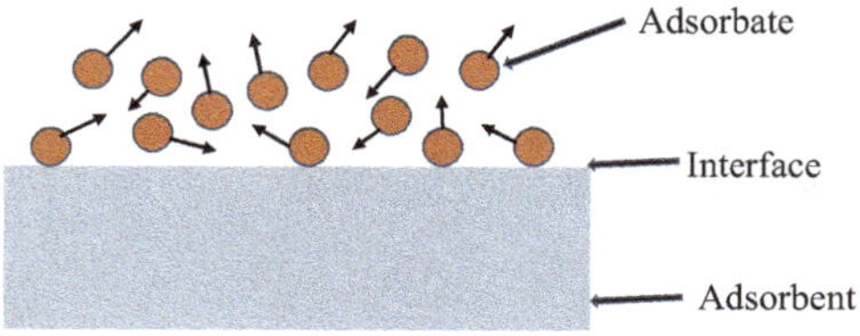

Figure 8.1: Adsorption of the adsorbate on the surface of the adsorbent.

8.2.2 Difference Between Adsorption and Absorption

Both adsorption and absorption involve interaction between two different phases. Most of the time, one phase is solid and another is liquid or gas. But adsorption is different than absorption, which is visualized and discussed in Figure 8.2 and Table 8.1, respectively. In adsorption, the molecules or particles of one phase adhere to the surface of another, but in the case of absorption, particles of one phase penetrate the bulk of another. Both processes are different in many ways.

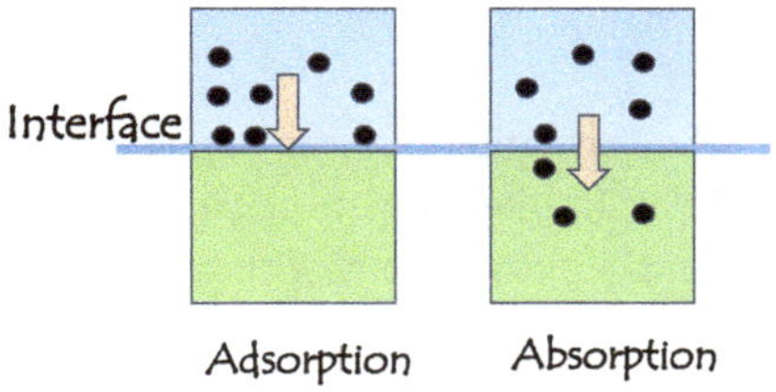

Figure 8.2: Surface adhesion (adsorption) versus bulk uptake (absorption).

Table 8.1: Difference between adsorption and absorption.

Adsorption	Absorption
Adsorption involves the adherence of molecules to the surface of one phase due to attractive forces.	In absorption, the molecules of one phase diffuse and penetrate through another phase.
It is an exothermic process because heat is released during bond formation.	It is an exothermic process.
It is a bulk phenomenon.	It is a surface phenomenon.
This happens at low temperatures.	There is no effect of temperature on absorption.
Pressure plays a very important role in assisting molecules of adsorbate to stay near the surface of the adsorbent.	Pressure has no role to play in absorption.
This is a reversible process.	This is an irreversible process.
Adsorption of O_2 on an activated carbon surface is an example of adsorption.	Absorption of water by salt or sugar is an example of absorption.

8.2.3 Physical Adsorption and Chemical Adsorption

8.2.3.1 Physical Adsorption (Physisorption)

Physical adsorption [1], or physisorption, arises due to weak van der Waals forces between the adsorbate and the adsorbent. Since these forces are weak, the process is reversible and does not alter the chemical structure of either the adsorbate or the adsorbent. Heat of adsorption in this process is low due to weak interaction and lies in the range of 20–40 kJ/mol. The nature of the adsorbate or adsorbent has very less to do with this process but still can affect the process. During adsorption, multiple layers of adsorbate may form on the surface of the adsorbent. When the temperature of the system is raised, the weak van der Waals forces become even weaker, which results in decreased extent of adsorption. Similarly, with decrease in temperature increases the extent of adsorption. Adsorption of oxygen or nitrogen on the surface of activated charcoal is an example of physisorption.

8.2.3.2 Chemical Adsorption (Chemisorption)

Chemical adsorption, or chemisorption, involves the formation of strong covalent or ionic bonds between the adsorbate and the adsorbent. Unlike physisorption, this process is irreversible due to the strength of the chemical bonds formed. The heat of adsorption is much higher than physisorption between 40 and 400 kJ/mol. In chemisorption, there is single-layer formation of the adsorbate on the surface of the adsorbent. Chemisorption is a highly selective process due to the involvement of strong bond formation. Adsorption of hydrogen gas on the surface of nickel and adsorption of oxygen on the surface of iron during rusting are chemisorptions. Chemisorption plays a key role in heterogeneous catalysis and surface reactions.

Thus, physical adsorption is different from chemical adsorption in many aspects, which is tabulated in Table 8.2.

Table 8.2: Differences between physisorption and chemisorption.

Physisorption	Chemisorption
Physisorption is caused by physical forces called van der Waals forces of attraction. No chemical bonds are formed here.	In chemisorption, chemical bonds are formed between the adsorbate and the adsorbent.
Physisorption is a weak phenomenon.	Because of the involvement of strong bonds like covalent or electrostatic bonds, this is a strong process.
In this process, adsorbate forms multiple layers on the surface of the adsorbent.	Chemisorption involves single-layer formation of adsorbate on the surface of the adsorbent.
Physisorption takes place all around the adsorbent and is not site-specific.	Chemisorption is site-specific, taking place only at certain locations on the adsorbent surface.
Factors affecting this process include surface area, temperature, pressure, and the type of adsorbent.	The extent of chemisorption depends on factors such as surface area, temperature, pressure, and the type of adsorbent.
Physisorption happens with a very low activation energy of the order 20–40 kJ/mol.	The energy of activation in this process is higher than that of physisorption and lies in the range 40–400 kJ/mol.

8.2.4 Factors Affecting Adsorption

The extent of adsorption is influenced by several factors, discussed as follows:

1. **Nature of the adsorbent:** Adsorbents with a larger surface area exhibit higher adsorption, as more surface is available for adsorbate interaction. Common examples include activated charcoal and silica gel.

2. **Nature of the adsorbate:** Molecules with higher polarity or easily liquefiable gases are more readily adsorbed due to stronger interactions with the adsorbent surface.
3. **Surface area:** The extent of adsorption increases with the surface area of the adsorbent, as greater surface availability enhances the interaction between adsorbate and adsorbent.
4. **Temperature:** The effect of temperature differs for physisorption and chemisorption. Being exothermic, physisorption decreases with increasing temperature, since adsorbate molecules gain enough kinetic energy to overcome weak van der Waals forces and desorb from the surface. Chemisorption, on the other hand, generally increases with temperature up to a certain limit, as activation energy is required to form chemical bonds. Beyond surface saturation or at very high temperatures, chemisorption also declines. This contrasting behavior of the two processes with respect to temperature is illustrated in Figure 8.3.
5. **Concentration:** For adsorption from solutions, the extent of adsorption increases with the concentration of the solute, since more adsorbate molecules are available to occupy the surface sites of the adsorbent. At equilibrium, a balance is established between the concentration of adsorbate in solution and on the surface.

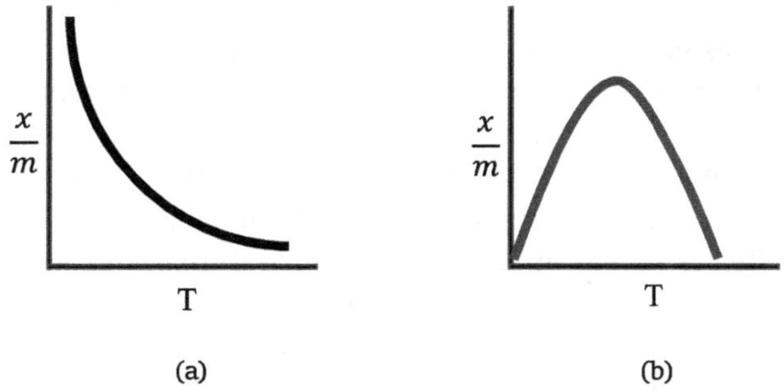

Figure 8.3: Variation of extent of adsorption with temperature in (a) physisorption and (b) chemisorption.

8.2.5 Adsorption of Gases on Solids

When gases are adsorbed on the surface of a metal or any other solid surface, the process is called adsorption. This is influenced by several factors, such as the nature of the adsorbent, surface area, temperature, pressure, and the type of gas. These factors determine the extent and effectiveness of adsorption:

1. **Nature of adsorbent:** The adsorption capacity largely depends on the properties of the adsorbent [2]. Substances with a highly porous structure show greater adsorption capacity. Common adsorbents include activated carbon, silica gel, alumina (Al_2O_3), metal oxides, and clays. The adsorption efficiency can be signifi-

cantly improved by activation, which increases the pore size and the number of active sites. For example, ordinary charcoal at 24 °C can adsorb only 0.011 g of a gas, whereas activated charcoal shows much higher adsorption.

2. **Activation of adsorbents:** Activation involves treating the adsorbent to increase its effectiveness. Typically, the adsorbent is heated with steam at around 1,500 °C, which removes impurities and increases the available surface area. Activation can be achieved in the following ways:
   (i) Roughening the surface of the adsorbent
   (ii) Heating the adsorbent in a vacuum to remove water vapor from its pores
   (iii) Treating with steam at high temperature (around 1,500 °C) to open pores and hence increase surface area
3. **Effects of surface area on adsorption:** An increase in the surface area of an adsorbent enhances the adsorption of gases. The extent of adsorption is influenced by two key factors:
   (i) **Surface area:** A large surface area provides more active sites for gas molecules, leading to higher adsorption capacity.
   (ii) **Porosity:** Materials that are finely divided and highly porous offer greater adsorption, making them highly effective adsorbents.
4. **Nature of gases**: The nature of a gas strongly affects its adsorption on a solid surface. Easily liquefiable gases such as HI, NH_3, Cl_2, and SO_2 are adsorbed more readily than permanent gases such as H_2, N_2, and O_2. Physical adsorption is nonspecific, so any gas can be adsorbed under suitable pressure and temperature conditions. Chemical adsorption occurs only when the gas can form a chemical bond with the surface. Adsorption of gases is influenced by two primary factors:
   (i) **The critical temperature**: The highest temperature at and above which a gas cannot be liquefied is known as the critical temperature. Gases with higher critical temperatures are more easily liquefied and, therefore, show stronger adsorption.
   (ii) **Stronger van der Waals forces**: These are found in gases that are easily liquefiable and hence responsible for adsorption.
5. **Heat of adsorption:** The heat evolved or absorbed when 1 g of a gas is adsorbed on a solid surface is called the heat of adsorption. Since adsorption is generally an exothermic process, the extent of adsorption decreases with increasing temperature. A rise in temperature provides gaseous molecules with enough kinetic energy to escape from the surface of the adsorbent.
6. **Pressure:** Adsorption increases with an increase in pressure, as higher pressure forces gas molecules to remain close to the adsorbent surface. At high pressures, adsorption reaches a saturation point when no further adsorption is possible because the surface is fully occupied.

8.2.6 Adsorption of Liquids

Solutions comprise two phases: solute (dissolved substance) and solvent (dissolving medium). Desorption in liquids refers to the process where a solute is adsorbed on the surface of a solid adsorbent from its solution. This technique is widely applied for the purification, separation, and decolorization of solutions. A common example is the adsorption of colored impurities from a solution by activated charcoal, leaving behind a colorless liquid. Similarly, charcoal can adsorb dissolved organic matter and acids, making it useful in water purification. The extent of adsorption from a solution depends on several factors:

1. **Concentration of the solute:** Higher solute concentration generally increases adsorption until the adsorbent surface is saturated.
2. **Nature of the adsorbent and adsorbate:** Different solids have varying affinities for different solutes. For example, silica gel, alumina, and activated carbon are commonly used in liquid-phase adsorption.
3. **Temperature:** Since adsorption is usually exothermic, the extent of adsorption decreases with increasing temperature.

8.3 Different Types of Adsorption Isotherms

To understand the adsorption of liquids more clearly, it is essential to describe the quantitative relationship between the amount of substance adsorbed and its equilibrium concentration in the solution. This relationship is expressed through adsorption isotherms, which provide valuable insights into the capacity of the adsorbent, the nature of the surface, and the strength of adsorbate-adsorbent interactions. Several models have been proposed to explain the adsorption behavior, among which the Freundlich and Langmuir isotherms are the most widely studied.

8.3.1 Freundlich Isotherm

The relationship between solute concentration and the extent of adsorption is quantitatively expressed by the Freundlich adsorption isotherm [3]. This empirical equation provides a simple yet effective way to describe how the amount of solute adsorbed varies with its equilibrium concentration in solution. It is particularly applicable to adsorption occurring on heterogeneous surfaces and offers a practical means to compare adsorption capacities of different adsorbents.

Let us understand the Freundlich adsorption isotherm in detail. Consider a solution of concentration C. If the amount of solute adsorbed on the surface of an adsorbent of mass m is x, then the adsorption capacity is expressed as x/m . According to

Freundlich, the extent of adsorption is related to the solute concentration by the following empirical equation:

$$\frac{x}{m}\ \alpha\ C^{1/n} \tag{8.1}$$

$$\frac{x}{m} = k_F C^{1/n} \tag{8.2}$$

$$\log\left(\frac{x}{m}\right) = \log k_F + \frac{1}{n}\log C \tag{8.3}$$

For a gaseous mixture, pressure (p) is considered instead of concentration. Hence, eq. (8.3) becomes:

$$\log\left(\frac{x}{m}\right) = \log k_F + \frac{1}{n}\log\ p \tag{8.4}$$

This equation can be compared with the equation of a straight line

$$y = mx + c$$

The plot of log (x/m) versus log p comes out to be a straight line with slope $1/n$ and intercept $\log k_F$ (Figure 8.4) at constant temperature.

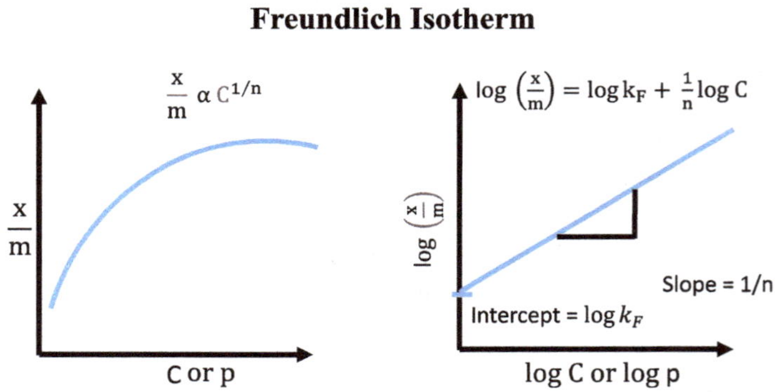

Figure 8.4: Plot of the Freundlich isotherm depicting the relationship between the extent of adsorption and concentration/pressure.

Here, k_F is a proportionality constant called as Freundlich coefficient. The value of k_F reflects the adsorption capacity of the adsorbent added to the solution. n is a constant related to the intensity of adsorption, known as the Freundlich constant. The value of $1/n$ serves as a correction factor and typically ranges from 0.7 to 1.0 at intermediate pressures.

The value of k_F depends on the nature of the solute, particle size of solute, temperature of solution, and nature of solvent. The variation of concentration C of solution with $\frac{x}{m}$ is parabolic in nature (Figure 8.4), and the variation of log $\left(\frac{x}{m}\right)$ with respect to $\log C$ comes out to be a straight line with slope $\frac{1}{n}$. The nature of the graph indicates that as the concentration of the solute rises, the adsorption increases up to a certain point, after that it becomes constant.

In eq. (8.5), at low pressure, the value of the correction factor n is almost equal to 1, and hence, the adsorption varies directly with pressure

$$\frac{x}{m}\ \alpha\ p^1 \tag{8.5}$$

At high pressure, the adsorption becomes independent of pressure as the value of n becomes very high, which leads to the following equation:

$$\frac{x}{m}\ \alpha\ p^0 \tag{8.6}$$

But, at intermediate pressure, n has value greater than 1 and hence shows the variation as shown in Figure 8.5.

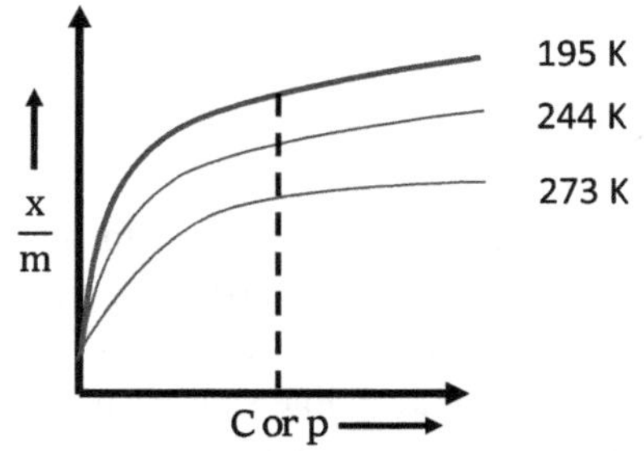

Figure 8.5: Variation of the extent of adsorption with concentration or pressure at different temperatures.

8.3.1.1 Drawbacks of the Freundlich Adsorption Isotherm

The Freundlich isotherm, despite its extensive application, is associated with several inherent limitations. It is derived on the premise that the solution acts as a homogeneous system and that adsorption sites are evenly distributed across the surface of the adsorbent. In practice, however, adsorbent surfaces are heterogeneous, with adsorption sites exhibiting different affinities for the adsorbate. Moreover, the applicability of the Freundlich equation is restricted to a limited pressure range. At pressures beyond saturation, adsorption attains a maximum value and remains constant, which the isotherm fails to represent. Another drawback is that the constants k_f and n are temperature-dependent, as evidenced by the variation in the slope of the x/m versus p plot with temperature. In liquid systems, the isotherm also shows deviations at very high solute concentrations, where the experimental data no longer fits the predicted relationship.

8.3.2 Langmuir Isotherm

Another important isotherm is the Langmuir isotherm. This isotherm also explains the variation of adsorption with pressure and was given by Irving Langmuir in 1916 [1]. The Langmuir isotherm is based on the following assumptions:

1. The surface of adsorbents has fixed adsorption sites with uniform adsorption energy.
2. The size and shape of all adsorption sites are the same. Each site adsorbs one molecule per site and releases a fixed amount of heat.
3. The heats of adsorption throughout the surface are the same and lie between 0 and 1.
4. The adsorbed gaseous molecules and free molecules above the adsorbed layer are continuously exchanging due to a dynamic equilibrium between them, called adsorption and desorption.
5. Only a single layer is formed as a result of adsorption (called a monolayer), which covers all possible binding sites, because all adsorbing sites are blocked after that.

8.3.2.1 Mathematical Derivation of the Langmuir Isotherm

The basis of the Langmuir isotherm equation [4] is the concept of monolayer formation, which represents maximum covering capacity by occupying all adsorbing sites and the binding affinity (called as adsorption constant). Let the fractional occupancy of the adsorption sites be q_e, which represents the fraction of available sites occupied by adsorbate molecules at pressure p_e or solute concentration C_e in equilibrium with the adsorbent surface:

$$q_e = \frac{(K_l \times p_e)}{1 + (K_l \times p_e)} \tag{8.7}$$

$$q_e = \frac{(K_l \times C_e)}{1 + (K_l \times C_e)} \tag{8.8}$$

Here, K_l is the Langmuir constant, which represents how strongly an adsorbate molecule interacts with a specific site on the surface of the adsorbent, p_e is the pressure of the gases when a gas is being adsorbed, and C_e is the concentration when the adsorption is taking place from liquid solutions. The Langmuir adsorption isotherm is graphically discussed in Figure 8.6.

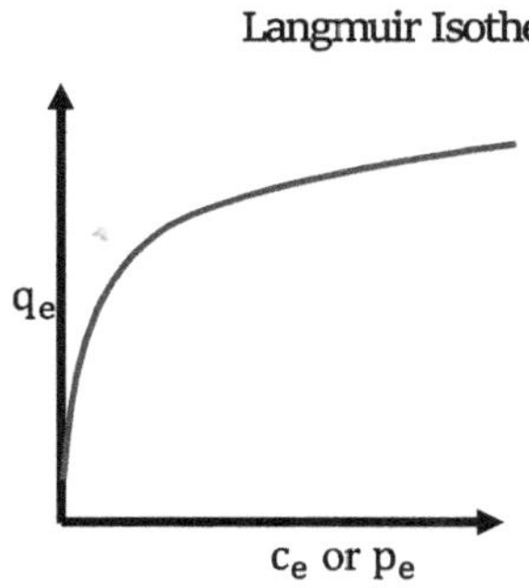

$$q_e = \frac{(K_l \times C_e)}{1+(K_l \times C_e)}$$

or

$$q_e = \frac{(K_l \times p_e)}{1+(K_l \times p_e)}$$

Figure 8.6: Langmuir isotherm showing variation of fractional binding site occupancy with concentration or pressure of adsorbate.

8.3.2.2 Significance of Langmuir Isotherm

The Langmuir isotherm holds great importance as it explains the influence of temperature, pressure, and concentration on the equilibrium of adsorption. It provides a theoretical basis for predicting adsorption capacity and surface coverage under different conditions. Such understanding is valuable in numerous practical fields, including heterogeneous catalysis, gas purification and separation, and the removal of contaminants from aqueous systems. For example, in the removal of dyes from wastewater, the Langmuir isotherm can be applied to determine the maximum adsorption capacity of activated carbon [5]. This helps in designing treatment systems that efficiently clean water by predicting how much dye can be adsorbed before the adsorbent surface becomes saturated.

8.3.2.3 Limitations of Langmuir Isotherm

The Langmuir isotherm is applicable under the assumption that gases and solutions exhibit ideal behavior. In practice, however, deviations from ideality occur, which restrict the validity of the isotherm to low pressures or dilute concentrations. The assumption of monolayer adsorption also holds true only under such conditions. At higher pressures, gas molecules accumulate near the adsorbent surface, leading to multilayer adsorption that contradicts the monolayer concept. Furthermore, the Langmuir model assumes a homogeneous adsorbent surface with uniformly distributed adsorption sites [1]. In reality, most adsorbent surfaces are heterogeneous, possessing sites with varying affinities toward the adsorbate. For example, in the adsorption of nitrogen gas on activated carbon, the Langmuir model fits well only at low pressures, where a monolayer is formed. At higher pressures, additional layers of nitrogen molecules build up on the first layer, producing multilayer adsorption. This behavior cannot be explained by the Langmuir isotherm but is better described by the Brunauer, Emmett, and Teller (BET) model.

8.3.3 BET Equation and Different Isotherms

The adsorption of gases on the surface of the adsorbent is monolayer only in ideal cases. The scenario changes when there is an increase in pressure and a decrease in temperature. At high pressures, the extent of adsorption of molecules increases, which is multilayered. Similarly, when the temperature falls, molecules do not get enough energy to escape or overcome the van der Waals forces of attraction between the adsorbed and unadsorbed molecules. This leads to multilayer adsorption. The formation of a multilayer is favored more when the pressure of the gas approaches the vapor pressure of the liquefied adsorbent at the given experimental temperature. Consequently, the shape of the isotherm varies from Figure 8.6 and turns out to be as shown in Figure 8.7 at different pressures (explained in the next section).

Multilayer isotherms were explained by Brunauer, Emmett, and Teller through an equation popularly known as BET equation or BET adsorption isotherm. The explanation for multilayer adsorption is similar to that of the Langmuir equation. In deriving the BET equation, interactions between adsorbed molecules are completely neglected. The process of multilayer adsorption can be represented by the following equations:

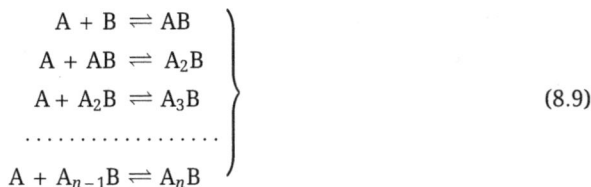

$$\left.\begin{array}{c} A + B \rightleftharpoons AB \\ A + AB \rightleftharpoons A_2B \\ A + A_2B \rightleftharpoons A_3B \\ \cdots\cdots\cdots\cdots \\ A + A_{n-1}B \rightleftharpoons A_nB \end{array}\right\} \tag{8.9}$$

where A, B, AB, A_2B, etc. represent the unadsorbed gaseous molecules, the vacant site in the adsorbent surface, single molecules adsorbed per site, two molecules adsorbed per vacant site, and so on. For the equilibrium represented by the set of eq. (8.9), the respective reaction constants are given as:

$$K'_1 = \frac{[AB]}{[A][B]} \tag{8.10}$$

$$K'_2 = \frac{[A_2B]}{[A][AB]} \tag{8.11}$$

$$K'_3 = \frac{[A_3B]}{[A][A_2B]} \tag{8.12}$$

Here, [A], [B], [AB], [A_2B], and [A_3B] represent the concentrations of the respective components. The concentrations of gases (here A, the adsorbate) are proportional to the pressure of the gas.

Hence, [A] $\propto$ pressure of the gas, or [A] $\propto p$; [B] $\propto$ fraction of the free surface, or [B] $\propto \theta_0$; [AB] $\propto$ fraction of surface covered with single-molecule adsorption or [AB] $\propto \theta_1$; [A_2B] $\propto$ fraction of surface covered with two-molecule adsorption, or [A_2B] $\propto \theta_2$; and so on for consecutive adsorption.

From the above equations, the equilibrium constants in terms of θ_0, θ_1, θ_2, etc. become:

$$\left.\begin{aligned} K_1 &= \frac{\theta_1}{p\theta_0} \\ K_2 &= \frac{\theta_2}{p\theta_1} \\ K_3 &= \frac{\theta_3}{p\theta_2} \end{aligned}\right\} \tag{8.13}$$

and so on.

K_1 has a higher value than consecutive equilibrium constants. This is because as the distance of adsorbate molecules increases from the adsorbent surface, the concentration of adsorption decreases and hence the equilibrium constant. But with every adsorbed layer after first layer, the difference between consecutive equilibrium constants becomes smaller, and hence all equilibrium constants, K_2, K_3, K_4, etc., are considered to be almost equal,

i.e., $K_2 \simeq K_3 \simeq K_4 \simeq K_5 \simeq \ldots \simeq K_L$

where K_L is the equilibrium constant corresponding to the saturated vapor-liquid equilibrium and, also

Saturated vapor $\leftrightharpoons$ Liquid

$$K_L = \frac{1}{p_0} \tag{8.14}$$

With the approximations given in eq. (8.14):

$$\left.\begin{aligned} K_1 &= \frac{\theta_1}{p\theta_0} \\ K_2 = K_L &= \frac{\theta_2}{p\theta_1} \\ K_3 = K_L &= \frac{\theta_3}{p\theta_2} \end{aligned}\right\} \tag{8.15}$$

and so on.

These equations can be further rearranged to get the values of θ_1, θ_2, θ_3, etc. as shown below:

$$\left.\begin{aligned} \theta_1 &= K_1 p\theta_0 \\ \theta_2 = K_L p\theta_1 &= \left(\frac{1}{p_0}\right) p(K_1 p\theta_0) = K_1 p\left(\frac{p}{p_0}\right)\theta_0 \\ \theta_3 = K_L p\theta_2 &= \left(\frac{1}{p_0}\right) p\left(K_1 p\frac{p}{p_0}\theta_0\right) = K_1 p\left(\frac{p}{p_0}\right)^2 \theta_0 \\ &\cdots\cdots\cdots\cdots\cdots\cdots \end{aligned}\right\} \tag{8.16}$$

When the adsorption process is complete and the adsorbent surface can no longer adsorb any adsorbent, the value of θ_{total} can be given as the sum of all other fractions of surface covered as shown below:

$$\theta_{\text{total}} = \theta_0 + \theta_1 + \theta_2 + \theta_3 + \cdots = 1$$

The set of eq. (8.16) is

$$\begin{aligned} \theta_{\text{total}} &= \theta_0 + K_1 p\theta_0 + K_1 p\left(\frac{p}{p_0}\right)\theta_0 + K_1 p\left(\frac{p}{p_0}\right)^2 \theta_0 + \cdots\cdots\cdots = 1 \\ &= \theta_0\left[1 + K_1 p\left\{1 + \left(\frac{p}{p_0}\right) + \left(\frac{p}{p_0}\right)^2 + \cdots\right\}\right] = 1 \end{aligned} \tag{8.17}$$

Since $p_0 \geq p$, $p/p_0 \leq 1$, eq. (8.17) becomes

$$\left\{1 + \left(\frac{p}{p_0}\right) + \left(\frac{p}{p_0}\right)^2 + \cdots\right\} = \left[1 - \left(\frac{p}{p_0}\right)\right]^{-1} = \frac{1}{1 - \left(\frac{p}{p_0}\right)} \tag{8.18}$$

Hence, from eq. (8.17)

$$\theta_{\text{total}} = \theta_0\left[1 + K_1 p\left[1 - \left(\frac{p}{p_0}\right)\right]^{-1}\right] = \theta_0\left[1 + \frac{K_1 p}{\left[1 - \left(\frac{p}{p_0}\right)\right]}\right] = 1 \tag{8.19}$$

or

$$\theta_0 = \frac{1}{\left[1 + \frac{K_1 p}{\left[1 - \left(\frac{p}{p_0}\right)\right]}\right]} = \frac{1 - \left(\frac{p}{p_0}\right)}{1 + K_1 p - \frac{p}{p_0}} \tag{8.20}$$

The total volume of the gas can be calculated as

$$V_{\text{total}} = V_{\text{mono}}(\theta_1 + 2\theta_2 + 3\theta_3 + \cdots) \tag{8.21}$$

Here, V_{mono} is the volume required for the monolayer adsorption. Substituting the values of $\theta_1, \theta_2, \theta_3, \ldots$ in eq. (8.21)

$$V_{total} = V_{mono}\ K_1 p \theta_0 \left(1 + 2\left(\frac{p}{p_0}\right) + 3\left(\frac{p}{p_0}\right)^2 + \cdots\right) = V_{mono}\ K_1 p\ \theta_0 \left[1 - \left(\frac{p}{p_0}\right)\right]^{-2} \tag{8.22}$$

Here,

$$\left[1 - \left(\frac{p}{p_0}\right)\right]^{-2} = \left(1 + 2\left(\frac{p}{p_0}\right) + 3\left(\frac{p}{p_0}\right)^2 + \cdots\right) \tag{8.23}$$

Substituting the value of θ_0 and rearranging, eq. (8.22) becomes

$$V_{total} = \frac{V_{mono}\ K_1 p}{\left(1 - \left(\frac{p}{p_0}\right)\right)\left(1 + K_1 p - \left(\frac{p}{p_0}\right)\right)} \tag{8.24}$$

The pressure p in the above expression may be replaced in terms of relative pressure p/p_0 as shown below:

$$p = p_0 \frac{p}{p_0} = \frac{1}{K_L}\left(\frac{p}{p_0}\right)$$

Hence, V_{total} from eq. (8.24) becomes

$$V_{total} = \frac{V_{mono} \frac{K_1}{K_L}\left(\frac{p}{p_0}\right)}{\left(1 - \left(\frac{p}{p_0}\right)\right)\left(1 + \frac{K_1}{K_L}\left(\frac{p}{p_0}\right) - \left(\frac{p}{p_0}\right)\right)} \tag{8.25}$$

Substituting $\frac{K_1}{K_L} = C$ in eq. (8.25)

$$V_{total} = \frac{V_{mono} C\left(\frac{p}{p_0}\right)}{\left(1 - \left(\frac{p}{p_0}\right)\right)\left(1 + C\left(\frac{p}{p_0}\right) - \left(\frac{p}{p_0}\right)\right)} \tag{8.26}$$

Equation (8.26) is known as the BET equation for multilayer adsorption of gaseous molecules.

Further rearrangement of eq. (8.26) gives the following expression for the BET equation:

$$\frac{p}{V_{total}(p - p_0)} = \frac{1}{V_{mono} C} + \frac{C - 1}{V_{mono} C} \frac{p}{p_0} \tag{8.27}$$

Comparing this equation with the equation of a straight line $y = mx + c$.

Hence, the plot of $\frac{p}{V_{total}(p - p_0)}$ versus $\frac{p}{p_0}$ should be a straight line with slope $= \frac{C - 1}{V_{mono} C}$ and intercept $\frac{1}{V_{mono} C}$.

The constants C and V_{mono} can thus be calculated, and hence the surface area of the adsorbent can be evaluated from the following equation:

$$V = \frac{4}{3}\pi r^3 = \frac{M}{N_A \rho} \tag{8.28}$$

$$A = \pi r^2 = \pi \left(\frac{3}{4} \frac{M}{N_A \rho} \right)^{2/3} \tag{8.29}$$

Here, A is the cross-sectional area of the molecules, ρ is the density, and $V = V_{\text{mono}}$ is the volume occupied by a single molecule.

8.3.3.1 The Thermodynamic Aspect of Constant *C* in the BET Equation

The standard free energy of any equilibrium can be given as:

$$\Delta G^\circ = \Delta H^\circ - T\Delta S^\circ = -RT \ln K_1^0 \tag{8.30}$$

Here, $K_1^0 = K_1 P^0$.

For adsorption, eq. (8.19) can be rearranged for K_1^0 as

$$K_1^0 = \exp(\Delta_{\text{ads}} S_1^0 / R)\ \exp(-\Delta_{\text{ads}} H_1^0 / RT) \tag{8.31}$$

$$= g_1 \exp(\Delta_{\text{des}} H_1^0 / RT) \tag{8.32}$$

Here g_1 is the entropy factor and $\Delta_{\text{des}} H_1^0$ is the standard entropy of desorption of monolayer formation.

Similarly, for the equilibrium constant K_L, the expression becomes

$$K_L^0 = g_L \exp(\Delta_{\text{vap}} H_L^0 / RT) \tag{8.33}$$

where $K_L^0 = K_L P^0$ and $\Delta_{\text{vap}} H_L^0$ is the standard entropy of vaporization of the liquid adsorbate.

Hence, the expression for C can be evaluated from the value of K_1^0 and K_L^0

$$C = \frac{K_1^0}{K_L^0} \tag{8.34}$$

$$= \frac{g_1 \exp(\Delta_{\text{des}} H_1^0 / RT)}{g_L \exp(\Delta_{\text{vap}} H_L^0 / RT)} \tag{8.35}$$

$$\simeq \exp\left(\frac{\Delta_{\text{des}} H_1^0 - \Delta_{\text{vap}} H_L^0}{RT} \right) \tag{8.36}$$

After rearranging eq. (8.27):

$$\frac{pp_0}{V_{\text{total}}(p-p_0)} = \frac{1}{V_{\text{mono}}K_1} + \frac{1}{V_{\text{mono}}}\frac{C-1}{C}p \tag{8.37}$$

If $C > 1$ and $p_0 \gg p$, then eq. (8.22) is modified to

$$\frac{p}{v_{\text{total}}} = \frac{1}{V_{\text{mono}}}\frac{1}{K_1} + \frac{p}{V_{\text{mono}}} \tag{8.38}$$

This equation is none other than the Langmuir equation.

8.3.4 The Different Adsorption Isotherm Curves

The different isotherms [6] obtained in Figure 8.7 can be explained using the BET equation:

1. **Isotherm I:** This type of curve is obtained when $p/p_0 \ll 1$ and $C \gg 1$. This is the case of monolayer adsorption and was explained by the Langmuir isotherm.
2. **Isotherm II:** This is the case when $C \ll 1$ or $\Delta_{\text{des}}H_1^0 < \Delta_{\text{vap}}H_{\text{L}}^0$. In such isotherms, only multilayer adsorption takes place since the beginning of the process.
3. **Isotherm III:** This adsorption isotherm is observed in cases where $C \gg 1$, or $\Delta_{\text{des}}H_1^0 > \Delta_{\text{vap}}H_{\text{L}}^0$. The constant value of x/m between points a and b represents monolayer adsorption within particular pressure range.
4. **Isotherm IV:** In low-pressure region, the monolayer adsorption takes place, which resembles isotherm II. In this case too, $\Delta_{\text{des}}H_1^0 < \Delta_{\text{vap}}H_{\text{L}}^0$ the shape of the curve when p approaches p^0 is different from that in isotherm III. Here, the adsorption approaches a limit when the pressure is close to the saturation vapor pressure. This kind of behavior of the isotherm can be explained by the multilayer formation along with the filling of capillary pores due to the condensation of the adsorbate below the saturation vapor pressure.
5. **Isotherm V:** In isotherm V, the behavior of the curve at lower pressure is similar to that in isotherm II, which again indicates $\Delta_{\text{des}}H_1^0 < \Delta_{\text{vap}}H_{\text{L}}^0$. As pressure rises, the curve becomes similar to curve IV, indicating saturation in adsorption for similar reason as mentioned in isotherm IV.

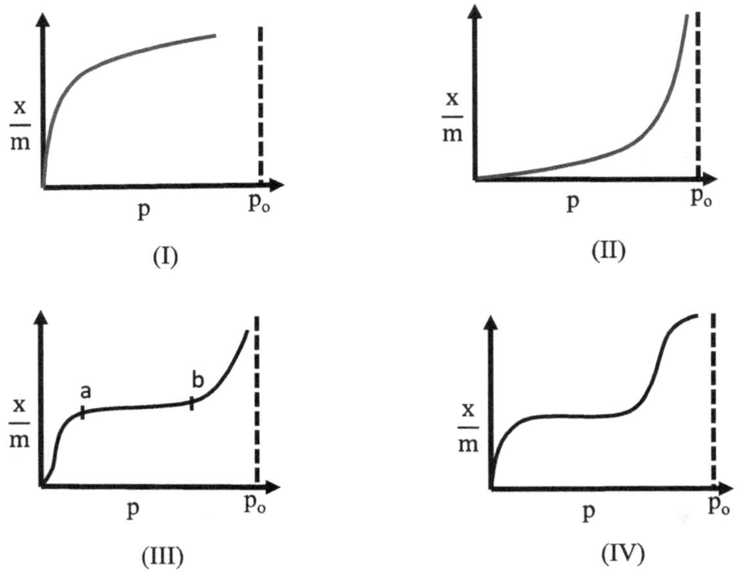

Figure 8.7: BET isotherms.

8.4 Applications of Adsorption

Adsorption is the process in which molecules accumulate on the surface of a solid or liquid rather than distributing throughout the bulk. Since it releases heat, adsorption is generally an exothermic phenomenon. The rate of adsorption is usually rapid in the beginning but gradually slows down as the surface becomes saturated, eventually reaching equilibrium. This property makes adsorption extremely useful in both industry and daily life. It is employed in air and water purification, heterogeneous catalysis, gas separation, chromatography, ion exchange processes, and the creation of high-vacuum conditions, as well as in solar refrigeration, pollution control, and biological systems such as enzyme activity. Practical examples of adsorption are seen in many situations: activated charcoal in gas masks removes toxic gases, silica gel packets absorb moisture to keep goods dry, and activated charcoal tablets are used in medicine to adsorb toxins from the digestive system. Adsorption also plays a vital role in energy storage, environmental protection, and healthcare, highlighting its broad scientific and practical importance.

8.4.1 Role of Adsorption in Catalysis

Adsorption plays a significant role in catalysis, particularly in heterogeneous catalytic reactions. A heterogeneous catalytic reaction is one in which the reactants and the catalysts are in different phases. In heterogeneous catalysis, usually the reactants are in the gaseous state while the catalyst is in the solid state. Adsorption facilitates the interaction between the solid catalyst and the gaseous reactions. As the concentration of reactant molecules at the catalyst surface increases, these are held either by van der Waals forces of attraction or chemical attractions, leading to an increase in the reaction rate, allowing the reaction to proceed quickly and efficiently toward completion. Let us understand it with some examples:

8.4.1.1 Haber-Bosch Process

In this process, nitrogen (N_2) and hydrogen (H_2) react in the presence of an iron catalyst to form ammonia (NH_3):

$$N_2(g) + 3H_2(g) \underset{\text{Mo}}{\overset{\text{Fe (s)}}{\longrightarrow}} 2NH_3$$

The basic mechanism of this reaction is shown below:

$$N_2(g) \rightarrow N_2(\text{adsorbed on solid catalyst}) \rightarrow 2N(\text{adsorbed})$$
$$H_2(g) \rightarrow H_2(\text{adsorbed on solid catalyst}) \rightarrow 2H(\text{adsorbed})$$
$$N(\text{adsorbed}) + 3H(\text{adsorbed}) \rightarrow NH_3(\text{adsorbed}) \rightarrow NH_3(g)$$

Step 1: In the first step, nitrogen and hydrogen molecules are adsorbed on the catalyst`s active sites.

Step 2: This involves the dissociation of the nitrogen-nitrogen and hydrogen-hydrogen bonds as these bonds get weakened, resulting in the formation of individual nitrogen and hydrogen atoms adsorbed on the catalyst surface.

Step 3: This involves the combination of nitrogen and hydrogen atoms to form ammonia molecules.

Step 4: Finally, these ammonia molecules undergo desorption and detach from the catalyst surface.

This ammonia, formed from the Haber-Bosch process, is used to manufacture fertilizers, explosives, textiles, chemical synthesis, plastics, etc.

8.4.1.2 Hydrogenation of Ethene

This reaction involves the hydrogenation of ethene to ethane using metal catalysts such as nickel, palladium, or platinum:

$$\underset{\text{Ethene}}{C_2H_4} + H_2 \xrightarrow{\text{Pd/Pt}} \underset{\text{Ethane}}{C_2H_6}$$

The reaction involves a series of steps, where ethene and hydrogen adsorb onto the surface of the catalyst, react, and the product is formed, which desorbs from the catalyst surface

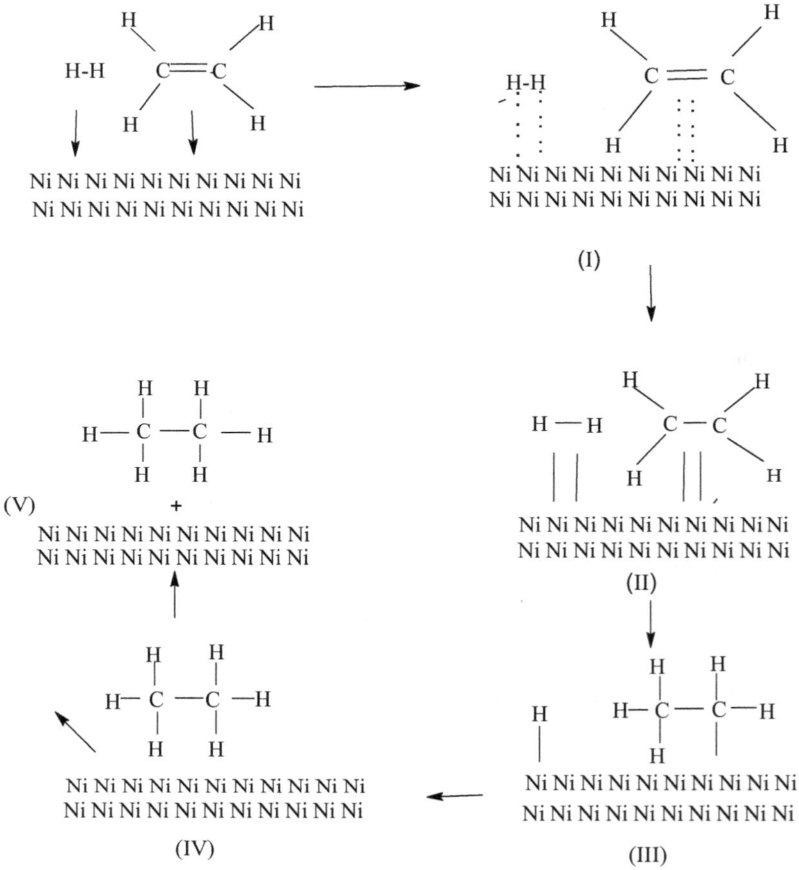

Step 1: Initially, the π-bonds of the ethene molecules are broken due to the adsorption of ethene on the nickel surface.

Step 2: Hydrogen molecules also adsorb on the Ni catalyst, breaking the H–H bond, and new bonds are formed between the catalyst and hydrogen atoms.

Step 3: The hydrogen atoms formed interact with the adsorbed ethene as these atoms migrate from the nickel surface, forming new carbon-hydrogen bonds leading to the formation of ethane.

Step 4: As ethane is formed, it desorbs from the surface of nickel.
This reaction holds prime importance in industries such as petrochemical industries and food industries, where liquid vegetable oils are converted into solid fats.

8.4.1.3 Catalytic Converters

Adsorption also plays a significant role in the working of catalytic converters [7]. These are installed in the engines to minimize the emissions of harmful substances in the air. Its pictorial representation is given in Figure 8.8. As they involve the adsorption of harmful exhaust gases such as CO, NO_x, and HC on the catalyst surface, usually platinum, palladium, and rhodium, which have a high surface area, they are converted into less harmful substances such as CO_2, H_2O, N_2, and NH_3. Usually, chemisorption is observed in the catalytic converters.

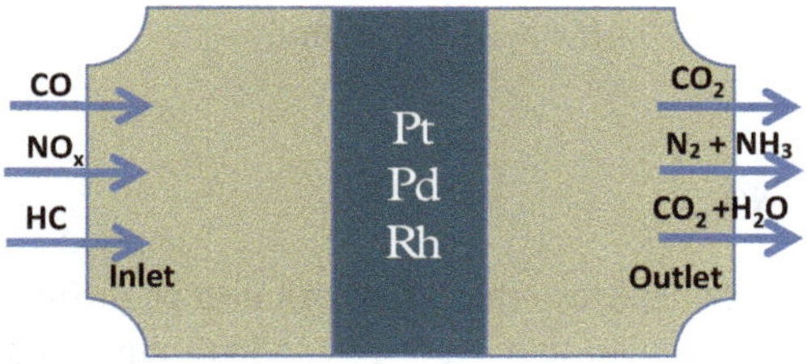

Figure 8.8: A diagrammatic representation of a catalytic converter.

Let us understand the mechanism by which a catalytic converter works:

Step 1: Catalytic converters usually use platinum and rhodium as a catalyst, as they have a higher surface area, due to which the harmful exhaust gases such as CO, NO_x, and HC are adsorbed on it.

Step 2: The adsorbed molecules undergo chemical reactions on the surface of the catalyst, i.e.,

$$CO + O_2 \xrightarrow{Pt/Pd} CO_2$$

$$HC + O_2 \xrightarrow{Pt/Pd} CO_2 + H_2O$$

Step 3: As the reaction gets completed, the products formed are desorbed from the catalyst`s surface, allowing the catalyst to become available for further reactions.

This mechanism is observed in gasoline-powered vehicles, diesel engines, etc., and holds great importance in reducing the pollution due to exhaust gases emitted from vehicles.

8.4.1.4 Ostwald's Process

It is one of the most significant methods used in industries for the production of nitric acid (HNO_3). This method is also used in the production of fertilizers. Usually, platinum in the form of platinum gauze is often used as a catalyst during this process.

The process involves the following reactions:

1. $4NH_3 + 5O_2 \xrightarrow{600-800\ ℃} 4NO + 6H_2O\ \Delta H = -24.8\,\text{kcal/mol}$
2. $2NO + O_2 \xrightarrow{150\ ℃} 2NO_2$
3. $3NO_2 + H_2O \rightarrow 2HNO_3 + NO$

The conditions of the reactions are to be optimized, as other gases can also be formed with NO. Even in this reaction, adsorption plays an important role, where ammonia (NH_3) and oxygen (O_2) adsorb onto the platinum catalyst surface, followed by the formation and desorption of nitric oxide (NO) and water (H_2O) from the surface. This reaction is exothermic.

Let us understand the mechanism of this reaction:

Step 1: It involves the adsorption of NH_3 and O_2 molecules onto the surface of the platinum catalyst, bringing them closer to the catalyst and thus initiating the reaction.

Step 2: These adsorbed molecules react on the catalyst surface, resulting in the formation of nitric oxide (NO) and water (H_2O).

Step 3: The nitric oxide and water molecules formed now desorb from the surface of the catalyst, which, in the presence of water, eventually converts to nitric acid.

8.4.1.5 Contact Process

This method is a unique and important method for the manufacturing of sulfuric acid. This method is preferred over many other methods as it is highly efficient and produces concentrated sulfuric acid. It involves the oxidation of sulfur dioxide to sulfur trioxide, and then this sulfur trioxide is absorbed in concentrated sulfuric acid to form oleum. The catalyst used in this process is vanadium pentaoxide, V_2O_5.

The reactions involved are:

$$1.\ SO_2(g) + O_2(g) \underset{\substack{400\text{-}450^{o}c \\ 1\text{-}2\ atm}}{\overset{V_2O_5}{\rightleftharpoons}} SO_3(g) \qquad \triangle H\text{=-}197KJ/mole$$

$$2.\ \underset{(catalyst)}{V_2O_5} + SO_2(g) \longrightarrow \underset{(unstable\ intermediate)}{V_2O_4 + SO_3(g)}$$

$$3.\ 2V_2O_4 + O_2(g) \longrightarrow V_2O_5$$

$$4.\ SO_3(g) + H_2SO_4(l) \longrightarrow H_2S_2O_7(l) \xrightarrow{diluted} 2H_2SO_4$$

The detailed mechanism of the contact process has been explained.

Step 1: It involves the formation of SO_2 from the elemental sulfur or pyrite ores.

Step 2: The sulfur dioxide and oxygen adsorb on the surface of V_2O_5 to form SO_3. This process is exothermic and reversible.

Step 3: The formed SO_3 is now desorbed from the catalyst surface, and the unstable intermediate formed in the presence of oxygen regenerates V_2O_5.

Step 4: The formed SO_3 absorbs in concentrated sulfuric acid (H_2SO_4) to form oleum ($H_2S_2O_7$), which is diluted to produce H_2SO_4.

The contact process holds great importance in industries such as chemical synthesis, fertilizers, mineral processing, and oil refining. There are other examples of adsorption useful in preventing corrosion of a metal surface (mild steel) in the literature. For example, the polymeric substances having heteroatoms such as S, O, and N have shown to adsorb on the surface of the metal via chemisorption, forming a protective layer on the metal surface and protecting it against corrosion even in aggressive acidic medium and environments.

8.4.2 Role of Adsorption in Chromatography

Adsorption plays a pivotal role in chromatography, a widely used separation technique in chemistry and biology. In chromatography, a mixture of molecules is separated based on their differential interactions with two phases: a stationary phase, which remains fixed in position, and a mobile phase, which moves through the stationary medium. The principle of separation lies in the varying degrees of adsorption of the components of the mixture onto the stationary phase. Molecules that are more strongly adsorbed move slowly, while those with weaker adsorption travel faster, leading to their effective separation. Chromatography is extensively applied in diverse fields such as environmental monitoring (air, water, and soil analysis), pharmaceuticals (drug formulation and purity testing), and the food industry (detection of pesticides, preservatives, and additives). It also has vital applications in clinical and forensic sciences, where biological samples such as blood, urine, and saliva are analyzed to detect drugs, poisons, or overdoses of medicines. To better understand the role of adsorption, let us examine some commonly used types of chromatography.

8.4.2.1 Paper Chromatography

Paper chromatography is one of the simplest and most commonly used techniques for separating components of a mixture. In this method, paper acts as the stationary phase, while a solvent (mobile phase) moves through it by capillary action [8]. The special properties of chromatographic paper, such as uniform texture, high wicking ability, and strong capillary action, make it suitable for separation. Typically, this paper is composed of cotton cellulose (98–99% α-cellulose and 0.3–1% β-cellulose), which ensures uniformity and effective adsorption. Paper chromatography is versatile and can be applied to both colored and colorless solutions. The separation process is governed by adsorption phenomena, where molecules of the mobile phase are adsorbed onto the surface of the paper (stationary phase) through intermolecular forces such as hydrogen bonding, dipole-dipole interactions, and van der Waals forces.

Mechanism of separation in paper chromatography, as shown in Figure 8.9:

(i) **Step 1:** The stationary phase is the thin film of water retained within the cellulose fibers of the paper. This water layer provides active sites for adsorption.

(ii) **Step 2:** The mobile phase, containing the sample mixture dissolved in a suitable solvent, moves up the paper by capillary action.

(iii) **Step 3:** As the solvent rises, the different components of the mixture are distributed between the mobile phase and the stationary water layer, depending on their ability to form intermolecular hydrogen bonds and other interactions.

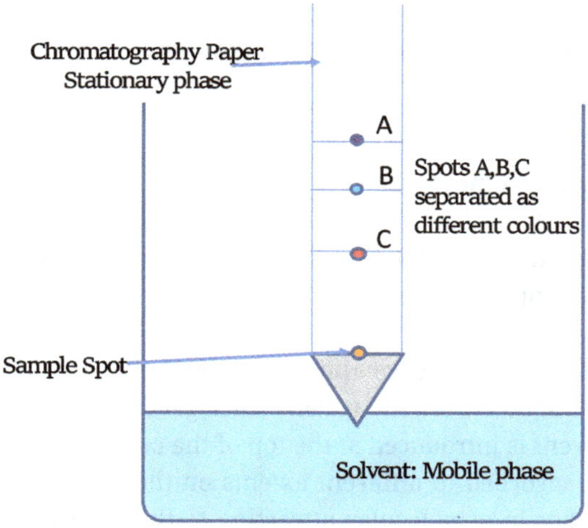

Figure 8.9: Separation of components using paper chromatography.

(iv) **Step 4:** Components that interact weakly with the stationary phase (low adsorption affinity) travel faster, while those with stronger interactions move more slowly. This differential movement leads to the separation of components into distinct spots or bands on the chromatographic paper.

Thus, adsorption governs the rate of movement of different molecules, making paper chromatography an effective and inexpensive technique for both qualitative and quantitative analyses.

8.4.2.2 Column Chromatography

Column chromatography is a separation technique in which the components of a mixture are separated using a solid stationary phase and a liquid mobile phase. A glass column is packed with a solid material such as silica gel or alumina (the stationary phase), while a suitable solvent or solvent mixture (mobile phase) carries the mixture through the column. This method is versatile and can separate molecules ranging from small organic compounds to large biomolecules, making it suitable for both small-scale laboratory uses and large-scale industrial applications [9].

8.4.2.2.1 Applications of Column Chromatography

(i) Chemical and pharmaceutical industries: Separation and purification of drugs, and removal of impurities.
(ii) Biochemical research: Isolation and purification of DNA, peptides, and proteins.

(iii) Food industry: Analysis of food components to ensure quality and standards.
(iv) Environmental studies: Detection and analysis of pollutants in environmental samples.

8.4.2.2.2 Role of Adsorption in Column Chromatography

Adsorption is central to this technique. Molecules in a mixture interact with the solid adsorbent surface through intermolecular forces [10]. The extent of adsorption determines the speed at which each component moves through the column, thus enabling separation.

Mechanism of adsorption in column chromatography, as shown in Figure 8.10:
Step 1: A glass column is packed with a solid adsorbent, usually silica gel or alumina.
Step 2: The mixture dissolved in a solvent is introduced at the top of the column.
Step 3: Molecules in the mixture are adsorbed to different extents on the surface of the adsorbent. Components with weaker intermolecular attraction to the adsorbent move down the column faster. Components with stronger attraction are retained longer.
Step 4: Fresh solvent is added to elute (wash out) the adsorbed molecules, which are collected separately as they leave the column.

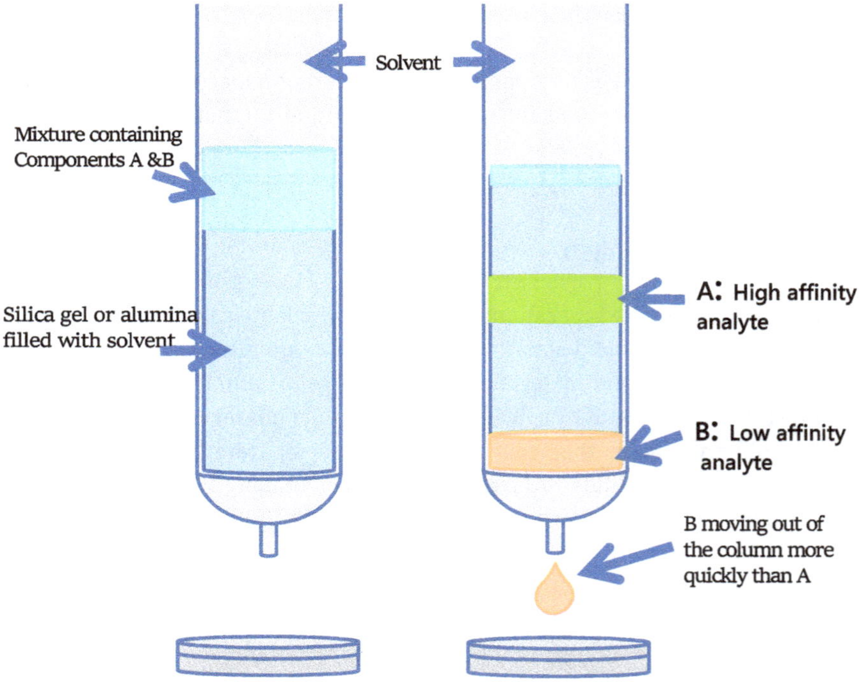

Figure 8.10: Extraction of components by column chromatography.

8.4.2.2.3 Factors Affecting Adsorption in Column Chromatography

1. **Nature of adsorbent:** Polar adsorbents tend to adsorb polar compounds more strongly.
2. **Nature of solvent:** Polar solvents can reduce the adsorption of polar compounds by competing for interaction with the adsorbent.
3. **Molecular properties:** Surface area, size, shape, and presence of functional groups influence the extent and rate of adsorption.

8.4.2.3 Thin-Layer Chromatography (TLC)

Thin-layer chromatography (TLC) is a simple, rapid, and widely used technique for separating the components of a mixture. Its principle is similar to column chromatography: the separation of components depends on their differential adsorption on a solid stationary phase and their solubility in a mobile phase [11].

In TLC, a plate made of glass, plastic, or aluminum foil is coated with a thin layer of adsorbent material, usually silica gel or alumina, which serves as the stationary phase. The mobile phase is a suitable solvent or solvent mixture that travels upward through the stationary phase by capillary action.

Principle of separation: When a mixture is spotted at the base of the TLC plate and the plate is placed in a chamber containing the mobile phase:

1. Molecules with a strong affinity for the stationary phase (adsorbent) are held back and move slowly.
2. Molecules with a weaker affinity travel faster with the mobile phase. This difference in adsorption and solubility leads to the separation of components.

The steps involved in TLC are as follows (Figure 8.11):

(i) A TLC plate is prepared with a thin coating of silica gel or alumina as the stationary phase.
(ii) A small drop of the sample mixture is spotted near the bottom of the plate using a capillary tube.
(iii) The plate is placed in a closed chamber containing a shallow layer of the solvent (mobile phase).
(iv) The solvent rises the plate by capillary action, carrying the sample components with it.
(v) Components separate into distinct spots based on their adsorption affinity and solubility in the solvent.
(vi) The separated spots are visualized under UV light or with suitable staining reagents.

Key features of TLC:

1) Requires only a small amount of sample.

2) Fast and inexpensive.
3) Useful for checking the purity of compounds, monitoring reactions, and preliminary identification of components in mixtures.

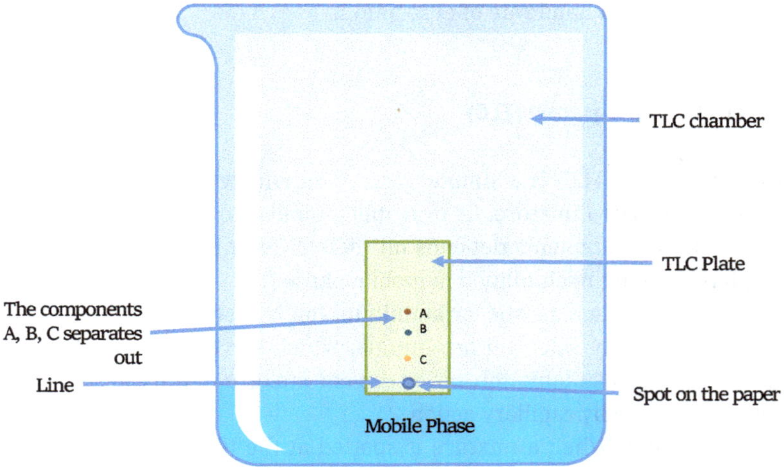

Figure 8.11: Thin-layer chromatography (TLC) showing the separation of components as distinct spots on a plate.

8.4.3 Role of Adsorption in the Removal of Pollutants

Adsorption plays an important role in controlling the levels of pollution. Various adsorption techniques have proved to be an effective and low-cost way for the removal of inorganic pollutants such as heavy metals and organic pollutants, too. The most commonly used adsorbents, such as activated carbon, nanomaterials, and natural adsorbents, have proved to be quite beneficial for controlling the release of these pollutants.

Let us understand the various ways by which adsorption helps in the control of two major types of pollution: Air pollution and water pollution.

8.4.3.1 Role of Adsorption in Controlling the Emissions Due to Air Pollution

The two main methods discussed are:
(i) Role of carbon canister in controlling automobile pollutants
(ii) Role of catalytic converters

(i) **Role of the carbon canister:** A carbon canister is a device installed in automobiles to decrease the amount of air pollution. It is usually a rectangular box that is near the fuel tank. It is usually filled with charcoal or carbon pellets, which act as an adsorbent. The canister has two ports, one is the inlet and the other is the outlet. The inlet is connected to the fuel tank's vent port, while the outlet is connected to the valve in the side of the vehicle's intake manifold. When the vehicle's engine is turned off, there is a pressure imbalance in the fuel tank, which produces hydrocarbons that appear in the form of vapors. The carbon canister traps these vapors, and the carbon adsorbs them rather than allowing them to move out of the engine, sending them back to the engine. When the engine starts again, it pulls all these vapors back and burns them in the engine, thereby stopping their emissions. These evaporative emissions or vapors are responsible for about 20% of all the harmful automobile pollutants. These canisters can be cleaned by splurging fresh air into them, thus allowing them to be reused many times. These are also capable of trapping odors like cigarette smoke.

(ii) **Role of catalytic converters:** As already discussed in the previous section, the role of adsorption in controlling the harmful emissions in the air is played by catalytic converters. This is accomplished by using the catalysts such as Pt, Pd, and Rh in the converters onto which the harmful exhaust gases such as CO, NO_x, and HC are adsorbed and converted into less harmful substances such as CO_2, H_2O, N_2, and NH_3.

8.4.3.2 Role of Adsorption in Controlling the Effluents Due to Water Pollution

Adsorption also helps in controlling water pollutants by adsorbing them onto the surface of a solid adsorbent and later removing them from water. Various adsorbents can be used, such as activated carbon, zeolites, silica gel, carbon nanotubes, and graphene oxide (these are the materials with advanced adsorption properties). These can effectively eliminate the heavy metals from water, pesticides from the crops, cleaning industrial wastewater water, and even purifying the drinking water.

Let us understand how the water can be made clean for drinking purposes by the process of water filtration through RO (reverse osmosis) using the concept of adsorption.

8.4.3.2.1 Working of RO

Adsorption plays an important role in the working of Aquaguards, where activated carbon is usually used as an adsorbent, which is a porous material with a large surface area, so that the contaminants in water can easily adhere to it and are held either by van der Waals forces of attraction or chemical attractions. The contaminants such as chlorine or certain organic compounds that cause bad taste or odor can be easily removed. Though these Aquaguards are less effective in removing heavy metals, nitrates, etc., the removal of heavy metals from the wastewater can be effectively achieved using nanomaterials as adsorbents, such as graphene, graphene oxide, and reduced graphene oxide, as they have a larger surface area and enhanced active sites, which makes them work effectively to adsorb heavy metals. For example, it has been found that if a hydrophilic group is added to graphene, the adsorption capacity of graphene toward chromium increases tremendously. The adsorption of chromium on graphene oxide has been found to be an endothermic and spontaneous reaction.

Several graphene composite materials have also been developed, which act as good adsorbents to remove heavy metals from wastewater. There is one more example of a natural biomaterial that has proved to be a good adsorber of lead from the wastewater.

Lavandula pubescens is a shrub usually found in the Middle East, and its oils contain flavonoids, which are mainly used in aromatherapy. Some studies have shown that it also acts as an adsorber for Pb(II) ions from the wastewater.

8.4.4 Role of Adsorption in Biological Systems

Adsorption is an important mechanism in our biological system and helps in the optimum working of biomaterials. It is an important mechanism not only for human bodies but also for microorganisms.

Let us now understand the following processes where the adsorption plays a significant role.

1. Adsorption in the digestive system
2. Adsorption in the microbial system
3. Adsorption of proteins on biomaterials
4. Adsorption between enzyme and substrate

8.4.4.1 Adsorption in the Digestive System

Adsorption in the digestive system helps in the following mechanisms in our body:

(i) Absorption of nutrients
(ii) Removal of waste materials from the body

(i) **Absorption of nutrients**

Our small intestine is lined with villi and microvilli, which are finger-like projections in the small intestine. Villi are larger and microvilli are smaller hair-like projections, which increase the surface area, make the adsorption of digested food molecules easy, and make the movement of nutrients such as glucose, amino acids, and electrolytes across the intestinal lining into the blood stream and lymphatic system easy.

(ii) **Removal of waste materials from the body**
Our body at times gets accumulated with certain toxins in the form of solid, liquid, and gaseous waste, which can be adsorbed by the lining of the intestine and then removed through faeces, preventing their absorption in the body. There is also an example of shellfish, which contains chitosan, a derivative of chitin, which can effectively adsorb heavy metals from the digestive tract and remove them from the body [12].

8.4.4.2 Adsorption in Microbial Systems

This process is usually referred to as biosorption, where these microorganisms bind (adsorb) the pollutants such as heavy metals and toxic pollutants onto their cell surfaces by physio-adsorption or chemical adsorption. The microorganisms usually used for this are bacteria, fungi, algae, yeast, etc. This is bioremediation process where contaminated environment is cleaned. Bioremediation involves the adsorption of contaminants by the microorganisms onto the cell surface and converting them into less harmful substances such as CO_2, water, or biomass. This process can be carried out directly at the contaminated site or by removing the contaminated material for treatment somewhere else. The adsorption capacity of these microorganisms can be increased by immobilizing these microorganisms on the support materials like activated carbon.

Though this process is quite promising in cleaning the environment, it still has some limitations, like:

(i) Adsorption rate is influenced by various factors such as temperature, pH, nutrient availability, and oxygen levels.
(ii) The process is time-consuming.

8.4.4.3 Adsorption of Proteins on Biomaterials

Adsorption of proteins on the biomaterials (usually implants), which are placed in the human body, plays a significant role in the overall process. Whenever any biomaterial (biomedical implants) is placed, the proteins from the body fluids, such as blood, adsorb onto the biomaterial surface, forming a protein layer through hydrophobic interactions, van der Waals forces, or hydrogen bonds. This adsorbed protein layer can impact how the body responds to the biomaterial. This protein adsorption also plays a role in different aspects of implant performance, including osseointegration, which is called bone formation around implants. There are various factors that influence the rate of adsorption, such as shape, size, charge, structural stability, and hydrophobicity of proteins. Protein adsorption is a molecular phenomenon occurring at the nanoscale, which is also driven by physicochemical properties of the surface. Therefore, we can say that protein adsorption is a phenomenal consideration in the understanding of a biological response to a foreign surface.

8.4.4.4 Adsorption Between the Enzyme and the Substrate

The enzymes, which are proteins, play an important role in the biological processes in our bodies. It would be very difficult to maintain a balance without these biological catalysts. Every enzyme operates through a specific mechanism. Every enzyme has active sites onto which the substrate adsorbs, forming an enzyme-substrate complex, including van der Waals forces, ionic bonds, and hydrogen bonds that are usually weaker and do not change the original structure of the enzyme.

It can be easily understood by looking at the reactions:

$$E + S \rightleftharpoons ES$$

Enzyme Substrate (enzyme substrate complex)

$$ES \rightarrow P + E$$

Complex Product Enzyme

Enzymes accelerate the rate of reactions by lowering the activation energy (E_a).
Let us understand this concept by taking the following examples:

1. Enzymes in the digestion process
2. Enzymes in biological washing powder
3. Enzymes in meat tenderizers

8.4.4.4.1 Enzymes in the Digestion Process

Enzymes such as amylase, lipase, and protease are mainly produced in the pancreas but are secreted by salivary glands, the lining of the stomach, and the small intestine. Every enzyme has an active site that adsorbs the substrate, initiating the chemical reactions that break the food into smaller components. These smaller components are absorbed through the walls of the intestines into the bloodstream and are transported to cells throughout the body. Deficiency of these digestive enzymes can pose problems to our health.

8.4.4.4.2 Enzymes in Biological Washing Powders

Enzymes such as proteases, lipases, and amylases act as a stain remover in washing powders. These enzymes work best between 30 to 40 °C. Enzymes work by allowing the substrates (stains) to bind onto them, forming an enzyme-substrate complex, which lowers the activation energy, allowing the stain to break down quickly. The products formed are usually water-soluble and are easily washed away with water. Also, the enzymes used in biological detergents are biodegradable and do not pose any threat to the environment. Proteases used in detergents are very effective in removing protein-based stains like blood or egg stains. Amylases work for the stains caused by potatoes, and lipases work for stains due to oils, grease, etc.

8.4.4.4.3 Enzymes in Meat Tenderizers

Enzymes are also used to make the meat softer and tender. Some enzymes that are already present in meat are called endogenous enzymes, such as cathepsins and calpains, and the enzymes that are added to meat from outside are exogenous enzymes, such as papain and bromelain.

These enzymes catalyze the hydrolysis of peptide bonds, which break down the large protein molecules to smaller molecules. This causes the breakdown of connective tissues, such as collagen, and muscle fiber proteins, such as myofibrillar proteins, which are responsible for the rough meat structure and stiffness, thereby making it soft and tender. Temperature, pH, and enzyme concentration all play a significant role in carrying out this breakdown of muscle and connective tissue proteins.

Practice Questions

1. What is adsorption? How is it different from absorption?

2. Differentiate between physisorption and chemisorption. Briefly explain the effect of pressure on the extent of adsorption in both cases with increasing pressure.

3. Explain the following:
   (I) Adsorbent
   (II) Adsorbate
   (III) Interface

4. How is desorption different than sorption?

5. Explain the following:
   (I) Adsorption is exothermic in nature.
   (II) Physisorption is reversible, while chemisorption is irreversible.
   (III) Effect of pressure on physisorption and chemisorption.

6. Explain the Freundlich adsorption isotherm and explain its significance.

7. Plot and explain the Langmuir adsorption isotherm.
8. Give any two applications of adsorption. Explain in brief.
9. Why are finely divided substances more effective as adsorbents?
10. What is the BET equation? What does it explain?
11. How is the BET adsorption isotherm different from the Langmuir adsorption isotherm?
12. Explain the nature of the adsorption isotherms obtained for $\Delta_{des}H_1^0 < \Delta_{vap}H_L^0$ and $\Delta_{des}H_1^0 > \Delta_{vap}H_L^0$, from BET equation.
13. Explain the role of adsorption in the Haber's Bosch process.
14. Explain in detail the working of catalytic converters.
15. Explain the mechanism of adsorption in chromatographic techniques.
16. Illustrate and explain the separation of components in TLC.
17. Explain the mechanism of adsorption in the working of carbon canisters.
18. Explain the working of an Aquaguard.
19. What is bioremediation?
20. What do you understand by protein adsorption?
21. How do the enzymes act as a stain remover in washing powders?
22. Explain, giving suitable examples, about the adsorption between the enzyme and substrate.

References

[1] Langmuir I. Surface chemistry. Chem Rev. 1933;13(2):147–191.

[2] Haghseresht F, Nouri S, Finnerty JJ, Lu GQ. Effects of surface chemistry on aromatic compound adsorption from dilute aqueous solutions by activated carbon. J Phys Chem B. 2002;106(42):10935–10943.

[3] Vigdorowitsch M, Pchelintsev A, Tsygankova L, Tanygina E. Freundlich isotherm: An adsorption model complete framework. Appl Sci. 2021;11(17):8078.

[4] Liu Y. Some consideration on the Langmuir isotherm equation. Colloids Surf a Physicochem Eng Asp. 2006;274(1–3):34–36.

[5] Heraldy E, Hidayat Y, Firdaus M. The Langmuir isotherm adsorption equation: the monolayer approach. IOP Conf Ser Mater Sci Eng. 2016;107(1):012067.

[6] Redlich O, Peterson DL. A useful adsorption isotherm. J Phys Chem. 1959;63(6):1024–4.

[7] Kritsanaviparkporn E, Baena-Moreno FM, Ramirez Reina T. Catalytic converters for vehicle exhaust: Fundamental aspects and technology overview for newcomers to the field. Chemistry. 2021;3:630–646. doi: 10.3390/chemistry3020044.

[8] Dubey R, Firdosi MK. Paper chromatography: A modern review of techniques and applications. Int J Novel Res Dev. 2024;9(10). Oct. ISSN:2456-4184.Published I.D IJNRD2410045, Registration ID:1JNRD_226922.

[9] Aveyard R, Haydon DA. An introduction to the principles of surface chemistry. Cambridge: Cambridge University Press; 1973.

[10] Ravindranath B. Principles and practice of chromatography. New York: John Wiley & Sons; 1989. ISBN: 0470213280.

[11] Geiss F. Fundamentals of thin layer chromatography. Heidelberg, New York; ISBN-3778508547.

[12] Helenius A, Simons K. Solubilization of membranes by detergents. Biochim Biophys Acta. 1975;415:29–79. doi: 10.1016/0304-4157(75)90016-7.

Index

https://doi.org/10.1515/9783112208236-009

www.ingramcontent.com/pod-product-compliance
Lightning Source LLC
Chambersburg PA
CBHW080926220326
41598CB00034B/5700
* 9 7 8 3 1 1 9 1 4 6 6 4 7 *